Lecture Notes in Physics

Edited by H. Araki, Kyoto, J. Ehlers, München, K. Hepp, Zürich
R. Kippenhahn, München, H.A. Weidenmüller, Heidelberg,
J. Wess, Karlsruhe and J. Zittartz, Köln
Managing Editor: W. Beiglböck

301

G. Ferenczi F. Beleznay (Eds.)

W0225763

New Developments in Semiconductor Physics

Proceedings of the Third Summer School
Held at Szeged, Hungary
August 31 – September 4, 1987

Springer-Verlag
Berlin Heidelberg GmbH

Editors

G. Ferenczi
F. Beleznay
Research Institute for Technical Physics
of the Hungarian Academy of Sciences
P.O. Box 76, H-1325 Budapest, Hungary

Organizing Committee

F. Beleznay   (Chairman)
G. Ferenczi   (Program Chairman)
B. Pődör   (secretary)
M. Török   (local arrangements)
É. Németh   (administrative secretary)
I. Tánczos   (administrative secretary)

Organized by

Semiconductor Branch of the Roland Eötvös Physical Society
Research Institute for Technical Physics of the HAS
Attila József University, Szeged

ISBN 978-3-662-13669-0   ISBN 978-3-540-39145-6 (eBook)
DOI 10.1007/978-3-540-39145-6

© Springer-Verlag Berlin Heidelberg 1988
Originally published by Springer-Verlag Berlin Heidelberg New York in 1988
Softcover reprint of the hardcover 1st edition 1988

2158/3140-543210

PREFACE

The semiconductor branch of the Hungarian Physical Society organized its third Summer School at Szeged, Hungary from August 31 to September 4, 1988 with the intention of discussing the most recent developments in semiconductor physics. (Proceedings of the previous meetings are published in the Lecture Notes in Physics series as Volume 122 and Volume 175.) 84 participants from 18 countries attended the meeting, which included 15 invited talks covering the areas of multilayer growth technology, theory of electron states, transport theory, defect related effects and structural properties of semiconductors. 33 contributed papers, most of them closely related to the invited talks, gave an exciting insight into the research in this field.

The present volume is a selection of the most interesting papers presented at the Summer School, and its format follows that of the meeting: invited papers are accompanied by related contributions. As the table of contents indicates, we found that transport theory and defect-related effects are the most widely researched subjects in contemporary semiconductor physics. Other topics are, nevertheless, well represented.

At the closing session it was suggested that, since the meeting had been so successful, similar events held at regular intervals would be welcome. First steps are being taken to organize future Schools as joint ventures of the Physical Societies of neighbouring countries.

The editors are grateful to Eva Nemeth for her expert help in preparing this volume.

Budapest, Hungary   George Ferenczi
January 1988   Program Chairman

TABLE OF CONTENT

INTEGER QUANTUM HALL EFFECT
- Present state of the theory -

J. Hajdu
Institut für Theoretische Physik
Universität zu Köln, D-5000 Köln

The basic experimental facts concerning the integer quantum Hall effect (QHE) are summarized and confronted with the prediction of standard transport theory. The different ideas and approaches to explain the
QHE are reviewed and commented upon. The phenomenological model which
simulates localization in additive disordered systems at zero temperature by bound states is considered for two system-geometries. For a finite cylinder a spectral stability condition for the QHE is formulated
and shown to be sufficient for current compensation. For a torus the
topological quantization of the Kubo Hall conductivity in mobility gaps
is outlined. Some problems facing current and future research are pointed
out.

1. Facts

At high magnetic fields (B~10 T) and low temperatures (T ~ 1 K) the
Hall resistance R_H of a silicon MOSFET as a function of the gate voltage
U_G shows characteristic plateaux. Since U_G is proportional to the carrier
density n this observation contradicts the established theory which
predicts $R_H \sim 1/n$.

In 1980 Klaus von Klitzing discovered that the plateau values of R_H
are entirely independent of the properties of the sample, and are given
by

$$R_H (\text{plateau}) = \frac{1}{k} \frac{h}{e^2} , \quad k = 1, 2, \cdots \qquad (1.1)$$

/1/. Furthermore, in the plateau regimes of R_H the (longitudinal) resistivity R practically vanishes. Shortly thereafter this integer quantum Hall effect (QHE) was also observed in GaAs heterostructures (in
this case n is constant and B is varied). The plateaux are centered
(approximately) around the corresponding integer values η = k of the
filling factor

$$\eta = (2\pi \ell^2) n \qquad (1.2)$$

Here l is the magnetic length, $l^2 = \hbar/eB$. Figs. 1 and 2 show some
typical experimental data. At present the experimental accuracy of the

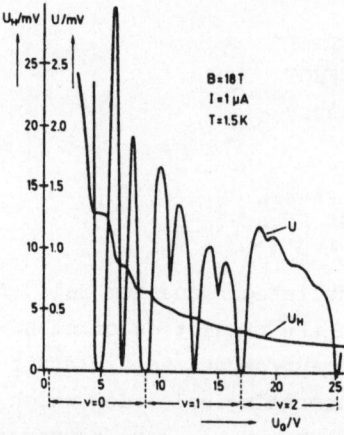

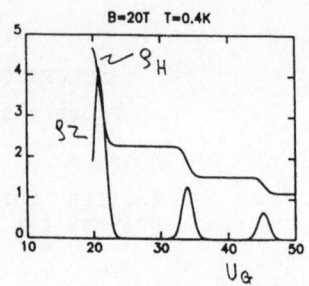

Fig. 2: QHE /2/. Both data
refer to Si MOSFETs.
$U_G \sim n$

Fig. 1: The quantum Hall
effect /1/

quantization (1.1) is better than 10^{-7}. Thus, the QHE provides a high precision measurement of the Sommerfeld fine structure constant e^2/hc.

The conductivity layers in MOSFETs and heterostructures behave like two-dimensional systems. At sufficiently weak currents the conduction properties of an homogeneous and isotropic rectangular system (with area $Ar = L_x L_y$) in a perpendicular magnetic field are described by the linear relations between currents and voltages,

$$U_x = R I_x + R_H I_y$$

$$U_y = - R_H I_x + R I_y \tag{1.3}$$

Introducing the electric field $\underline{E}$, $E_x = U_x/L_x$, $E_y = U_y/L_y$ and the current density $\underline{j}$, $j_x = I_x/L_y$, $j_y = I_y/L_x$ we get

$$E_x = \rho j_x + \rho_H j_y$$

$$E_y = -\rho_H j_x + \rho j_y \tag{1.4}$$

with resistivities

$$\rho = R(L_y/L_x), \quad \rho_H = R_H \tag{1.5}$$

Notice that in two dimensions R_H is independent of the size of the sample. Inverting (1.4)

$$j_x = \sigma E_x - \sigma_H E_y$$

$$j_y = \sigma_H E_x + \sigma E_y \tag{1.6}$$

with conductivities

$$\sigma = \frac{\rho}{\rho^2 + \rho_H^2} \quad , \quad \sigma_H = \frac{\rho_H}{\rho^2 + \rho_H^2} \qquad (1.7)$$

Notice that (if $\rho_H \neq 0$) σ and ρ vanish simultaneously. In the usual Hall measurement $I_y = 0$. Since in the plateau regimes $R = 0$ ($U_x < 10^{-14}$V) there $\sigma = 0$ and

$$\sigma(\text{plateau}) = k\frac{e^2}{h} \quad , \quad k = 0, 1, 2, \cdots \qquad (1.8)$$

In 2d the physical dimension of the conductivity is

$$[\sigma] = \frac{[\dot{\rho}]}{[E]} = \frac{\text{charge} \cdot \text{velocity/area}}{\text{voltage/length}} \qquad (1.9)$$

$$= \frac{\text{charge/time}}{\text{energy/charge}} = \frac{(\text{charge})^2}{\text{action}}$$

- in accordance with (1.8) and (1.1). The plateau values of σ_H are integer multiples of the atomic unit e^2/h of the 2d conductivity.

2. Hints

According to the simple classical kinetic model the Hall conductivity of free electrons is

$$\sigma_H^0 = \frac{en}{B} \qquad (2.1)$$

or with (1.2)

$$\sigma_H^0 = \eta\frac{e^2}{h} \qquad (2.2)$$

Notice that, at integer filling, $\eta = k$, the measured values of the Hall conductivity of (rather complex) real 2d systems coincide with the corresponding values calculated for the (fictitious) free electron system by using the most simple classical model (cf. Fig. 3). In fact this is even true for the Hall conductivity of disordered systems calculated by perturbation-theoretical evaluation of the Kubo formula (cf. Fig. 4). The stability of the Hall conductivity at integer filling ($\sigma_H = k\, e^2/h$) is, perhaps, not so surprising because the states corresponding to completely filled Landau levels are highly resistent against perturbation.

Due to the disorder the free electron Landau levels

$$\mathcal{E}_\alpha^0 = \mathcal{E}_\nu = \hbar\omega_c(\nu + \tfrac{1}{2}) \qquad (2.3)$$

$\omega_c = eB/m$, $\nu = 0, 1, 2, ..$, are broadened to energy bands. In lowest order

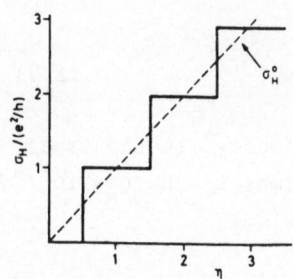

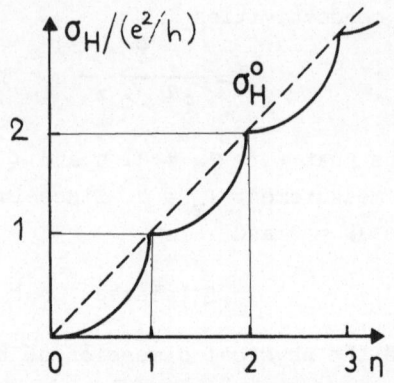

Fig. 3: QHE - extrapolated to
zero temperature. At
integer filling, η = k,
σ_H and σ_H^0 coincide,
both being k e^2/h .

Fig. 4: σ_H calculated by perturbation theory /3/
(schematic).

cumulant approximation for a white-noise correlated random potential,
the free electron density of states

$$n^0(\varepsilon) = \frac{1}{2\pi\ell^2} \sum_\nu \delta(\varepsilon - \varepsilon_\nu) \qquad (2.4)$$

is replaced by

$$n(\varepsilon) = \frac{1}{2\pi\ell^2} \frac{\sqrt{2/\pi}}{\Gamma} exp\left[-2\left(\frac{\varepsilon - \varepsilon_\nu}{\Gamma}\right)^2\right] \qquad (2.5)$$

with

$$\Gamma = \frac{2}{\pi} \hbar\omega_c \hbar/\tau \qquad (2.6)$$

where τ is the B = 0 relaxation time (in Born approximation) /4/. Each
band contains $1/2\pi\,\ell^2$ states per unit area (the degree of degeneracy of
the Landau levels). For high mobility samples and sufficiently high
magnetic fields, such that $\omega_c\tau \gg 1$, the overlapping of bands is
insignificant. The density of states can be used to express the chemical
potential ξ in terms of n,B and T,

$$n = \frac{1}{A\tau} \sum_\alpha f_\alpha = \int f(\varepsilon) n(\varepsilon) d\varepsilon \qquad (2.7)$$

where f_α = f(ε_α) is the Fermi distribution function

$$f(\varepsilon) = \left\{ exp\left[(\varepsilon - \xi)/k_B T\right] + 1 \right\} \qquad (2.8)$$

For free electrons, with (2.4)

$$n = \frac{1}{2\pi \ell^2} \sum_{\nu} f(\varepsilon_\nu) \tag{2.9}$$

and, with (2.2),

$$\sigma_H^0(\zeta) = \frac{e^2}{h} \sum_{\nu} f(\varepsilon_\nu) \tag{2.10}$$

In the limit $T \to 0$ $\zeta = \varepsilon_F$ and

$$\sigma_H^0(\varepsilon_F) = \frac{e^2}{h} \sum_{\nu} \theta(\varepsilon_F - \varepsilon_\nu) \tag{2.11}$$

Thus $\sigma_H^0(n)$ is a straight line but $\sigma_H^0(\varepsilon_F)$ is a step function (Fig. 5). (This is all right since $n = n(\varepsilon_F)$ is a step function as well.) In contrast to this, the observed Hall conductivity extrapolated to $T = 0$ (shown in Fig. 3), is a step function on the n scale. Assuming, however,

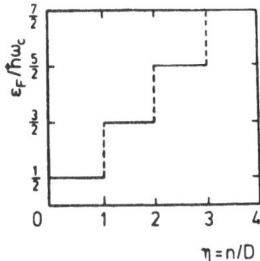

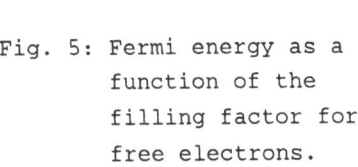

$\eta = n/D$

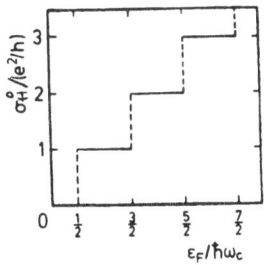

$\varepsilon_F/\hbar\omega_c$

Fig. 5: Fermi energy as a
function of the
filling factor for
free electrons.
$D = 1/2\pi \, 1^2$

Fig. 6: Hall conductivity
as a function of
the Fermi energy
for free electrons

that in a real system $n = n(\varepsilon_F)$ is a smooth function (as indicated by model calculations) we can conclude that on the ε_F scale the extrapolated values of the measured Hall conductivity is exactly the same as for free electrons, given by (2.11). Thus, a way to characterize the QHE effect is to say that, at $T = 0$, $\sigma_H(\varepsilon_F)$ is the same as for an ideal free electron gas - inspite of the broadening of the Landau levels to energy bands. This behaviour is obviously radically different from that predicted by the traditional transport theory for disordered systems (perturbation-theoretical evaluation of the Kubo formula).

3. Ideas and Approaches

Shortly after the discovery of the QHE several ideas were developed
to explain this suprising phenomenon. Aoki and Ando /5/ pointed out that
the QHE may be brought about by <u>localization</u> of electrons in a 2d random potential. In fact, the vanishing longitudinal conductivity in the
plateau regimes of σ_H seems to be an obvious indication of localization.
Furthermore, localization was shown to occur in the tails of the Landau
bands /6/ and, in the limit of very high magnetic fields, everywhere
outside of small ranges around the band centers ($\varepsilon = \varepsilon_\nu$) /7/. Since
an operative high field transport theory which incorporates localiza-
tion was - and still is - not available, Aoki and Ando simulated localization by assuming bound states in which the expectation value of the
velocity vanishes,

$$\langle \alpha | \underline{v} | \alpha \rangle = 0 \quad \text{for bound states} \tag{3.1}$$

Connecting this phenomenological description of localization with the
Kubo formula they demonstrated that, at T = 0, the longitudinal conductivity vanishes and the Hall conductivity σ_H keeps constant as long
as the Fermi energy varies within a regime of bound states. Unfortunately their proof, showing the plateau values of σ_H to coincide with
the quantized ones (1.8), seen in experiment turned out to be incomplete. At first glance the proof can be supplemented by reference to
perturbation theory /8/ or to the Streda formula /9/ both of which lead,
for $\eta = k$, to $\sigma_H = k\ e^2/h$. However both ways of fixing the plateaux
at the observed quantized values are rather unsatisfactory. The perturbation theory definitely tails for $\eta \neq k$. The Streda formula which requires spectral gaps between the Landau bands in order to yield the
quantized values for integer filling, seems to hold for a confined system only /10/. For such systems, however, due to edge states no band
gaps exist. As we shall see (cf. Section 4) considerably more effort is
needed to prove quantization within the phenomenological description of
localization /10/. In any case, numerical analysis by Ando /11/ has confirmed that the QHE can be explained in terms of independent electrons
moving in a random potential.

If the QHE is due to localization, the loss of current (caused by
localization) must be compensated by an additive acceleration in the
delocalized states - in order to maintain $\sigma_H = k\ e^2/h$ for $\eta = k$.
Prange /12/ and subsequently other authors /13,14/ have attempted to
demonstrate compensation for some model systems. An elegant way of
proof utilizes Levinson's theorem. In the usual formulation this theorem

relates the scattering phase shift to the number of bound states brought about by a certain potential. We shall see (cf. Section 4) that the conditions for the QHE in the phenomenological description of localization are sufficient to prove conpensation (and to derive Levinson's theorem) /15/. This is an important fact because localization cannot adequately be explained by potential scattering.

A simple model which illustrates localization and explains qualitatively the QHE is a system of independent electrons moving in a slowly varying random potential V(x,y) and a strong magnetic field /16,17/. Introducing center and relative coordinates with respect to the cyclotron motion,

$$x = X + v_y/\omega_c \, , \, y = Y - v_x/\omega_c \tag{3.2}$$

(v_x, v_y) and (X,Y) are pairs of conjugate variables,

$$\left[v_x, v_y \right] = i \left(\ell \, \omega_c \right)^2 \tag{3.3}$$

$$\left[X, Y \right] = i \ell^2 \tag{3.4}$$

The velocity components v_x and v_y are bounded; the expectation values of the relative coordinates in an energy eigenstate is proportional to l, i.e. to 1/B. Consequently, for sufficiently large B we can approximate V(x,y) by V(X,Y). Furthermore, since (according to (2.4)) the limit $B \rightarrow \infty$ is equivalent to the classical limit $h \rightarrow 0$ we can approximate the quantum dynamics by the corresponding classical one. Replacing the kinetic energy $mv^2/2$ by its eigenvalue ε^0_α , we get the slow motion Hamiltonian

$$H = \varepsilon^0_\alpha + V(X,Y) \tag{3.5}$$

The equations of motion

$$\dot{X} = \frac{1}{eB} \frac{\partial V}{\partial Y} \, , \, \dot{Y} = -\frac{1}{eB} \frac{\partial V}{\partial X} \tag{3.6}$$

describe a reversible 1d motion along equipotential lines V(X,Y) = const. If the space average of V vanishes then, according to percolation theory, in the thermodynamic limit, all equipotential lines in the bulk with energy $\varepsilon \neq \varepsilon^0_\nu$ are closed (localized states), and open equipotential lines (delocalized states) exist only at $\varepsilon = \varepsilon_\nu$ (i.e. at the centers of the Landau bands).

To calculate the current we have to add to V the potential energy e ϕ due to a driving electric field $\underline{E}$ = -grad ϕ . Obviously, the motion along closed equipotential lines does not contribute to the net current

$$\dot{j}_y = \frac{-e}{A_r} \sum_v \iint \frac{dX\,dY}{2\pi \ell^2} \, f\left(-\frac{1}{eB}\frac{\partial U}{\partial X}\right) \qquad (3.7)$$

$A_r = L_x L_y$, $U = V + e$. Assuming that all channels of open (percolating) equipotential lines with $\mathcal{E}_v \leqslant \mathcal{E} \leqslant \mathcal{E}_{v+1}$ are occupied we get for the current at T = 0

$$\dot{j}_y = \frac{e^2}{h} \sum_v \theta(\mathcal{E}_F - \mathcal{E}_v) \sum_\ell \Delta U_\ell / eL_x \qquad (3.8)$$

where $\Delta U_1/e$ is the potential drop across the 1th channel. Since $\sum_\ell \Delta U_1/e$ is nothing else but the total potential drop across the sample, (3.8) is equivalent to

$$\sigma_H = k \frac{e^2}{h} \qquad (3.9)$$

where k is the largest integer for which k $\leqslant \eta$. Unfortunately, this impressingly simple derivation of the desired result (3.9) (Iordansky /16/) does not even exclude corrections of the order 1/L. The formulation based on linear response theory (Kubo formula) /17,18/ is subject to the same limitation. Since in experiments, $1/L \approx 10^{-4}$ and the accuracy of quantization is 10^{-8}, the required accuracy of the proof of (3.9) is at least $(1/L)^2$. This may have motivated the aim to combine the high field percolation model with the gauge argument /13/ (see below). The problem, however, is not to prove the stability of (3.9) within the leading order high field model (which is guaranteed by the topologically distinct nature of closed and open equipotential lines) but to determine the accuracy of the model as such. For the relation of the high field percolation model to percolation theory, cf. Trugman /19/. The onset of dissipation is investigated in /20,21/. The high field model also provides an explanation of observed thermoelectric effects analogous to the QHE /22/.

The gauge argument asserts that the QHE is due to a particular symmetry property: for a 2d system on the surface of a cylinder, the change of axial flux by a unit flux quantum, together with the transfer of a unit charge from one edge of the cylinder to the other (which are at different potentials) is a symmetry transformation of the system if, at T = 0, the Fermi energy lies in a mobility gap (Laughlin /23/) or, more generally, the ground state of the system is non-degenerate and separated from the rest of the energy spectrum /24/. The gauge argument requires some interpretation. This is provided, for instance, by the topological approach to the QHE /25,26/ which was initiated by the observation that for an electron in an ideal 2d lattice the Hall conductivity defined by the Kubo formula is topologically quantized and equal to an integer multiple of e^2/h if, at T = 0, the Fermi energy lies in an energy gap /27/. (The same result follows also from the Streda formula /28/). The characteristic feature of the topological approach is a double-periodic Hamiltonian (2d system on a torus). As we shall see (cf. Section 4) the Kubo Hall conductivity for such a system can be proved to be topologically quantized if, at T = 0, the Fermi energy lies in a mobility gap - the localization being simulated by bound states /29/.

Of course, a theory of the QHE as a localization phenomenon is only acceptable if it explains rather than assumes localization. Confronted with this requirement, the state of the art in QHE theory is rather unsatisfactory. Following the lines of the B = 0 self-consistent localization theory Ono /30/ obtained for the high field longitudinal conductivity exponential delocalization at the band centers ($\varepsilon = \varepsilon_\nu$). Perturbation theory yields qualitatively the same result /31/. Similar investigations for the Hall conductivity have not yet been reported. For the time being the only localization theory which treats σ and σ_H on equal grounds is the field theory by Levine, Libby and Pruisken /32/. In this theory the relevant long range modes are described by the Lagrangian

$$\mathcal{L} = \sigma^{(0)} \mathcal{L}_1 + \sigma_H^{(0)} \mathcal{L}_2 \tag{3.10}$$

where $\mathcal{L}_2$ is due to the axial symmetry breaking by the magnetic field and the coupling constants are the mean field values of the longitudinal and the Hall conductivity respectively. For finite action field configurations (instantons) $\mathcal{L}_2$ is a topological invariant /33/. COnsequently, the two parameter scaling space decomposes into equivalent sectors. Levine et al /32/ argue that for $T \to 0$, $\sigma^{(0)}$ and $\sigma_H^{(0)}$ are renormalized to 0 and $k\, e^2/h$ respectively. Although this seems rather plausible a formal proof (solution of renormalization group equations) is still missing.

Still not clarified is the role played by the Coulomb interaction between the electrons. As pointed out recently the Hall conductivity of a 2d free electron system depends strongly both on the potential distribution of the driving force (<u>E</u>) /34/ and on the system size (in the direction of the field) /35/. It is believed that the Coulomb interaction

substantially reduces these effects and, thus, acts to set up the classical free electron Hall conductivity (2.1).

4. Spectral stability

4.1 Finite cylinder

We consider a 2d system of independent electrons in a random potential V(x,y) on a rectangular part of the (x,y)-plane in a perpendicular
homogeneous magnetic field B. In the x direction the electrons are assumed to be confined by a potential $V_c(x)$ to an interval of finite
length L_x. In order to get a non-vanishing current, say in the y direction, we impose the periodic boundary condition

$$\Psi(x,y+L_y) = \Psi(x,y) \tag{4.1}$$

on the wave functions. By (4.1) the geometry of the system is - with
respect to connectivity - equivalent to the surface of a cylinder
R x S'. Unfortunately, the usual coordinate representation of quantum
mechanics cannot be implemented on S'. The reason is that (4.1) prevents
both the existence of a global multiplication operator corresponding to
y and the unitary equivalence of the operators $-i\hbar\,\partial/\partial y$ and $-i\hbar\,\partial/\partial y\,+$
const. One way to overcome this difficulty is to apply the U(1) bundle
theory. Another, essentially equivalent, way is to repeat the system
periodically in the y direction

$$V(x,y+L_y) = V(x,y) \tag{4.2}$$

and apply the usual coordinate representation

$$H = -\frac{\hbar^2}{2m}\frac{\partial^2}{\partial x^2} + \frac{1}{2m}\left(-i\hbar\frac{\partial}{\partial y} + eBx\right)^2 + V(x,y) + V_c(x) \tag{4.3}$$

Owing to (4.2) the Bloch theorem applies in the y direction

$$H\phi_n(x,y;\vartheta) = \mathcal{E}_n(\vartheta)\,\phi_n(x,y;\vartheta) \tag{4.4}$$

$$\phi_n(x,y;\vartheta) = e^{i\vartheta y/\hbar}\,u_n(x,y;\vartheta) \tag{4.5}$$

$$u_n(x,y+L_y,\vartheta) = u_n(x,y;\vartheta) \tag{4.6}$$

being the Bloch wave number. The Bloch factors u_n satisfy

$$H(\vartheta)u_n(x,y;\vartheta) = \mathcal{E}_n(\vartheta)\,u_n(x,y;\vartheta) \tag{4.7}$$

with

$$H(\vartheta) = -\frac{\hbar^2}{2m}\frac{\partial^2}{\partial x^2} + \frac{1}{2m}\left(-i\hbar\frac{\partial}{\partial y} + \hbar\vartheta + eBx\right)^2 + V(x,y) + V_c(x) \quad (4.8)$$

Thus, $u_n(x,y;\vartheta)$ can be interpreted as the wave function $\psi_n(\vartheta)$ corresponding to a representation

$$p_y : p_y(\vartheta) = -i\hbar\frac{\partial}{\partial y} + \hbar\vartheta, \quad \vartheta \in \frac{2\pi}{L_y}(\mathbb{R}/\mathbb{Z}) \quad (4.9)$$

and restricted to $y \in \left[0, L_y\right[$.

By restricting ϑ to the first Brillouin zone, $\vartheta = \vartheta_o \in \left[-\frac{\pi}{L_y}, \frac{\pi}{L_y}\right[$ we get the usual <u>energy bands</u> $\mathcal{E}_n(\vartheta_o)$ labeled by the band index n. In the case of free electrons $(V = 0)$ with Dirichlet boundary conditions $\psi_n = 0$ for $x = \pm L_x/2$ replacing the confinement potential (Teller-model) the energy spectrum of $H(\vartheta = 0)$ depends on the eigen values $\hbar k_n$, $k_n = \frac{2\pi}{L_y} n$, $n \in \mathbb{Z}$ of the conserved momentum $p_y = -i\hbar\,\partial/\partial y$ as well as on the Landau quantum number $\nu = 0,1,2,\ldots$ (Fig. 1). The energy bands $\mathcal{E}_n(\vartheta_o)$ are related to the spectrum $\mathcal{E}_\nu(k_n)$ by

$$\mathcal{E}_n(\vartheta_o) = \mathcal{E}_\nu(k_{n'} + \vartheta_o) \quad (4.10)$$

and can be joined together to form continuous <u>energy branches</u>

$$\mathcal{E}_\nu(\vartheta) = \mathcal{E}_\nu(k_n + \vartheta_o), \quad \vartheta \in \frac{2\pi}{L_y}\mathbb{R} \quad (4.11)$$

In the Landau model $(V_c = 0, L_x \to \infty)$ these energy branches are the familiar equidistant degenerate Landau levels $\mathcal{E}_\nu(\vartheta) = \mathcal{E}_\nu = \hbar\omega_c(\nu + \frac{1}{2})$ In the Teller model the energy branches depend on ϑ. For sufficiently large values of L_x/l the lifting of degeneracy is significant at the edges $\vartheta \simeq \pm L_x/2l^2$ only. For the Hall effect in the Teller model, cf. the recent work by Ono and Kramer /36/.

In a disordered system $(V \neq 0)$ both extended and localized states may exist. In the phenomenological model /5/ to be adopted in this and in the following section, the localization is simulated by bound states in which the expectation value of the velocity $v_y = \frac{i}{\hbar}[H,y]$

$$\langle \phi_n(\vartheta)|v_y|\phi_n(\vartheta)\rangle = \langle \psi_n(\vartheta)|v_y(\vartheta)|\psi_n(\vartheta)\rangle = \frac{1}{\hbar}\frac{\partial \mathcal{E}_n(\vartheta)}{\partial \vartheta} \quad (4.12)$$

with

$$v_y(\vartheta) = \frac{1}{\hbar}\frac{\partial H}{\partial \vartheta} \quad (4.13)$$

vanishes in the first Brillouin zone,

$$\frac{\partial \mathcal{E}_n(\vartheta)}{\partial \vartheta} = 0 \,,\, \vartheta \in \left[-\frac{\pi}{L_y}, \frac{\pi}{L_y}\right[\tag{4.14}$$

(cf. (3.1)). If the random potential is sufficiently weak, those energy eigenvalues $\mathcal{E}_n(\vartheta)$ which correspond to extended states in the first Brillouin zone can be connected to smooth energy branches $\mathcal{E}_\nu(\vartheta)$. The construction procedure is similar to that for free electrons in the Teller model - except for the fact that each localized state gives rise to a missing interval $\Delta\vartheta = \frac{2\pi}{L_y}$ (i.e. a Brillouin zone). Recent numerical studies by Ohtzuki and Ono /37/ indicate the existence of continuous energy branches corresponding to extended states. The co-existence of localized and delocalized states at the same energy has been related to the fractal dimension of the wave function /38/.

4.2. QHE

In the presence of an external homogeneous electric field along the x axis, the Hamiltonian of the system is changed to $H' = H(\vartheta) + eEx$ with eigenvalues $\mathcal{E}_\alpha$ and eigenstates $|\alpha\rangle$. The average Hall current is given by

$$\dot{J}_y = \frac{-e}{A_T} \mathrm{Tr}\left(\hat{f}\, v_y\right) \tag{4.15}$$

A density matrix which describes a stationary state with a homogeneous charge distribution is

$$f_{\alpha\alpha'} = f\left(\mathcal{E}_n(\vartheta)\right)\delta_{\alpha\alpha'} \tag{4.16}$$

In such a state

$$\dot{J}_y = \frac{-e}{A_T} \sum_\alpha f\left(\mathcal{E}_n(\vartheta)\right)\frac{1}{\hbar}\frac{\partial \mathcal{E}_\alpha}{\partial \vartheta} \tag{4.17}$$

Since to first order in E

$$\mathcal{E}_\alpha = \mathcal{E}_n(\vartheta) + eE\, X_n(\vartheta) \tag{4.18}$$

where

$$X_n(\vartheta) = \langle\Psi_n(\vartheta)|x|\Psi_n(\vartheta)\rangle \tag{4.19}$$

(4.17) yields for the Hall conductivity $\sigma_H = (d \mathfrak{z}_y / dE)_{E=0}$

$$\sigma_H = -\frac{e^2}{h} \sum_\nu \frac{1}{L_x} \int_{BZ} d\vartheta f(\varepsilon_n(\vartheta)) \frac{\partial X_n(\vartheta)}{\partial \vartheta} \tag{4.20}$$

This expression can also rigorously be derived from the Kubo Hall conductivity formula. Notice that for bound states $\partial X_n(\vartheta)/\partial \vartheta = 0$.

In the phenomenological localization approach the quantization (1.8)
for T = 0 is furnished by a sufficient condition on the spectrum and
and the interpretation pf L_x as an effective system length. The spectral
condition asserts that for extended states $|\Psi_\nu(\vartheta)\rangle$ the expectation
value $X_\nu(\vartheta)$ of X is globally independent of disorder. In particular,
if ε is the smallest constant for which the equation $\varepsilon_\nu(\vartheta) = \varepsilon$ has
two solutions ϑ_1 and ϑ_2, the effective system length

$$\tilde{L}_x \equiv X_\nu(\vartheta_1) - X(\vartheta_2) \tag{4.21}$$

(for each fixed ν) must be the same as for free electrons (Teller model). Integrating in

$$\sigma_H = -\frac{e^2}{h} \sum_\nu \frac{1}{L_x} \int d\vartheta f(\varepsilon_\nu(\vartheta)) \frac{\partial X_\nu(\vartheta)}{\partial \vartheta} \tag{4.22}$$

by parts, using (4.21) and setting $L_x = \tilde{L}_x$ we get for T = 0 (1.8).
Since in a disordered system the number of extended states and, therefore, $\vartheta_2 - \vartheta_1$ is much less than in the Teller model (for each bound
state an interval $\Delta \vartheta = 2\pi/L_y$ is missing) the assumed global behaviour of $X_\nu(\vartheta)$ is in fact a (far reaching) stability condition
for the quantization (1.8) to be concluded. For a finite number of
point impurities this stability condition is fulfilled - as long as the
number of bound states is less than the number of extended ones. Nothing is known about more realistic systems.

4.3. Compensation

In a disordered system the free electron value of the Hall conductivity (2.2) is maintained for integer filling, η = k since the loss
of current due to bound states is exactly compensated by an excess current carried by extended states. This compensation can be proved to
follow from the stability condition formulated above without any reference to scattering theory. On the other hand this condition proves to

be sufficient to derive the energy shift formula and the Levinson theorem of potential scattering theory, both of which were used to demonstrate the current compensation in the scattering theoretical approach to the QHE.

In linear order in E the energy of extended states is given by

$$\mathcal{E}'_\nu(\vartheta) = \mathcal{E}_\nu(\vartheta) + eEX_\nu(\vartheta) \qquad (4.23)$$

For E = 0 the total current density carried by the extended states of each energy band must vanish,

$$\int_{\vartheta_1}^{\vartheta_2} d\vartheta\, \sigma_{y\nu}(\vartheta) = \left[\mathcal{E}_\nu(\vartheta_2) - \mathcal{E}_\nu(\vartheta_1)\right]/\hbar = 0 \qquad (4.24)$$

By (4.23), (4.24) the condition (4.21) with $\tilde{L}_x = L_x$ turns out to be a stability condition for $\mathcal{E}'_\nu(\vartheta)$:

$$\overline{\mathcal{E}} - \underline{\mathcal{E}} \equiv \mathcal{E}'_\nu(\vartheta_1) - \mathcal{E}'_\nu(\vartheta_2) = eEL_x \qquad (4.25)$$

For free electrons the expectation value of the velocity in each bulk state $|\Psi_\nu(\vartheta)\rangle$, $\vartheta \in [\vartheta_1, \vartheta_2]$ is the drift velocity $\nu_D = -E/B$. Thus, the total velocity of bulk states belonging to each energy branch is $N\nu_D$ where $N = (L_y/2\pi)$ ($\vartheta_2^0 - \vartheta_1^0$) is the corresponding number of states (quantities marked by the subscript o refer to free electrons). The expectation value of the velocity in a disordered system can be related to ν_D by introducing the inverse function $\theta(\vartheta)$ of $\mathcal{E}'_\nu(\vartheta)$,

$$\mathcal{E}'_\nu(\vartheta) = \mathcal{E}'^0_\nu(\theta_\nu(\vartheta)) \qquad (4.26)$$

which is well-defined for $\vartheta \in [\vartheta_1, \vartheta_2]$ (since in this interval $\mathcal{E}'^0_\nu = \hbar\nu_D\vartheta + \mathcal{E}^0_\nu$ is a monotonic function with values in $[\underline{\mathcal{E}}, \overline{\mathcal{E}}]$). Due to the stability condition (4.25)

$$\theta_\nu(\vartheta_1) = \vartheta_1^0, \quad \theta_\nu(\vartheta_2) = \vartheta_2^0 \qquad (4.27)$$

With $\nu_y(\vartheta) = \partial H'/\partial\hbar\vartheta$ and (4.26)

$$\nu_{y\nu}(\vartheta) = \sigma_D\, d\theta_\nu(\vartheta)/d\vartheta \qquad (4.28)$$

and, thus, the total velocity of bulk states belonging to a single energy branch is

$$\frac{L_y}{2\pi} \int_{\vartheta_1}^{\vartheta_2} \upsilon_{yv}(\vartheta) \, d\vartheta = \frac{L_y}{2\pi} \upsilon_D \left[\theta_v(\vartheta_2) - \theta(\vartheta_1) \right] = N \upsilon_D \qquad (4.29)$$

Since this is the same as for free electrons, the current compensation is established.

5. Topological quantization

5.1. Torus

Repeating the rectangular system considered in Sect. 4 both in x and in y direction,

$$V(x + L_x, y) = V(x, y + L_y) = V(x, y) \qquad (5.1)$$

is - with respect to connectivity - equivalent to assuming the geometry of a torus, $S^1 \times S^1$. The Hamiltonian (4.3) with (5.1) is now invariant under the discrete two-parameter group of magnetic translations. This group is, for arbitrary values of a perpendicular magnetic field, non-Abelian. For the discrete values

$$B = \frac{k}{Ar} \frac{h}{e}, \qquad k \text{ integer} \qquad (5.2)$$

of the magnetic field which correspond to the rational values $\eta = N/k$ of the filling factor, however, the magnetic translational group is isomorphic with the usual space translation group and, thus, Abelian. For the values (5.2) of the magnetic field the Bloch theorem applies,

$$H \phi_n(x, y; \vartheta) = \mathcal{E}_n \phi_n(x, y; \vartheta) \qquad (5.3)$$

$$\phi_n(x, y; \vartheta) = e^{i(\vartheta_1 x + \vartheta_2 y)} u_n(x, y; \vartheta) \qquad (5.4)$$

$$u_n(x + \alpha L_x, y + \beta L_y; \vartheta) = u_n(x, y; \vartheta) \qquad (5.5)$$

$$H(\vartheta) u_n = \mathcal{E}_n(\vartheta) u_n \qquad (5.6)$$

$\vartheta = (\vartheta_1, \vartheta_2)$. $\mathcal{E}_n(\vartheta)$ and $u_n(\vartheta)$ have the following properties:

$$\mathcal{E}_n(\vartheta + g) = \mathcal{E}_n(\vartheta) \qquad (5.7)$$

$$u_n(x,y;\vartheta+g) = e^{-i(g_1 x + g_2 y)} u_n(x,y;\vartheta) \tag{5.8}$$

where $\quad g = (g_1, g_2) ; \quad g_i \in (2\pi/L_i)\mathbb{Z} , \quad i = 1,2 \,\widehat{=}\, x,y \tag{5.9}$

In the generic case, $\varepsilon_n(\vartheta)$ is, for fixed values of n, an analytic function of ϑ (energy bands). "Generic" means that no degeneracy occurs - neither intrinsically (brought about by the random potential) nor accidentally (in the sense of the Wigner-von Neumann theorem).

The velocity operators acting on the Bloch factors are

$$v_i(\vartheta) = \partial_i H(\vartheta)/\hbar \tag{5.10}$$

$$\partial_i \equiv \partial/\partial\vartheta_i$$

5.2 QHE

For an additive system on a torus the Kubo conductivity formula can be written as

$$\sigma_{k\ell} = \iint_{BZ} d\vartheta_1 d\vartheta_2 \, \sigma_{k\ell}(\vartheta) \tag{5.11}$$

$$\sigma_{k\ell}(\vartheta) = i e^2 \hbar \sum_{n \neq m} f(\varepsilon_n(\vartheta)) \frac{(v_\ell(\vartheta))_{nm} (v_k(\vartheta))_{mn} - (v_k(\vartheta))_{nm} (v_\ell(\vartheta))_{mn}}{(\varepsilon_n(\vartheta) - \varepsilon_m(\vartheta))^2}$$

In our model no dissipation occurs, i.e. the longitudinal conductivity σ_{kk} vanishes identically.

For T = 0, $f(\varepsilon) = \theta(\varepsilon_F - \varepsilon)$ and two cases can be distinguished: If the Fermi energy ε_F lies in a mobility gap (simulated by bound states with property (3.1)), (5.11) can be transformed to give

$$\sigma_{yx} = \frac{e^2}{h} \sum_n{}' \left\{ \frac{i}{2\pi} \int \langle du_n | du_n \rangle \right\} \tag{5.12}$$

where $|du_n(\vartheta)\rangle = |\partial_1 u_n(\vartheta)\rangle d\vartheta_1 + |\partial_2 u_n(\vartheta)\rangle d\vartheta_2$ and the sum goes over all occupied bands of extended states. Since the quantity in the curly bracket is an integer (Chern number) (5.12) asserts topological quantization of the Hall conductivity as a function of the

Fermi energy: as long as ε_F varies in a mobility gap $\sigma_{yx}(\varepsilon_F)$ remains constant taking one of the quantized values k e^2/h.
If, on the other hand, the Fermi energy lies in a regime of delocalized states the Kubo Hall conductivity is not quantized in the above
sense (even if the many-body ground state is non-degerate). For sufficiently weak disorder the relation between the integer k and the filling
factor can be established by perturbation theory, the topological quantization assures the QHE.

6. Problems

Does the Hall conductivity depend, in the thermodynamical limit, on
the geometry (boundary conditions, topology) of the system? If the an-
swer is yes, we have to ask for the physical reason. In the case of
the torus, does a condition exist similar to the spectral stability
condition which holds for a finite cylinder? And: does the topological
quantization hold for the cylinder as well? Is the high field limit a
model for the phenomenological localization approach? If so, what is
the reason that the correction of the order $1/L_x$ vanishes? I believe
that these questions will be answered in the near future. Also, the
role of the Coulomb interaction will be clarified soon (at least for
a simple model, perhaps numerically). A much more difficult problem is,
however, that of developing a theory of the localization-delocalization
phase transition in strong magnetic fields. All attempts to extend
the self consistent localization theory to the Hall conductivity have
failed so far. A presumably easier task is to solve the two-parameter
renormalization group equations of the non-linear σ model (in the di-
lute instanton approximation). Only very little is known about finite
size effects, finite temperature transport (onset of dissipation),
finite frequency response, as well as the local current and field distribution.
There is still much to be done in the theory of the integer quantum
Hall effect...

References

/1/ K. von Klitzing, G. Dorda, M. Pepper, Phys. Rev. Lett 45, 494 (1980)

/2/ T. Englert, Lecture Notes in Physics 177, 87 (1983)

/3/ T. Ando, Y. Matsumoto, Y. Uemura, J. Phys. Soc. Japan 39, 279 (1975)

/4/ R. Gerhardts, Z. Phys. B21, 275; 285 (1975); Surf. Sci. 58, 227 (1976)

/5/ H. Aoki, T. Ando, Solid State Comm. 38, 1079 (1981)

/6/ H. Aoki, H. Kamimura, Solid State Comm. 21, 45 (1977)

/7/ M. Tsukada, J. Phys. Soc. Japan 41, 1466 (1976)

/8/ D.J. Thouless, J. Phys. C 14, 3475 (1981)

/9/ P. Středa, J. Phys. C15, L717 (1982). Cf. also
    L. Smrčka, J. Phys. C17, L63 (1984)

/10/ J. Hajdu, M. Janßen, O. Viehweger, Z.Phys. B66, 433 (1987)

/11/ T. Ando, in Anderson Localization, ed. by Y. Nagaoka and
     H. Fukuyama (Sol. St. Sci. 39) Springer, 1982, p. 176

/12/ R.E. Prange, Phys. Rev. B 23, 4802 (1981)

/13/ R. Joynt, R.E. Prange, Phys. Rev. B 29, 3303 (1984)

/14/ W. Brenig, Z. Phys. B 50, 305 (1983); W. Brenig, K. Wysokinski,
     Z. Phys. B 63, 149 (1986)

/15/ M. Janßen, J. Hajdu, Z. Phys. B in print

/16/ S.V. Iordansky, Solid State Comm. 48, 1 (1982)

/17/ Y. Ono, in op. cit. 11, p. 207

/18/ S.M. Apenko, Yu. E. Lozovik, J. Phys. C18, 1197 (1985);
     Sov. Phys. JETP 62, 328 (1985)

/19/ S.A. Trugman, Phys. Rev. B27, 7539 (1983)

/20/ Y. Ono, J. Phys. Soc. Japan 51, 237 (1982)

/21/ B. Shapiro, Phys. Rev. B33, 8447 (1986)

/22/ A. Grunwald, J. Hajdu, Solid State Comm. 63, 289 (1987)

/23/ R.B. Laughlin, Phys. Rev. B23, 5632 (1981)

/24/ R. Tao, Y.S. Wu, Phys. Rev. B30, 1097 (1984)

/25/ Q. Niu, D.J. Thouless, J. Phys. A17, 2453 (1984);
     Q. Niu, D.J. Thouless, Y.S. Wu, Phys. Rev. B31, 3372 (1985)

/26(J. Avron, R. Seiler, Phys. Rev. Lett. 54, 259 (1985)

/27/ D.J. Thouless, M. Kohmoto, M.P. Nightingale, M. den Nijs,
     Phys. Rev. Lett. 49, 405 (1982)

/28/ P. Středa, J. Phys. C15, L 1299 (1982)

/29/ W. Pook, J. Hajdu, Z. Phys. B 66, 427 (1987)

/30/ Y. Ono, J. Phys. Soc. Japan 51, 2055, 3544 (1982); 53, 2342 (1984);
     Prog. Theor. Phys. Suppl. No. 84, 138 (1985)

/31/ S. Hikami, Phys. Rev. B29, 3726 (1984)

/32/ H. Levine, S.B. Libby, A.M.M. Pruisken, Nucl. Phys. B240, 30, 49,
     71 (1984)

/33/ M. Janßen, U. Phys. in press

/34/ R. Johnston, L. Schweitzer, Z. Phys. in press

/35/ Y. Ono, T. Ohtusuki, Z. Phys. in press

/36/ Y. Ono, B. Kramer, Z. Phys. B67, 341 (1987)

/37/ T. Ohtsuki, Y. Ono, Solid State Comm. in press; preprint Köln,
     1987

/38/ B. Kramer, Y. Ono, T. Ohtsuki, in: Proceedings EP2DS-VII, in press

THEORY OF THE ENERGY LOSS RATE OF HOT ELECTRONS IN 2D SYSTEMS

Erich Vass

Institut für Experimentalphysik, University of Innsbruck
Technikerstrasse 25, A-6020 Innsbruck, Austria

A novel general expression is derived for the average energy loss rate
<ELR> of hot 2D electrons due to their interaction with phonons. The
presented loss rate formula allows to explain quantitatively the tempe-
rature- and concentration dependence of the acoustic phonon limited
<ELR> found in previous experimental studies.

1. INTRODUCTION

The acoustic- [1-7] as well as the polar optical [8-10] phonon limited
<ELR> of hot 2D electrons have been investigated extensively in the past.
Shubnikov-deHaas measurements [2,3] performed at given values of the
input power and carrier concentration n show that the acoustic phonon
limited <ELR> of warm charge carriers in quantized Si inversion layers
increases $\propto T^x(T_e-T)$ if the lattice temperature T<10K whereby the actual
value of x depends on the mode type involved in the scattering process
and the surface orientation of the layer. In [6] it was found that the
static screened <ELR> of hot electrons in GaAs heterostructures should
be $\propto (T_e^x - T^x)n^{-3/2}$ where x=7 (5) if they are scattered by the deformation (piezoelectric) potential of acoustic phonons. Far infrared (FIR)
intensity measurements on n-Si [11] as well as n-GaAs [12] inversion
layers imply that the ratio of the electron heating temperature $\Delta T_e = T_e - T$
to the square root of the input power $e\mu E^2$ is independent on the carrier
concentration. It was shown [13] that this ratio is completely determined
by fundamental parameters of the lattice and is due to the n-independence
of the <ELR>. Recent investigations [14] of <ELR> substantiate this n-
independence which cannot be understood by the currently [1,2,7] existing energy loss rate calculations predicting the <ELR> $\propto n^{-3/2}$. A satis-
factory explanation of this apparent discrepancy is still missing and
will be presented for the first time in this paper. For this purpose
let us make the following assumptions:
i) The subband structure of the 2DEG is parabolic, ii) The electron-
electron collisions take place more frequently than those between the
electrons and phonons, iii) The density of the 2DEG is homogenious.

2. THEORY

If an electron with wave vector $\vec{k}$ interacts with phonons, the time
rate of change of its energy $\varepsilon_{\vec{k}}$ can be written in general as [15]

$$<-\frac{d\varepsilon_{\vec{k}}}{dt}> = \sum_{\vec{k}} \varepsilon_{\vec{k}} (\frac{\partial f_{\vec{k}}}{\partial t})_c / \sum_{\vec{k}} f_{\vec{k}} \tag{1}$$

where the collision term is given by

$$(\frac{\partial f_{\vec{k}}}{\partial t})_c = \sum_{\vec{k}'} [W_{\vec{k}\vec{k}'} f_{\vec{k}'} (1-f_{\vec{k}}) - W_{\vec{k}'\vec{k}} f_{\vec{k}} (1-f_{\vec{k}'})] \tag{2}$$

$W_{\vec{k}\vec{k}'}$ is the transition probability per unit time for the transition between the initial state $\vec{k}$ and the final state $\vec{k}'$. The occupation numbers
of these states are denoted by $f_{\vec{k}}$ and $f_{\vec{k}'}$ which are replaced in the following by a hot drifted Fermi-Dirac function. Inserting (2) into (1) and
using the fact that the value of the above sums cannot depend on the
notation of the summation indices we may replace $\vec{k}$ by $\vec{k}'$ and vice versa.
In addition we may change the order of the $\vec{k}'$- and $\vec{k}$-summations to get

$$<-\frac{d\varepsilon_{\vec{k}}}{dt}> = [\sum_{\vec{k}} f_{\underset{\sim}{\vec{k}}} (-\frac{d\varepsilon_{\vec{k}}}{dt})]/\sum_{\vec{k}} f_{\underset{\sim}{\vec{k}}} \qquad (\underset{\sim}{\vec{k}} = \vec{k} - \vec{k}_d) \tag{3}$$

$\vec{k}_d$ is the electron drift wave vector and

$$-\frac{d\varepsilon_{\vec{k}}}{dt} = \sum_{\vec{k}'} (\varepsilon_{\vec{k}} - \varepsilon_{\vec{k}'}) W_{\vec{k}\vec{k}'} (1-f_{\vec{k}'}) \qquad (\underset{\sim}{\vec{k}'} = \vec{k}' - \vec{k}_d) \tag{4}$$

On the other hand the energy- and momentum conservation laws provide
$\varepsilon_{\vec{k}'} = \varepsilon_{\vec{k}} \pm \hbar\omega_{\vec{q}j}$, $\vec{k}' = \vec{k} \mp \vec{q}$ with $\omega_{\vec{q}j}$ being the angular frequency of a phonon
with wave vector $\vec{q}$ belonging to the j-th phonon mode. Using this and

$$\sum_{\vec{k}'} \to \sum_{\vec{q},j} \quad , \quad n\,A = g_s g_v \sum_{\vec{k}} f_{\underset{\sim}{\vec{k}}} \qquad \text{(g_s, g_v: spin- and valley degeneracy,}$$
$$\text{A: area of the 2DEG)}$$

we obtain from (3)

$$<-\frac{d\varepsilon_{\vec{k}}}{dt}> = \frac{1}{n\,A} \sum_{\vec{q},j} \hbar\omega_{\vec{q}j} \{g_s g_v \sum_{\vec{k}} [W_{\vec{k}\vec{k}-\vec{q}} f_{\underset{\sim}{\vec{k}}} (1-f_{\underset{\sim}{\vec{k}-\vec{q}}}) - W_{\vec{k}\vec{k}+\vec{q}} f_{\underset{\sim}{\vec{k}}} (1-f_{\underset{\sim}{\vec{k}+\vec{q}}})]\} \tag{5}$$

where the factor $\{...\}$ can be interpreted as the number of phonons
emitted per unit time [14]. To evaluate (5) let us assume that

$$W_{\vec{k}\vec{k}\pm\vec{q}} = \frac{2\pi}{\hbar} |M^j_{\vec{k}\pm\vec{q}\vec{k}}|^2 (n_{\vec{q}j} + \frac{1}{2} \mp \frac{1}{2}) \delta(\varepsilon_{\vec{k}\pm\vec{q}} \varepsilon_{\vec{k}} \mp \hbar\omega_{\vec{q}j}) \tag{6}$$

with $M^j_{\vec{k}\pm\vec{q}\vec{k}}$ being the electron-phonon matrixelement. $n_{\vec{q}j}$ is the phonon
occupation number which can be different from the Planck-function.

To symmetrize (6) we introduce the center of mass position vector $\vec{k}^*$ by the transformation $\vec{k} = \vec{k}^* + \vec{k}_d \pm \vec{q}/2$. After that we replace in (5) the sums over $\vec{q}$ and $\vec{k}$ by integrals and integrate over the δ-function of (6) with respect to the angle between $\vec{q}$ and $\vec{k}^*$. As a result $\langle ELR \rangle$ of surface phonons having the wave vector $\vec{q}$ parallel to the 2DEG is found to be

$$\left\langle -\frac{d\varepsilon_{\vec{k}}}{dt} \right\rangle = \sum_j \frac{g_s g_v}{2\sqrt{2}\pi^3} \frac{m_0^{*3/2}(k_B T_e)^{1/2}}{\hbar^3 n} \int_0^\infty dq \int_0^{2\pi} d\varphi \; \omega_{\vec{q}j} \; A|M_{\vec{q}j}|^2 \; (n^e_{\vec{q}j} - n_{\vec{q}j}) \; I_{\vec{q}j} \quad (7)$$

where the hot carrier distribution function enters into the integral

$$I_{\vec{q}j} = \int_0^\infty du \, [f(u,n_-) - f(u,n_+)] \quad , \quad f(u,n_\pm) = [\exp(u^2 - n_\pm) + 1]^{-1} \quad , \text{ with}$$

$$n_\mp^+ = n - a_j[\cosh(s) \mp 1] \mp 2\cos\varphi\{a_j x_d[\cosh(s) \mp 1]\}^{1/2} - x_d \cos^2\varphi \; .$$

The normalized chemical potential $\eta = \zeta/k_B T_e = \ln[\exp(T_F/T_e) - 1]$ with T_F being the Fermi-temperature of the 2DEG. $a_j = \hbar\omega_{\vec{q}j}/2k_B T_e$, $x_d = \hbar^2 k_d^2/2m_0^* k_B T_e$,

$$n^e_{\vec{q}j} = [\exp(2a_j - 2\sqrt{2}a_j x_d \, e^{s/2} \cos\varphi) - 1]^{-1} \quad , \quad s = \ln(\hbar q^2/2m_0^* \omega_{\vec{q}j}) \text{ with } m_0^* \text{ be-}$$

ing the effective electron mass. Eq. (7) is valid for arbitrary carrier statistics and scattering mechanisms. For bulk phonons $\int dq$ must be replaced by $\int dq (d/2\pi) \int dq_z$ with q_z being the phonon wave vector component perpendicular to the 2DEG. The parameters T_e and k_d have to be determined from the balance equations [2] for the energy and momentum.

In the following (7) is evaluated for acoustic phonons whereby we assume that the drift energy $\hbar^2 k_d^2/2m_0^* \ll k_B T_e$. The wave function of the electrons in the QW having the thickness d is specified by

$$\Psi_{\vec{k}}(\vec{r},z) = A^{-1/2} \exp(i\vec{k}\,\vec{r})\xi(z) \quad , \quad \xi(z) = \sqrt{\frac{2}{d}}\sin\left(\frac{\pi}{d}z\right)$$

where $\vec{r}$ and $\xi(z)$ denote the position vector and the envelope function. The matrixelements for the interaction with the deformation- and piezo-electric potential of acoustic surface- and bulk phonons are then given by

$$A|M_{\vec{q}j}|^2 = \begin{cases} \dfrac{\Xi_{ac}^2 \hbar \gamma_1^2 q^2}{2\rho c_R J} \left[1 + \left(\dfrac{e_{14}e}{\Xi_{ac}\varepsilon_s q}\right)^2 \sin^2\varphi\cos^2\varphi\right] F\{S_1 \, e^{-\gamma_1 qz} + S_2 \, e^{-\gamma_2 qz}\} \\[4mm] \dfrac{\Xi_{ac}^2 \hbar q}{2\rho d} \left[\left(\dfrac{1}{c_\ell} + \dfrac{2}{c_t}\right) + \left(\dfrac{e_{14}e}{\Xi_{ac}\varepsilon_s q}\right)^2 \left(\dfrac{K_\ell^2}{c_\ell} + \dfrac{2K_t^2}{c_t}\right)\right] F\{e^{i q_z z}\} \end{cases} \quad (8)$$

where Ξ_{ac}, e_{14}, ρ and ε_s denote the deformation potential, the piezo-electric constant, the mass density and the static dielectric constant of the semiconductor. The coefficients γ_1, γ_2, S_1, S_2 and J can be deduced from [17] and are determined by the ratios c_R/c_ℓ with c_R, c_ℓ and c_t

being the group-velocities of the Rayleigh-, the longitudinal- and transverse sound wave. $K_\ell = \frac{3}{2} \sin 2\varphi (\cos \vartheta - \cos^3 \vartheta)$, $K_t = \frac{1}{2}(1 - 2\sin^2 \varphi)\sin 2\vartheta$ are the anisotropy factors of the piezoelectric interaction. The formfactor

$$F\{\Phi\} = \left| \int_0^d dz\, \xi^*(z)\, \Phi\, \xi(z) \right|^2 \tag{9}$$

describes the influence of $\xi(z)$ onto the electron-phonon coupling. The above matrixelements can be written as

$$A|M_{\vec{q}j}|^2 = B_j\, q^\nu\, F\{\Phi\}$$

which will be used in the following to calculate (7) for the non-degenerate and degenerate 2DEG.

1) Energy loss rate of the non-degenerate 2DEG: If the chemical potential $\eta < 0$, the distribution function of the charge carriers can be approximated by a hot Maxwellian. In this case (7) gives

$$\left\langle -\frac{d\varepsilon_{\vec{k}}}{dt} \right\rangle = \frac{2^{\nu+1}}{\sqrt{\pi}}\, B_j\, \frac{k_{th}^{\nu+3}}{k_B T_e}\, c_j \int_0^\infty dy\, y^{\nu+1}\, (n_{\vec{q}j}^e - n_{\vec{q}j})\sinh(a_j)\exp\left(-y^2 - \frac{T_{oj}}{4T_e}\right) F\{\Phi\} \tag{10}$$

where $T_{oj} = 2m_o^* c_j^2 / k_B$ is the minimum electron temperature to emit the j-th acoustic phonon mode. $y = q/2k_{th}$ with $k_{th}^2 = 2m_o^* k_B T_e / \hbar^2$.

2) Energy loss rate of the degenerate 2DEG: If $\eta \gg 1$, in (7) the integral $I_{qj} \approx a_j/\sqrt{\eta - a_j}\cosh(s)$. According as $a_j \ll 1$ or $a_j \gg 1$, we have to consider two different temperature ranges which are of interest for the experiment.

i) Bloch-range ($a_j \gg 1$): In this range the occupation numbers $n_{\vec{q}j}^e$ and $n_{\vec{q}j}$ exponentially converge to zero with increasing values of q. Therefore it is reasonable to cut off in (7) the integration over q at the maximum value $k_B T_e/\hbar c_j$ and $k_B T/\hbar c_j$. This cut off procedure leads to

$$\left\langle -\frac{d\varepsilon_{\vec{k}}}{dt} \right\rangle = \frac{1}{\pi}\, B_j\, \frac{m_o^{*2} k_B^{\nu+3} T_e}{\hbar^{\nu+6} c_j^{\nu+1} k_F^3}\, [L_\nu(T_e) T_e^{\nu+2} - L_\nu(T) T^{\nu+2}] \tag{11}$$

where the function $L_\nu(T_e)$ is defined by

$$L_\nu(T_e) = \int_0^1 dx\, x^{\nu+2}\, e^{-x} \left[1 - \frac{T_e^2}{4T_{oj}T_F}\left(x^2 + \frac{T_{oj}^2}{T_e^2}\right)\right]^{-1/2} F\{\Phi\}$$

with $k_F = 4\pi n / g_s g_v$ being the Fermi-radius of the 2DEG. The above expression is valid in the temperature range $T_e \ll \sqrt{T_{oj}T_F}$.

ii) Equipartition range ($a_j \ll 1$): In this range the phonons are fully excited so that $n_{\vec{q}j}^e - n_{\vec{q}j} \approx k_B \Delta T_e / \hbar \omega_{\vec{q}j}$. Since the electron system is assumed to be degenerate, the upper limit for the q-integration in (7) is given by $q = 2k_F$. Phonons with $q > 2k_F$ cannot contribute to the loss rate due to the energy- and momentum conservation laws. As a result we obtain

$$<-\frac{d\varepsilon_{\vec{k}}}{dt}> = \frac{2^{\nu+2}}{\pi} B_j \frac{m_o^{*2}k_B\Delta T_e}{\hbar^4} c_j k_F^{\nu-1} \frac{1}{\int_0} dw\, w^{\nu+1} [1-(w^2+\frac{T_{oj}}{2T_F})]^{-1/2} F\{\Phi\} \qquad (12)$$

where $w=q/2k_F$. The above expression is valid in the temperature range $\sqrt{T_{oj}T_F}<<T_e<T_F$. Eq.(11) and (12) imply that the n-dependence of <ELR> of the degenerate 2DEG is completely different in the Bloch-Grüneisen- and the equipartition range. In previous calculations [4,6,7] a rigorous distinction between these two range has not been done.

In order to show that (12) is nearly independent of n if the interaction is dominated by the deformation potential of acoustic bulk phonons ($\nu=1$) we multiply with $(d/2\pi)\int dq_z$ to take into account the dissipation of the hot carrier energy by phonons propagating perpendicular to the 2DEG or having $q_z\neq 0$. The upper limit of the q_z-integration is cutted off at the value $q_z^{max} \approx k_B T_e/\hbar c_j$ since the mean free path of an acoustic phonon is much larger than the width d of the QW. Eq.(12) then gives

$$<-\frac{d\varepsilon_{\vec{k}}}{dt}> = \frac{\Xi_{ac}^2 m_o^{*2} k_B^2 \Delta T_e(\Delta T_e+T)}{\pi^2 \rho\hbar^4 c_j} \frac{1}{\int_0} dw\, w^2[1-(w^2+\frac{T_{oj}}{2T_F})]^{-1/2} F\{\Phi\} \qquad (13)$$

In real 2D systems $T_{oj}<<2T_F$ implying $\int dw...=\pi/4$ if $F\{\Phi\}=1$. Using this and inserting (13) into the power balance equation we find

$$\frac{\Delta T_e}{\sqrt{e\mu E^2}} = (\frac{4\pi\rho\hbar^4 c_j}{\Xi_{ac}^2 m_o^{*2} k_B^2})^{1/2} \quad , \quad \Delta T_e>>T \qquad (14)$$

where μ and E denote the mobility and the heating field of the electrons. In agreement with experimental data [11,12] the ratio (14) is independent on the carrier concentration n and is completely determined by fundamental parameters of the lattice.

3. NUMERICAL RESULTS AND DISCUSSION

In fig.1a the energy loss rate (7) is plotted versus T_e for different acoustic phonon modes at constant values of n and T. Fig.1b shows the corresponding loss rates as a function of n for given values T_e and T. These curves were calculated with the following numerical values of the material parameters: $\rho=5.31 gcm^{-3}$, $m_o^*=0.067m_o$, $\Xi_{ac}= 10$ eV, $c_\ell=5200$ ms^{-1}, $c_t=3200$ ms^{-1}, $c_R=0.9c_t$, $e_{14}=2.1\times10^9$ V/m, $d=100$ Å. The experimental data have been deduced from the energy loss measurements of [14] and are indicated by a hatched data bar.

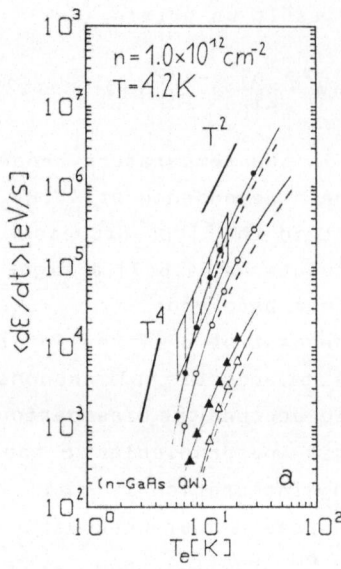
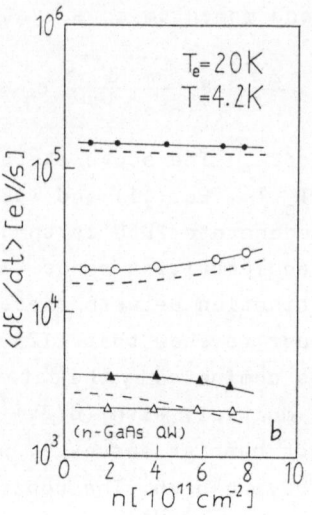

Fig.1: Unscreened (——) and screened (---) average energy loss rate of hot 2D electrons in a square QW as a function of the temperature T_e and the concentration n of the charge carriers for their interaction with the deformation- and piezoelectric potential of acoustic bulk- and surface phonon modes. ● (o): energy loss rate limited by the bulk- (surface-) deformation potential. ▲ (Δ): energy loss rate limited by the bulk- (sur- face-) piezoelectric potential.

The T_e-dependence of the presented loss rate curves can be understood as follows: If the average thermal energy of an electron is comparable with the excitation energy $\hbar\omega_{\vec{q}j}$ of a phonon, the electron mainly inter- acts with phonons having momenta $\hbar q \lesssim k_B T_e / c_j$ since the phonon occupation number $n_{\vec{q}j}$ converges exponentially to zero if $\hbar\omega_{\vec{q}j} \to \infty$. As a consequence the loss rate (7) turns out to be proportional to a power law of the form $T_e(T_e^x - T^x)$ in agreement with (11) provided the condition $T_e \ll \sqrt{T_{oj} T_F}$ is fulfilled whereby the exponent x depends on the kind of coupling as well as the mode type involved in the scattering. In the warm electron range $T_e \gtrsim T$ the above power law can be approximated by $T^{x-1}(T_e-T)$ im- plying that the electron heating temperature $T_e - T$ increases $\propto E^2$ if the mobility of the electrons is completely determined by elastic scatter- ing mechanisms (e.g. ionized impurities). Mobility calculations show [16] that at He-temperatures this assumption is reasonable for n-GaAs QW's. The <ELR> limited by the deformation potential stronger increases with T_e than the corresponding loss rate limited by the piezoelectric poten- tial. In addition this increase is also stronger for surface modes than for bulk modes as indicated in fig.1 by full straight lines. This be- havior can be attributed to the different q-dependence of the matrix-

elements (8) which imply that the deformation- (piezoelectric-) potential dissipates the energy of fast (slowly) moving electrons more
effectively. The energy dissipation via the piezoelectric potential
is therefore only essential in the temperature range $T_e \lesssim 2K$ and is reduced by a factor 2-3 if static screening [6] of the bare interaction
is taken into account. The loss rate associated with the deformation
potential coupling is weaker reduced by the screening effect since this
potential is of short range order. The loss rates for finite QW-width
d are found to be smaller than those for vanishing width. This observation can be attributed to the formfactor $F\{\Phi\}$ which is equal to 1
for d=0. If $\hbar\omega_{\vec{q}j} \ll k_B T_e$ and $T_e < T_F$, phonons with $q \approx 2k_F$ are mainly involved in the scattering. Consequently $\langle ELR \rangle \propto T_e \Delta T_e$ according to (13)
and is found to be independent of n if the interaction takes place via
the bulk-deformation potential. According to (10) the $\langle ELR \rangle$ of the
non-degenerate 2DEG is also independent of n since the chemical potential chancels out in (7) if $T_e \gg T_F$. In this limit the electron heating temperature ΔT_e increases proportional to E^2 in agreement with previous theoretical studies.

Acknowledgement: This work was supported by the "Fonds zur Förderung
der Wissenschaftlichen Forschung" in Austria, Project No. P6128P.

REFERENCES

[1] H. Ezawa, S. Kawaji and K. Nakamura, Jpn. J. Appl. Phys. 13,
126 (1974)
[2] K. Hess, T. Englert, T. Neugebauer, G. Landwehr and G. Dorda,
Phys. Rev. B16, 3652 (1977)
[3] T. Neugebauer and G. Landwehr, Phys. Rev. B21, 702 (1980)
[4] Y. Shinba, K. Nakamura, M. Fukuchi and M. Sakata, J. Phys. Soc.
Jpn. 51, 157 (1982)
[5] W. Hönlein and G. Landwehr, Surf. Sci. 113, 260 (1982)
[6] P.J. Price, J. Appl. Phys. 53, 6863 (1982)
[7] S.J. Manion, M. Artaki, M.A. Emanuel, J.J. Coleman and K. Hess,
Phys. Rev. B35, 9203 (1987)
[8] E. Vass and K. Hess, Z. Phys. B25, 323 (1976)
[9] D.K. Ferry, Surf. Sci. 75, 86 (1978)
[10] E. Vass, Z. Phys. B, (1987) in press
[11] R.A. Höpfel, E. Vass and E. Gornik, Solid State Commun. 49,
501 (1984)
[12] R.A. Höpfel and G. Weimann, J. Appl. Phys. Lett. 46, 291 (1985)
[13] E. Vass, Solid State Commun. 55, 847 (1985)
[14] K. Hirakawa and H. Sakaki, Appl. Phys. Lett. 49, 889 (1986)
[15] K. Seeger, Semiconductor Physics, p.179, Springer Verlag (1973)
[16] X.L. Lei, J.L. Birman and C.S. Ting, J. Appl. Phys. 58, 2270
(1985)
[17] E. Vass, Solid State Commun. 61, 127 (1987)

THE TRANSPORT PROBLEM

S.Nettel
Rensselaer Polytechnik Institute, Department of Physics
Troy, N.Y. 12180-3590, USA

1.Introduction

Art and science have been entwined as much in semiconductor as
in most other technologies. The technical developments in the early
fifties were accompanied by basic studies of electron dynamics, for
example, by Haynes and Westphal at the Bell Telephone Laboratories in
New Yersey. [1] According to the Naynes-Shockley picture established
at this time, electrons with a square-wave spatial distribution in
the density injected at one end of a semiconductor diffuse as a well-
defined spreading Gaussian, while drifting in the direction of an
applied electric field. The Einstein relation, $D = \mu kT$, relating the
diffusion constant D to the electric mobility μ, is closely satisfied [2].

This picture, while aparently reliable, was soon subject to basic
questions of theory. It was recognized that the Boltzmann equation,
which supports the model, is heuristic only. In 1957 Van Hove analysed
electric conductivity using higher order perturbation theory.[3] In
this work he was obliged to set $\lambda^2 t$, λ^2 the electron-phonon coupling
constant, t the elapsed time, equal to a finite constant, the so-
called "van Hove" limit. Many workers have used the Kubo formula [4].
Here an ad hoc dissipation term proportional to a parameter E is
introduced, and, subsequently, to find the conductivity, a selection
of Feynmon graphs must be made. This leads to a convergent series
in (λ^2/E), which latter parameter must be considered as bound[5].
Among recent promising work using the Kubo formula we mention that
of Götze[6].

Going back somewhat, we remember that Feynman and his co-workers
pointed out in relation to the Boltzmann equation that at anything
but vanishing coupling strength the interaction of the electron with
its scattering agents can not be resolved into single scattering
event.

An attempt to deal with the problem in terms of the path-integral
quantum mechanics of Feynman was published by Feynman, Hellwarth,
Iddings and Platzman in 1962 [7]. Feynman at al eliminated the lattice

mode coordinates with which a semiconducting electron is interacting, thereby obtaining a representation for the electron density as a double path-integral. A considerable effort has gone into obtaining reliable information from this exact reformulation[8]. Dr. F. Brosens will be reporting here at this institute on the work of Prof.Devreese's group which has been very active in this field.

Semiconductor technology has recently advanced in many directions, making the need for theoretical understanding more urgent. Devices are manifesting quantum effects simply by becoming smaller. A new direction in semiconductor technology is the use of amorphous materials, leading to the problem of strong coupling between electron and its scatters.

Our own work, I should like particularly to mention as collegue Hans Beck in Neuchatel, Switzerland, is also a derivative of the FHIP formulation. We were, first, interested in using this exact development to directly provide an understanding of the basic Haynes-Shockley picture.[9] Later we found a way of incorporating static crystal disorder into FHIP, which was developed to deal with scattering by phonons [10]. This has led us to seek to extend our work to the stronger coupling situations that apply in amorphous semiconductors.

FHIP belongs the "polaron" tradition. The polaron is a single electron accompanied by a cloud of phonons which it itself excites. Now, the quintessence of the polaron problem is a nonMarkovian character; the electron influences its host lattice in a manner depending on its dynamic history, i.e. development of its position and velocity in time: and the back reaction of the lattice then depends on this past "path" of the electron. How come, than, that the Einstein relation, a hall mark of Markovian propagation, holds? On the other hand, how come that in some selected disordered materials the dynamics, although also leading to systematic results, appears to be quite different?

Our analytic development is outlined in the next section. Just as the Feynman path-integral for a wave-function can be converted to the Schrödinger equation, so too can the double-path integral in FHIP for the density matrix be converted into an integral-differential equation of motion (i.-d. equation). We convert this equation into the coordinates $\vec{r}$ and $\vec{u}$, space and velocity, using the prescription of Wigner, and insert as ansatz a solution inspred by classical Grownian motion. We solve the equation by the method of Fourier transforms, using an aproximation which involves expanding both sides of the equation in Taylor series in the wave vectors, and equating coefficients.

In the third and last section we discuss the solution to the resulting time-dependent subsidiary equations for both an idealised model (cut-off model), and for real situations corresponding to the relatively weak coupling in crystalline semiconductors, and the stronger coupling of amorphous semiconductors.

2. Obtaining The Time-Dependent Equations

In 1932 E.Wigner introduced a formalism which gives quantum mechanics a more classical physics appearance [11]. He suggested rewritting a density matrix $\rho(R,R')$ by first defining variables

$$\vec{r} = (\vec{R} + \vec{R}')/2 \qquad \vec{y} = (\vec{R} - \vec{R}') \qquad \text{(1A,B)}$$

and then transforming $\rho(R,R')$ to $\rho_w(r,u)$

$$\rho_w(\vec{r},\vec{u}) \equiv (m/h)^3 \int_{-\infty}^{\infty} e^{\frac{im}{\hbar}(\vec{y}\cdot\vec{u})} \rho(\vec{r} + \vec{y}/2, \vec{r} - \vec{y}/2) d^3y \quad \text{(1C)}$$

Although ρ_w has no direct phyisical meaning, one can easily show that $\int \rho_w(\vec{r},\vec{u}) d^3r$ gives the distribution in $\vec{u}$, and $\int \rho_w(\vec{r},\vec{u}) d^3u$ in $\vec{r}$.

The FHIP double path integral can likewise formally be converted to the variable r and u . We shall refer to the integral of this double path integral as ρ_w^{open}, and to the path integral itself as ρ_w^{closed}. The relationship between these quantities is:

$$\rho_w^{closed} \equiv \rho_w(\vec{r},\vec{u}; \vec{r}_o,\vec{u}_o; t) =$$

$$\int_{-\infty}^{\infty} \rho_w^{open}(\vec{r}_N,\vec{u}_N \ldots \vec{r}_I,\vec{u}_I \ldots \vec{r}_o,\vec{u}_o; t) d^3r_{N-1} d^3u_{N-1} \ldots d^3r_o d^3u_1 \quad \text{(2)}$$

Here we have divided our time interval 0 to t into N intervals, ending respectively at times $t_1 \ldots t_I \ldots t_N$: $t_N = t$. A path starts at u_o, r_o at time t_o, goes thorugh r_I, u_I at t_I, and ends at time t at r_N, u_N. As we have said in the introduction the FHIP integral can be reduced to an i.d. equation for the density matrix which the integral represents. This equation, when converted to $\vec{r}$ and $\vec{u}$ etc. is of the form:

$$\frac{\partial \rho_w^{open}}{\partial t} = \lim_{N\to\infty} \sum_{I=1}^{N} \int V(\vec{r},\vec{u}; \vec{r}_I,\vec{u}_I; t - t_I) \rho_w^{open} d^3u_1 \ldots dr_{N-1} \quad \text{(3)}$$

It, thus, requires ρ_w open at at least one intermediate time. Just as with Green's functions we can set up a hiearchy of equations, relating ρ open at one intermediate time to ρ_{open} at two times, and so on. We shall find it convenient to represent ρ^{open}_{open}, and work with it.

It has also been found convenient to use Fourier representation. If we are interested only in the macroscopic spatial aspect of the problem, for instance the motion of electrons in an electric field over macroscopic times comprising many relaxation times, we can confirm our attention to small wave-length components. If the electron's spatial distribution satisfies a Gaussian distribution

$$\exp -\left[(\vec{r}'-\vec{r})^2/(4Dt)\right]$$

we shall want Fourier wave vectors of the order of $1/\sqrt{\Omega_\infty t}$, where Ω_∞ is a relaxation time, in crystalline materials $\sim 10^{-13}$ sec. at room temperature. We, accordingly, expand both sides of our i.d. equation of motion in Taylor series in wave vectors k and $\vec{w}$, conjugate to variable $\vec{u}$ and $\vec{r}$, respectively, and stop at quadratic order. Our equation for the distribution function $g_w(\vec{r},\vec{u};t)$ will ultimately have the form.

$$\partial g_w(\vec{r},\vec{u};t)/\partial t = \left[\text{quadratic form in } \vec{k}+\vec{w}\right] g \qquad (4)$$

with the solution of the form $\rho_w \propto \exp\left[\int(\text{form})\,dt\right]$, a Gaussian. We further expect that the spatial distribution $\int g_w \, d^3u$ will in some way resemble the mentioned distribution $\exp -\left[(\vec{r}-\vec{r}')^2/(4Dt)\right]$, which is what experiments yield for electrons in crystalline semi-conductors. We have the further input that the Einstein relation is closely satisfied, and one also rather expects a Maxwellian velocity distribution.

Now, as it happens all these features, the Gaussian spatial distribution, the velocity distribution, the Einstein relation, are found for a colloid particle in a liquid-Brownian motion. Let us write down as a Fourier representation the well known[12] distribution function $W(\vec{r},\vec{u};t)$ of Brownian motion:

$$W(\vec{r},\vec{u};t)=(2\pi)^{-6}\iint d^3k\,d^3w\,e^{i\vec{k}(\vec{u}-\vec{u}_0 e^{\beta t})}\,e^{i\omega(\vec{r}+\vec{u}/\beta-\vec{r}_0-\vec{u}_0/\beta)}\,e^{-V(\vec{k},\vec{w};t)/2} \qquad (5A)$$

$$V(\vec{k},\vec{w};t) = \left[b(t)k^2 - 2h(t)\vec{k}\cdot\vec{w} + a(t)w^2\right] \qquad (5B)$$

$$a(t) = \frac{2q}{\beta^2}\int_0^t dt = \frac{2q}{\beta^2}t \qquad (5C)$$

$$b(t) = 2q e^{-2\beta t} \int_0^t e^{2\beta t} dt = \frac{q}{\beta}\left(1 - e^{-2\beta t}\right) \tag{5D}$$

$$h(t) = -2q/\beta \, e^{-\beta t} \int_0^t e^{\beta t} dt = -\frac{2q}{\beta^2}\left(1 - e^{-\beta t}\right) \tag{5E}$$

Here q is a parameter which measures the velocity fluctuations, and β the viscous drag on the colloid particle. At long times, in real phase space $W(\vec{r}, \vec{u}; t)$ becomes:

$$\lim_{t \to \infty} W(\vec{r}, \vec{u}; t) \propto \exp\left[-(\vec{r} - \vec{r}_0)^2/(4Dt)\right] \exp\left[-mu^2/(2kT)\right] \tag{6}$$

where $D = q/\beta^2$ and $q/\beta = kT/m$. The form of a(t), b(t), h(t) in Eqs. (5) as integrals over time can be traced back to their being the solution of a diffusion equation. Our ultimate aim is to use $W(r, u:t)$ to inspire us to find the appropriate ansatz for ρ_w in our quantum mechanical problem of the scattered electron. Since we shall need ρ_w^{open}, we inquire if there is naturally such a thing as W^{open}?

Physically, a Markovian process is one in which the time evolution of a particle depends only on the present, and not the past. Mathematically, a Markovian propagator must satisfy:

$$W(\vec{r}, \vec{u}; \vec{r}_0, \vec{u}_0; t) = \int W(\vec{r}, \vec{u}; \vec{r}_I, \vec{u}_I; t - t_I) W(\vec{r}_I, \vec{u}_I; \vec{r}_0, \vec{u}_0; t_I) \, d^3\vec{r}_I \, d^3 u_I \tag{7}$$

The Fokker-Planck equation, of which W is the solution, is derived on the assumption of Markovian propagation. We, therefore, expect W to satisfy Eq.(7), and, indeed, the Fourier representation of W as given by Eqs.(5) allows us to verify this rather easily. As a matter of fact it will be convenient to generalize the property expressed by Eq.(7) by writing down W^{open} directly, i.e. opening everywhere:

$$W^{\text{open}}(\vec{r}_N, \vec{u}_N, \dots \vec{r}_I, \vec{u}_I, \dots \vec{r}_0, \vec{u}_0; t) = \frac{1}{(2\pi)^{6N}} \int_{-\infty}^{\infty} g_1^3 \, dk_L \cdots g_N^3 \, dk_N \int_{-\infty}^{\infty} d^3 w_1 \cdots d^3 w_{N-1}$$

$$\times e^{-V_0/2} \prod_{I=1}^{N} \frac{1}{\pi} e^{i \vec{k}_I \cdot (\vec{u}_I g_I - \vec{u}_{I-1} g_{I-1})} \prod_{I=1}^{N} e^{i \vec{w} \cdot (\vec{r}_I + \vec{u}_I/\beta - \vec{r}_{I-1} - \vec{u}_{I-1}/\beta)} \tag{8A}$$

$$V_0 = \sum_{I=1}^{N} V_I \tag{8B}$$

$$V_I = b_I k_I^2 - 2h_I \vec{k}_I \cdot \vec{w}_I + a_I w_I^2 \tag{8C}$$

$$a_I = (2q/\beta^2)(t_I - t_{I-1}) \tag{8D}$$

$$b_I = 2q \, e^{-2t\beta} \int_{t_{I-1}}^{t_I} e^{2\beta t} \, dt \tag{8E}$$

$$h_I = -2q/\beta \, e^{+t\beta} \int_{t_{I-1}}^{t_I} e^{\beta t} \, dt \tag{8F}$$

$$g_I = e^{-\beta(t-t_I)} \tag{8G}$$

If we now "close" Eq. (8A) at t_z, by integrating over u_1 and r_1, we find an integral

$$(2\pi)^{-3} \int e^{-i\vec{k}_2 \vec{u}_1 g_1} \, e^{i\vec{k}_1 (\vec{u}_1 g_1 - \vec{u}_0 g_0)} \, d^3 u_1 = \delta^3 (g_1 \vec{k}_2 - g_1 \vec{k}_1) e^{-i\vec{k}_2 g_0 \vec{u}_0} \tag{9}$$

Similarly, we shall ultimately find that $w_2 = w_1$.
Clearly, if we close Eq. (8A) over all intermediate variables, $I = 1,2 \ldots N-1$, we get back Eqs. (5). If we close over all intermediate variables except I, viz over $I' = 1,2 \ldots I-1, I+1, \ldots N-1$, we get exactly Eq. (7), thereby confirming the Markovian character of W. W^{open} corresponds to the kernel on the right of Eq. (7) when we generalize to an equation where we close explicitly over all intermeadiate variables.

We return to our quantum mechanical problem of an electron scattered by dynamic and static disorder. The quantum analogue of Eqs. (8), to be used as ansatz is:

$$g_W^{open} (\vec{r}_N, \vec{u}_N, \ldots \vec{r}_I, \vec{u}_I \ldots \vec{r}_0, \vec{u}_0; t) = (2\pi)^{-6N} \int_{-\infty}^{\infty} f_1 \, d^3 k_1 \cdots f_N \, d^3 k_N$$

$$\times \int_{-\infty}^{\infty} d^3 w_1 \ldots \int d^3 w_N \, e^{-F/2} \prod_{I=1}^{N} e^{i\vec{k}_I (\vec{u}_I f_I - \vec{u}_{I-1} f_{I-1})} \prod_{I=1}^{N} e^{i\vec{w}_I (\vec{r}_I + \chi_I \vec{u}_I - \vec{r}_{I-1} - \chi_{I-1} \vec{u}_{I-1})} \tag{10A}$$

$$F = \sum_{\substack{J=1 \\ I > J}}^{N-1} (A_{IJ} \vec{w}_I \vec{w}_J - \overset{o}{H}_{IJ} \vec{k}_I \vec{w}_I - \tilde{H}_{IJ} \vec{k}_I \vec{w}_J + B_{IJ} \vec{k}_I \vec{k}_J) \tag{10B}$$

In contradictinction to Eqs (8), Eq. (10) allows for nonMarkovian correlation between the present and the past, viz in the nondiagonal components of Eq. (10B). At the same time χ_J, which replaces $1/\beta$ in Eqs (8) can depend on the time t_I, and the functions f_I need not be exponentials. In fact they are ratios of functions of s and t, see below.

Various stratagems can be employed to find all the elements $A_{IJ}, \overset{o}{H}_{IJ}, \tilde{H}_{IJ}, B_{IJ}$. However, if we do not want to work explicitly with p open at various times, some simplifying principle or aproximation must be found. The elements $\overset{o}{H}_{IJ}, \tilde{H}_{IJ}, B_{IJ}$ are to be regarded as already compensated for the scaling in time of u_I, as built into

Eq.(10A). (The A_{IJ} need not be modified, and the uncompensated $\overset{\circ}{H}_{IJ}$, H_{IJ} are equal, see **II**.) . We suggest that the uncompensated elements be taken to be functions of only $(t_I - t_J)$. The original FHIP double path integral which we are representing is made up of elements which themselves depend only on time differences, viz. here $t_I - t_J$. Discussion of whether a similar time dependence for our elements is to be viewed as an aproximation, or as a necessity is beyond the scope of this lecture.

Our final equations will be given in terms of three functions of t and s, rather than in terms of the individual elements. We shall often use s to denote a typical time such that $t > s > 0$. Thus, $t_J = s$ below, in defining the functions:

$$A_1(t,s) = \sum_{I=1}^{\mathcal{J}} \sum_{I'=1}^{I} A_{I I'} \tag{11A}$$

$$A_2(t,s) = \sum_{I=\mathcal{J}+1}^{N} \sum_{I'=\mathcal{J}+1}^{I} A_{I I'} \tag{11B}$$

$$A_3(t,s) = \sum_{I=\mathcal{J}+1}^{N} \sum_{I'=1}^{\mathcal{J}} A_{I I'} \tag{11C}$$

$$A(t) = A_1(t,s) + A_2(t,s) + A_3(t,s) \tag{11D}$$

Similarly, we shall have functions $\tilde{H}_1(t,s)$, $\tilde{H}_2(t,s)$, $\tilde{H}_3(t,s)$ etc. We note that it is the functions A_3, B_3, H_3 which are the instruments of the nonMarkovian correlations between past and present.

We substitute the ansatz in Eqs.(10) for ρ^{open} into our i.d. equation of motion, close, (i.e. integrate over all intermediate variables $\vec{u}_I$, $\vec{r}_i$) and systematically expand both sides of the equation in Taylor series in the wave vectors $\vec{k}$ and $\vec{w}$, up to quadratic order. With constant terms vanishing, we are left with five equations from the k^2, $\vec{k}.\vec{w}$, w^2, $\vec{k}$ and $\vec{w}$. coefficients respectfully, for the five unknown functions A(t), B(t), H(t), n(t), X(t), where H(t) is $(\overset{\circ}{H} + \tilde{H})/2$. The equations are:

$$\frac{\dot{B}}{2} + \Omega B = Z + M/x_t \tag{12-I}$$

$$Z = G_2 \left[H_2 + x_t B_2 \, v(t,s) \right] - G_1 \left[n(s)/n(t) \right] \tag{12-IA}$$

$$M = \tfrac{1}{2} G_2 \left[B_3 \, x_t^2 \, v(t,s) + x_t \tilde{H}_3 \right] \tag{12-IB}$$

$$f_t(s) = n(s)/n(t) \tag{12-IC}$$

$$v(t,s) = (x_s/x_t)/f_t(s) \tag{12-ID}$$

$$\Omega(t) = G_2\left[X_t\left(v(t,s)-1\right)\right] \tag{12-IE}$$

$$\dot{H} + \Omega H + \dot{X}B + B - X_t\,\Omega B = Y - X_t Z + [N-M] $$

or, using Eq.(V)

$$\dot{H} + \Omega H = Y - X_t Z + [N-M] \tag{12-II}$$

$$Y = G_1[X_s] + G_2\left[A_2 + X_t H_2\, v(t,s)\right] \tag{12-IIA}$$

$$N = \tfrac{1}{2}G_2\left[X_t H_3\, v(t,s) + A_3\right] \tag{12-IIB}$$

$$H = (\tilde{H} + \overset{\circ}{H})/2 \tag{12-IIC}$$

And,

$$-\frac{\dot{A}}{2} - H - \dot{X}_t H + \Omega x H = X_t Y + X_t N$$

or, using Eq.(V)

$$-\frac{\dot{A}}{2} = X_t Y + X_t N \tag{12-III}$$

We also have

$$\dot{n}(t)/n(t) = \Omega(t) \tag{12-IV}$$

and

$$1 + \dot{X}_t = \Omega(t)\, X_t \tag{12-V}$$

$$G_1[\quad] \text{ means } \int_0^t ds \sum_k G_{1k}(t-s)[\quad] \tag{13A}$$

$$G_2[\quad] \text{ means } \int_0^t ds \sum_k G_{2k}(t-s)[\quad] \tag{13B}$$

i.e. integration on time s, summation over all wave-vectors $\vec{K}$ of the phonon modes and the static disorder. These wave-vectors are dimension-less quantitites, measured in units of 1^{-1}, $1 = v_{\text{thermal}}/\beta$ $\beta = \Omega(\infty)$. Further,

$$G_{1k}(t-s) = \frac{2 l^2 k^2}{3}\left[S_{1k}(t-s)\sin \Lambda_k - S_{2k}(t-s)\cos \Lambda_k\right]\exp\left[-k^2 R(t,s)/2\right] \tag{13C}$$

$$G_{2k}(t-s) = \frac{2L^2k^2}{3} \left[S_{1k}(t-s)\cos\Lambda_k + S_{2k}(t-s)\sin\Lambda_k \right] \exp\left[-k^2 R(t,s)/2 \right] \quad (13D)$$

$$R(t,s) = A_2 + 2H_2 + B_2 + (1-n(t)/n(s))(H_3+B_3) + (1-n(t)/n(s))^2 B_1 \quad (13E)$$

$$L = \hbar\beta/(2k_BT) \quad (13F)$$

$$\Lambda_k = Lk^2(1-n(s)/n(t)) \quad (13G)$$

Finally,

$$(S_{1k}, S_{2k}) = \frac{2}{\hbar^2} |C_k|^2 \left(\sin\omega_k(t-s), \frac{\xi_k}{i}\cos\omega_k(t-s) + \hat{S}_k \right) \quad (13H, I)$$

$$\xi_{\mathbf{R}} = 2\bar{n}_{\vec{R}} + 1, \quad \bar{n}_{\vec{R}} = \text{no. of phonons is mode } \vec{K} \quad (13J)$$

$$\hat{S}_k = (2\hbar^2 \Omega_o \Omega)^{-1} |\hat{\phi}(k)|^2 S_F(k) \quad (13K)$$

where $S_F(K)$ is the structure factor, and $\hat{\phi}(K)$ the Fourier transform of the electron-ion potential, and $\Omega_o = \Omega/N$ is the atomic volume. Incidentally, the units of S_{1K}, S_{2K} are (reciprocal time)^2.

These equations must be consistent with physical continuity. In the Wigner scheme the usual equation $\partial g/\partial t + \vec{\nabla} \cdot \vec{\mathcal{S}}$ equals zero, where $\vec{\mathcal{S}}$ is the flux, takes the form

$$\frac{\partial}{\partial t} \int g_w(\vec{r}, \vec{u}; t) \, d^3u + \int \vec{u} \cdot \vec{\nabla}_r g_w(\vec{r}, \vec{u}; t) \, d^3u = 0 \quad (14)$$

We substitute from Eq. (10A), expand in wave vectors, and close. Integrating over $\vec{u}$ yields $\vec{k} = -\vec{w}x_t$. Equating quadratic coefficients one gets.

$$\dot{H} + \frac{\dot{B}x_t}{2} + \frac{A}{2x_t} + \Omega(x_t B + H) = 0 \quad (15)$$

and from the linear coefficients:

$$-x_t \frac{\dot{n}}{n} + (1 + \dot{x}_t) = 0 \quad (16)$$

Eq. (15) can be readily obtained by combining Eqs. (12-I), (12-II) and (12-III), and Eq. (16) by combining Eqs. (12-IV) and (12-V).

Finally, applying an electric field, we find, as in I, that with $\vec{K} = e\vec{E}/m$, we must replace everywhere variables

$$\vec{r} + \vec{u}x_t \quad \text{by} \quad \vec{r} + \vec{u}x_t - K\int x_t dt \quad (17A)$$

$$\vec{u} \quad by \quad \vec{u} \; - \; \frac{\vec{K}}{n(t)} \int_0^t ds \, n(s) \tag{17B}$$

thereby identifying the electric mobility as

2. Discussion

The key to solving Equations (12) is the function $\Omega(t)$, and $\Omega(t)$ is largely governed by R/t,s) in the exponents of $G_{1K}(t-s)$, $G_{2k}(t-s)$, Eqs. (13-C,D). For a Markovian situation $exp-\left[\frac{KR(t,s)}{2}\right]$ turns out to be very nearly

$$exp-\left[\frac{(K\ell)^2}{2}(t-s)^2 \, \Omega^2(\infty)\right],$$

where l is the Einstein length as already given, and, again, K is the phonon or disorder wave vector. $\Omega(t)$ is presumed to go to a definite limit $\Omega(\infty)$. We need to juxtapose this exponential with n(t)/n(s) as found in Eqs. (12-ID), (12-IE), which define $\Omega(t)$. Again, in the Markovian situation n(t)/n(s) will turn out to be $exp\left[\Omega(\infty)(t-s)\right]$. The wave numbers K diminish from some maximum number, we shall call K_B, to zero. Let us consider a model, which we shall refer to as the cut-off model, in which we exlude all wave-numbers less than $K_0 \equiv \sqrt{2}/l$, so we have $K_B > K > K_0$. In this model $\left(exp-[R(t,s)]\right)$ x $\left(n(t)/n(s)\right) < 1$, and $\Omega(t)$ goes to a finite limit as $t \to \infty$, even for an exponentially increasing function as choice for n(t).

The cut-off model is not entirely academic. For a typical crystalline semi-conductor at room temperature $K_0/K_B \lesssim 10^{-2}$, so that only about 10^{-6} or less of the Brillouin zone is excluded. However, in disordered situations the Einstein length is much shorter. Even if we assume that the influence on the mobility of the long wave scattering modes is less than proportional to the volume in reciprocal space they accupy, still we see that the cut-off model must become progressively less representative as the disorder increases.

Limiting ourselves to the cut-off model we have found using Eqs. (12-IE,-IE), (13D), that with an exponentially decaying f(s) = n(s)/n(t), $\Omega(t)$ repidly increases from 0 to its asymptotic value $\Omega(\infty)$ with t. In our calculations in I, and in our discussion here, we ignore the initial transient build up and set

$$\frac{\dot{n}(t)}{n(t)} = \Omega(\infty) \tag{18.A}$$

The transient region would be of special importance in considering so-called ballistic transport.

Eqs.(12-I) to (12-V) now have as their unique solutions see II, exactly the same functional form for A(t), B(t), H(t), f(t) and χ as we encountered with classical Brownian motion. Letting $\Omega(\infty) \equiv \beta$, we get from Eq.(12-V) that $\chi(t)$ is a constant equal to β^{-1}, from Eq.(18A) that $n(t) = e^{\beta t}$, and from Eqs.(12) that

$$B(t) = (Z/\beta)(1 - e^{-2\beta t})$$

$$H(t) = (Y/\beta - Z/\beta^2)(1 - e^{-\beta t})$$

$$A(t) = -2Yt/\beta$$

and $B_3 = H_3 = A_3 = N = M = 0$

These solutions agree fully with the simple Haynes-Shockley picture as observed in the early '50's. We have the Gaussian diffusion, the Maxwellian velocity distribution, the Johnson noise spectrum. The details of the formula for the electron mobility $e/(\Omega(\infty)m)$, agree closely with the standard golden-rule result, see III, and the fomula was found to give in I very good results as regards both absolute value and temperature dependence. Lastly, computer calculations showed in I that the Einstein result $D/\mu = a_0 kT$ is forthcoming right down to the observed discrepancy of $a_0 \sim 1.02^{(2)}$. (a_0 can be calculated rather well, because details yartaining to a model tend to cancel out).

For completeness and to get at stronger coupling we must consider how to include scattering by waves with lengths beyond the cut-off. In this case, combining Eqs.(12-IC,ID,IE) with Eq.(12-IV), we have, instead of Eq.(18A)

$$\dot{n}^{-1}(t) = -G_2 \left[\chi_s/n(s) - \chi_t/n(t) \right] \tag{18B}$$

which does not contain any divergences.[*] Instead n(t) will increase slightly more slowly than its cut-off $\exp[\beta t]$ behavior, leading again to a well defined $\Omega(\infty)$ limit. Departures in n(t) from exponential have the further consequence that Y and Z now become

---

[*]Eq(18B) tends itself to Laplace transform techniques since G_2 is calling for integration over s, while introducing only functions of (t-s).

functions of t. As a result the functions A,B,H (A determining
diffusion in space, B the noise spectrum, H assuring physical
continuity) loose their Markovian character, and B_3, H_3, A_3 no
longer vanish. At weak coupling these effects are not expected
to have tangible results.

What happens as the coupling becomes stronger? Montroll and
Sher have identified a dynamic situation which is observed in
many disordered crystals[13]. In these one finds that the ratio
of distance gone in an electric field to the width of the spatial
distribution:

$$\vec{eE}/m \int_0^t \mu(t)dt / [D(t)t]^{1/2}$$

D a possible time dependent diffusion constant, is constant rather
than increasing as $t^{1/2}$ as it does in crystalline situations
where D does not depend on t.

The time dependence of the non-Markovian contributions N and
M, driving terms in Eqs. (I), (II) and (III), becomes of critical
interest. When do these terms become constants and linear functions of t, rather than diminishing functions, as observations
imply at weak coupling? As the coupling becomes stronger the relative contribution to $\Omega(t)$ of the modes with $K > K_o$, which
yields a well defined constant as $t \to \infty$ diminishes. If one
believes that for the situation considered by Montroll and Sher
there is still a well defined mobility $\chi(\infty)$, then the picture
one gets of $\Omega(t)$ is that of a more volatile quantity, oscillating, perhaps, indefinitely about $\chi^{-1}(\infty)$. This behaviour would
be reflected in a greater deviation of n(t) from an exponential,
and enhance the nonMarkovian terms N and M. In particular, a
linear time dependent component in N(t) will lead to the observed
$t^{1/2}$ time dependence of D.

In view of the above discussion one rather expects that at
still greater coupling $\Omega(t)$ assumes some unstable behavior as
$t \to \infty$. Thus, as K_o/K_B aproaches unity, the region of the
macroscopic aproach suggested have probably comes to an end. In
conclusion, our analysis suggests that the Markovian dynamics,
which seems to heavily predominate in good single crystals, is
easy of access. At stronger coupling, as the situation becomes
strongly nonMarkovian, the phenomena become as varied and interesting as they are difficult to calculate.

ACKNOWLEDGEMENT - It is a pleasure to thank my co-worker, Prof. Hans Beek at the University of Neuchatel, for many very helpful discussions.

REFERENCES

1. J.R.Haynes and W.C.Westphal
   Phys.Rev. $\underline{85}$, 680, (1952)

2. Transistor Teachers Summer School
   Phys.Rev. $\underline{88}$, 1368 (1952)

3. L.van Hove,
   Physica $\underline{21}$, 517 (1955)

4. R.Kubo,
   J.Phys.Soc.Jpn. $\underline{12}$, 570, (1957)

5. M.Huberman and G.V.Chester,
   Adv.in Physics $\underline{24}$, 489, (1973)

6. W.Götze,
   Philos,Mag. $\underline{B43}$, 219, (1981)

7. R.P.Feynman, R.W.Hellwarth, C.K.Iddings, and P.M.Platzman
   Phys.Rev. $\underline{127}$, 1004, (1962)

8. J.T.Devreese and R.Evrard
   Phys.Status Solidi $\underline{B78}$, 85 (1976)

9. S.Nettel and S.Anlage
   Phys.Rev. $\underline{B26}$, 2076 (1982) hereafter referred to as I, and
   S.J.Nettel and H.Beck, Phys.Rev, $\underline{B28}$, 4535 (1983), hereafter
   referred to as II in the text.

10.H.Beck and S.Nettel
   Phys.Lett. $\underline{105A}$, 319 (1984)

11.E.Wigner,
   Phys.Rev.$\underline{40}$, 749 (1932)

12.S.Chandrasekhar,
   Rev.Mod.Phys. $\underline{15}$, 1 (1943)

13.E.Montroll and R.Sher
   Phys.Rev. $\underline{B12}$, 2455 (1975)

CYCLOTRON RESONANCE OF QUASI-TWO-DIMENSIONAL POLARONS

J.T. Devreese* and F.M. Peeters[0]
University of Antwerp (U.I.A.), Department of Physics
Universiteitsplein 1, B-2610 Antwerp, Belgium

A theorerical survey will be given of the cyclotron resonance of *quasi-two-dimensional polarons* in heterostructures. One polaron and many-polaron effects will be discussed. The cyclotron resonance absorption spectrum of a quasi-two-dimensional many-polaron system is calculated with inclusion of the full *dynamical screening* of the electron-phonon interaction. Experimental results of several groups on the cyclotron mass in $GaAs/Al_xGa_{1-x}As$ heterostructures are well described. We found that for electron densities larger than $1.4 \times 10^{11} cm^{-2}$ the polaron mass renormalization is smaller than for the corresponding bulk GaAs system.

1. Introduction

Recently there has been increasing interest in the study of 2D quantum systems[1]. New effects have been discovered like e.g. the quantum Hall effect[2] and the fractional quantum Hall effect[3], the Haas-van-Alphen effect[4], ... Thanks to novel sample growth techniques (e.g. molecular beam epitaxy(MBE)) enormous improvements have been achieved in increasing the scattering time and in mastering the art of *band engineering*. This led already to a number of applications in the field of optical communications, opto-electrons, new fast transitors (e.g. HEMT),...

Many of the 2D structures are made out of III-V compound semiconducting materials which are weakly polar in nature. Examples of such systems are $GaAs-Al_xGa_{1-x}As$ heterostructures and superlattices, p-InSb inversion layers, $Ga_xIn_{1-x}As$ heterostructures, InAs-GaSb systems, ... In those weakly polar semiconductors electron-LO-phonon interaction is important and polaron effects may be observable.

Spatial confinement of electrons to a quasi-two-dimensional(Q2D) layer[1] will influence the electron-phonon interaction. Transport measurements (mobility, hot electron relaxation,...) gives information on *dissipation* as due to the electron-phonon interaction and is theoretically related to the imaginary part of the self-energy. On the other hand cyclotron resonance experiments, in particular the position of the cyclotron resonance peak, gives information on *non-dissipative* aspects of this interaction like polaron mass renormalization and coupling between Landau levels. This aspect of the interaction is often described by the real part of the self-energy. In this lecture we will mainly discuss the non-dissipative aspects of the electron-phonon interaction when a magnetic field is present and when the electrons are confined to a two dimensional(2D) layer with a thickness d.

One polaron theories[5,6] predict that polaron effects should be larger in confined systems than in the equivalent bulk material. This is for example evident from the approximate *scaling relations*[7,8] for the polaron ground state energy

$$E_{2D}(\alpha) = \frac{2}{3}E_{3D}\left(\frac{3\pi}{4}\alpha\right) \quad , \tag{1}$$

and the polaron mass

$$m_{2D}^*(\alpha) = m_{3D}^* \left(\frac{3\pi}{4} \alpha \right) \quad . \tag{2}$$

From these relations the 2D-result can be obtained from the corresponding 3D-result. These scaling relations are exact up to second-order perturbation theory and are approximately valid for arbitrary electron-phonon coupling strength α. The Feynman polaron approximation satisfies exactly Eqs. (1) and (2) (note that these equations are derived in the limit of zero magnetic field). The trends as given by the one polaron theory[9-11] seems to be confirmed by cyclotron resonance experiments[12,13] in InSb-inversion layers. But in recent studies[14-18] it was found that many-particle aspects of the system are very important and will reduce the polaron effects considerably.

This is different from 3D systems where the one particle approximation in e.g. bulk GaAs[19] is a very good approximation. The electron density is typically of the order of $n_e \sim 10^{13} cm^{-3}$ which leads to a Fermi energy $E_F \ll k_b T$ and consequently Boltzmann statistics is valid. In typical GaAs-heterostructures the electron density is $n_e \sim 10^{11} - 10^{12} cm^{-2}$ which for a typical electron layer thickness of $100 \mathring{A}$ results in an effective 3D electron density which is 4-5 orders of magnitude larger than for typical 3D systems. The resulting Fermi energy satisfies $E_F \gg k_b T$ and consequently the electrons will obey Fermi-Dirac statistics. The occupation probabilities of the Landau states is important and screening of the electron-phonon interaction may influence the results.

Often[9-11,20-21] the cyclotron resonance mass is obtained from the position of the Landau levels(E_n) in the following way: $m^* = m_b \omega_c / \omega_c^*$ (m_b is the electron band mass and $\omega_c = eH/m_b c$ is the cyclotron resonance frequency for a non-interacting electron) where the cyclotron resonance frequency is determined by $\omega_c^* = (E_1 - E_0)/\hbar = \omega_c - (\Delta E_1 - \Delta E_0)/\hbar$ with ΔE_n the electron-phonon correction to the position of the nth Landau level.

In this lecture we will mainly advocate a different approach and calculate the *magneto-optical absorption spectrum* which, in a cyclotron resonance experiment, is the experimental measured quantity. A full *dynamical screening* theory will be presented and applied to the GaAs-heterostructure system. No phenomenological parameters are introduced in the calculation.

2. The Landau levels of a 2D polaron

First we will study the more simple situation where *one* electron is restricted to move in a plane and a magnetic field is applied perpendicular to it. The energy of such an electron will be quantized into Landau levels $E_n = \hbar \omega_c(n + \frac{1}{2})$ with $n = 0, 1, 2, ...$ and $\omega_c = eH/m_b c$ is the cyclotron resonance frequency of a band electron with mass m_b. Due to the weak interaction of the electron with the LO phonons the landau levels will be shifted by ΔE_n. Within second-order perturbation theory the energy shift of the nth Landau level is given by

$$\Delta E_n = - \sum_{m=0}^{\infty} \sum_{q} \frac{|M_{n,m}(\mathbf{q})|^2}{D_{n,m}} \quad , \tag{3}$$

where

$$M_{n,m}(\mathbf{q}) = < m | \oplus < \mathbf{q} | H_I | 0 > \oplus | n > \quad , \tag{4}$$

is the matrix element of the electron-phonon interaction operator H_I between the quantized electron states $|n>$ and the single phonon states $|\mathbf{q}>$. The electron-phonon interaction term in the Hamiltonian is

$$H_I = \sum_{\mathbf{q}} \left(V_q a_q e^{i\mathbf{q}\cdot\mathbf{r}} + V_q^* a_q^+ e^{-i\mathbf{q}\cdot\mathbf{r}} \right) \quad , \tag{5}$$

with $a_q(a_q^+)$ the annihilation (creation) operator of an LO phonon with momentum $\hbar\mathbf{q}$ and energy $\hbar\omega_{LO}$. The interaction coefficient is

$$|V_q|^2 = \frac{1}{q}\frac{\sqrt{2}\pi\alpha}{A} \quad , \tag{6}$$

in 2D where we used units such that $\hbar = m_b = \omega_{LO} = 1$ and A is the surface area of the system. The energy dominator in Eq.(3) is given by

$$D_{n,m} = \hbar\omega_{LO} - \Delta_n + \hbar\omega_c(m - n) \quad , \tag{7}$$

where the choice: (1) $\Delta_n = 0$ leads to Rayleigh-Schrödinger perturbation theory (RSPT) which gives accurate results for ΔE_n if $\omega_c \ll \omega_{LO}$; (2) $\Delta_n = \Delta E_n$ results in Wigner-Brillouin perturbation theory (WBPT) which can account for the *splitting* of the degenerate energy levels; and (3) $\Delta_n = \Delta E_n - \Delta E_n^{RSPT}$ gives an "improved" Wigner-Brillouin theory (IWBPT) as was developed in Ref.22 and further discussed in Ref.19. IWBPT gives the correct *pinning* behavior for small α.

In Eq.(3) one has to sum over an *infinite* number of intermediate states m. This sum converges very slowly as is indicated in Fig. 1 for the RSPT approximation in the case of $\alpha = 0.07$ and for a relative small magnetic field $\omega_c/\omega_{LO} = 0.1$. For larger magnetic field values similar results are obtained. Note that about 10^4 terms are needed in order to obtain reliable results for the groundstate energy. This clearly shows that it is not sufficient to limit ourselves to the first few intermediate Landau states as is often done. We also give the results in the case of a non-parabolic conduction band. As an example we took an electron in a Kane band where the non-perturbed Landau levels are given by

$$E_n = -\frac{E_g}{2} + \sqrt{\left(\frac{E_g}{2}\right)^2 + E_g\hbar\omega_c(n + \frac{1}{2})} \quad , \tag{8}$$

with E_g the band gap which for GaAs is equal to $E_g/\hbar\omega_{LO} = 41.4$. For $n \to \infty$ the non-parabolic energy increases like $n^{1/2}$ which is different from the parabolic case where a linear increase in n

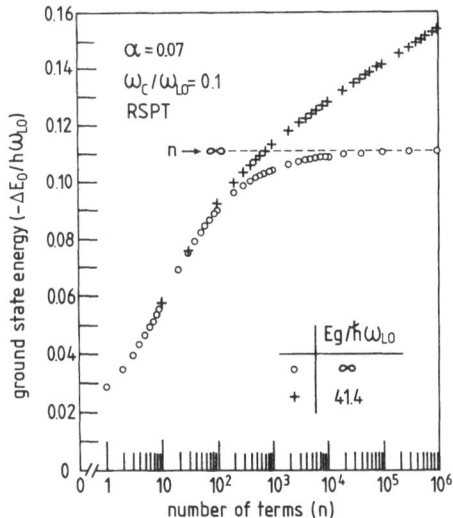

Fig. 1: The polaron ground state energy as function of the number of intermediate states (n) for a parabolic ('0') and a non-parabolic ('+') conduction band.

is found. This has the unfortunate consequence that the electron-phonon energy correction to the Landau levels is divergent! Of course in practical systems the sum over the intermediate states has to be cutoff at a certain m_{max} which reflects the finite size of the first Brillouin zone in real systems.

Fortunately the situation is much better if one looks at differences in the energy of the Landau levels. Note also that experimentally only differences in energies can be measured. The result is shown in Fig. 2 for the difference between the polaron correction of Landau level $n = 0$ and Landau level $n = 1$. Note that already 100 intermediate Landau levels are sufficient to obtain the $n \to \infty$ result with a reliable accuracy. Furthermore the non-parabolic result is finite but in this case more terms are needed in the sum in order to obtain the same accuracy as compared to the parabolic case.

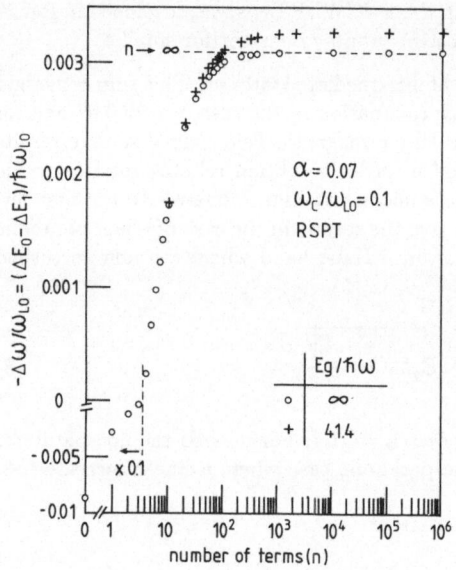

Fig. 2: Same as Fig. 1 but now for the difference in the polaron correction between the ground state and the first Landau level.

For a parabolic conduction band it is possible to performe the summation over the intermediate Landau levels explicitly. In the ideal 2D case a closed analytic expression can be obtained which has been calculated in Ref.11. The result is

$$\Delta E_n = -\frac{\pi}{2}\alpha\frac{1}{\sqrt{\omega_c}} \sum_{m=0}^{n} \binom{n}{m} \left(\frac{(2m-1)!!}{2^m}\right)^2 \frac{1}{m!} \cdot \frac{\Gamma((1-\Delta_n)/\omega_c - m)}{\Gamma((1-\Delta_n)/\omega_c + 1/2)} \quad , \tag{9}$$

where $\binom{n}{m}$ is the well-known combinatorics, $(2m-1)!! = (2m-1)(2m-3)...3\cdot1$ with $(2m-1)!! = 1$ if $m = 0$ and $\Gamma(x)$ is the gamma function. For $\Delta_n = 0$ Eq.(9) was also obtained by Larsen[10] for $n = 0, 1, 2$. In Fig. 3 the first three Landau levels are plotted as function of the magnetic field for an electron-phonon coupling constant $\alpha = 0.1$. The thin-dashed lines represent the unperturbed energy levels. The Landau levels for the 2D-polaron (thick, full curves) are compared with the corresponding 3D-polaron (thick, dashed curves) result. From Fig. 1 it is apparent that the polaron energy levels are: (1) shifted to lower energies, (2) at small magnetic fields they are bended downward due to a mass renormalization, (3) at $n\omega_c \simeq \omega_{LO}$ the nth Landau levels do not cross the LO-phonon+(E_0) level, and (4) for $\omega_c \to \infty$ all the Landau levels become pinned to $\hbar\omega_{LO} + \Delta E_0^{RSPT}$. When one compares the 2D result with the equivalent 3D result one notes that

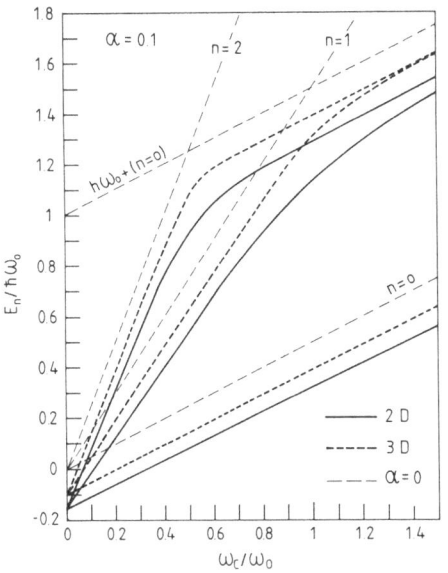

Fig. 3: The first three Landau levels as function of the magnetic field for the unperturbed (thin dashed curves), the 2D polaron (thick solid curves) and the 3D polaron (thick dashed curves).

in 2D the energy shift, the mass renormalization, and the level splitting at $n\omega_c \simeq \omega_{LO}$ are more pronounced than for the 3D polaron.

We will consider the limiting behaviour of the polaron correction for certain values of the magnetic field. First for small magnetic fields (i.e. $n\omega_c \ll \omega_{LO}$) the energy difference between two successive Landau levels is

$$\Delta E_{n+1} - \Delta E_n = \alpha\frac{\pi}{8}\omega_c\left(1 + \frac{9}{8}(n+1)\omega_c + \frac{5}{128}(30n^2 + 60n + 29)\omega_c^2 + ...\right) \quad, \qquad (10)$$

which results in the cyclotron mass

$$\frac{m^*}{m_b} = \frac{1}{1 - \alpha\frac{\pi}{8}(1 + \frac{9}{8}\omega_c + \frac{145}{128}\omega_c^2 + ...)} \quad. \qquad (11)$$

For $n\omega_c = \omega_{LO}$ two unperturbed levels will cross but because of the presence of the electron-phonon interaction they will split up due to the anti-crossing theorem of von Neumann. The splitting is found to be[11] equal to twice

$$|\Delta E_n| = \sqrt{\alpha}\sqrt{\sqrt{\frac{\pi}{n}}\frac{(2n-1)!!}{2^{n+1}n!}} \quad, \qquad (12)$$

which has the characteristic $\alpha^{1/2}$-dependence which is different from the $\alpha^{2/3}$-dependence as found for the 3D polaron.

In the limit of $\omega_c \to \infty$ the Landau levels become pinned. We found[11] the following limiting behaviour

$$E_n = E_0^{RSPT} + 1 + \frac{1}{2}\omega_c - \frac{\alpha}{\sqrt{\omega_c}}\frac{\sqrt{\pi}(2n-1)!!}{2^{n+1}n!} \quad. \qquad (13)$$

3. The cyclotron resonance absorption spectrum

Theoretically there are two different viewpoints in calculating the cyclotron resonance frequency. First, and the most often used in polaron physics because of its apparent simplicity, relies on a calculation of the position of the Landau levels. This approach was elaborated on in the forgoing section. An alternative approach which we want to advocate is by calculating the *cyclotron resonance absorption spectrum* itself. It is this quantity which is measured experimentally. The position of the cyclotron resonance peak then gives the cyclotron resonance frequency.

Within linear response theory the cyclotron resonance absorption spectrum in the Faraday active mode configuration is proportional to the real part of the conductivity tensor

$$Re\sigma_+(\omega) \sim \frac{Im\Sigma(\omega)}{(\omega - \omega_c - Re\Sigma(\omega))^2 + (Im\Sigma(\omega))^2} \quad , \tag{14}$$

where $\Sigma(\omega)$ is the memory function which is a function of the cyclotron resonance frequency ω_c, of the electron-phonon coupling constant α and also of the width of the 2D electron layer. The memory function is in essence a force-force correlation function where the time evolution is restricted to the Liouville space which is perpendicular to the velocity operator $\dot{x}+i\dot{y}$. In order to calculate the memory function we use second-order perturbation theory and i) decouple the electron-phonon system, ii) the phonons are taken at thermal equilibrium and iii) the electron system will be calculated within different approximations. To first order in the electron-phonon coupling constant the memory function has the form

$$\Sigma(\omega) = \frac{1}{\omega} \int_0^\infty dt(1 - e^{i\omega t})ImF(t) \quad , \tag{15}$$

with

$$F(t) = -\sum_{\mathbf{k}} \frac{k_\parallel^2}{n_e m_b \hbar} |V_{\mathbf{k}}|^2 \left[i\, D(k_\parallel, t) + i\, n(\omega_{LO})D^R(k_\parallel, t) \right] e^{-i\omega_{LO} t} \quad , \tag{16}$$

where $n(\omega) = (e^{\beta\hbar\omega}-1)^{-1}$ is the occupation number. $D(k_\parallel, t)$ ($D^R(k_\parallel, t)$) is the electron (retarded) density-density correlation function[23]. $V_{\mathbf{k}}$ is the interaction coefficient of the electron-phonon interaction and is given by

$$V_{\mathbf{k}} = i\omega_{LO} \left(\frac{4\pi\alpha}{Vk^2} \right)^{1/2} \left(\frac{\hbar}{2m_b\omega_{LO}} \right)^{1/4} \langle\psi_0|e^{ik_z z}|\psi_0\rangle \quad . \tag{17}$$

if we include the non-zero thickness of the 2D electron layer. $\psi_0(z) = (b^3/2)^{1/2}ze^{-bz/2}$ is the variational wave function of the electron in the direction normal to the 2D electron layer[1]. The variational parameter b is taken as $b = (48\pi N m_b e^2/\hbar^2\epsilon_0)^{1/3}$, where $N = n_d + (11/32)n_e$ and n_d and n_e are the depletion and carrier charge densities respectively. In Eq.(17) only the lowest subband is included and all higher electric subbands are neglected. This approximation should be valid for systems with not too high electron density ($n_e < 10^{12}cm^{-2}$) and for systems at low temperature so that no electrons are present in higher electric subbands.

At low temperatures one has $Im\Sigma(\omega) \simeq 0$ and consequently the cyclotron resonance frequency is given by the non-linear equation

$$\omega - \omega_c - Re\Sigma(\omega) = 0 \quad , \tag{18}$$

From the solution $\omega = \omega_c^*$ the cyclotron resonance mass is defined as $m^* = m_b\omega_c/\omega_c^*$. The physical meaning of Eq.(2) is apparent: $Re\Sigma(\omega_c^*)$ is the shift in the cyclotron resonance peak due to the electron-phonon interaction. Note also that theoretically the cyclotron mass is always referred to the electron band mass which is not a directly meausurable quantity.

3.1 One polaron theory

We will consider different approximations for the electron density-density correlation function. In the *one* electron approximation Eq.(15) may also be written as

$$F(t) = -\sum_{k} \frac{k_{\parallel}^2}{m_b \hbar} |V_k|^2 < e^{ik \cdot r(t)} e^{-ik \cdot r(0)} > e^{-i\omega_{LO} t} \quad , \tag{19}$$

where for simplicity we have limited ourselves to the zero temperature case. The electron density-density correlation function is given by

$$I(\mathbf{k}, t) = < e^{ik \cdot r(t)} e^{-ik \cdot r(0)} >$$

$$= e^{-\Gamma^2 t^2 / 4} \exp\left[-\frac{\hbar k^2}{2m_b \omega_c} (1 - e^{-i\omega_c t}) \right] \quad , \tag{20}$$

where a Gaussian broadening parameter Γ is introduced which mimicks the broadening of the delta-function like peaks in the electron unperturbed density of states. If Eq.(20) is inserted into Eq.(19) one finds for the memory function (15) a real part

$$Re\Sigma(\omega) = \sum_{n=0}^{\infty} \frac{B_n}{\omega \Gamma} \left[2D(\epsilon_n / \Gamma) - D((\epsilon_n + \omega)/\Gamma) - D((\epsilon_n - \omega)/\Gamma) \right] \quad , \tag{21a}$$

and an imaginary part

$$Im\Sigma(\omega) = \sum_{n=0}^{\infty} \frac{\sqrt{\pi} B_n}{2\omega \Gamma} \left[exp\left(-(\epsilon_n + \omega)^2 / \Gamma^2 \right) - exp\left(-(\epsilon_n - \omega)^2 / \Gamma^2 \right) \right] \quad , \tag{21b}$$

with

$$B_n = \frac{\eta}{n!} \int_0^{\infty} dx f(x, b_0) x^{2(n+1)} e^{-x^2} \quad ,$$

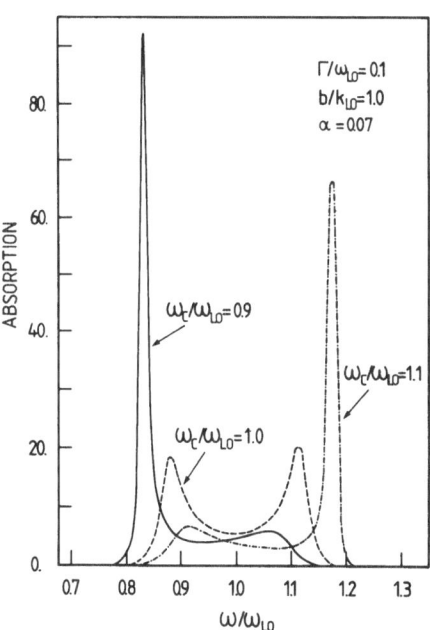

Fig. 4: The cyclotron resonance absorption spectrum of the Q2D system as due to the interaction of one electron with LO-phonons for different values of the magnetic field around the resonant condition.

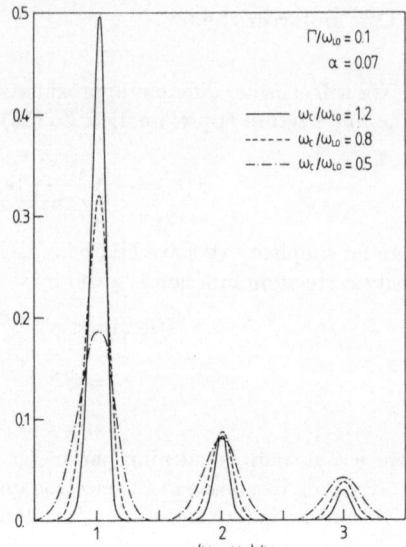

Fig. 5: The LO-phonon-assisted harmonics of the absorption spectrum for an ideal 2D system and for different values of the magnetic field.

where $f(x, b) = (8b^3 + 9b^2k + 3bk^2)/[8(b + k)^3]$ is a form factor resulting from the nonzero width of the 2D layer. In Eqs.(21) we defined $\epsilon_n = \omega_{LO} + n\omega_c$, $\eta = 2\alpha\omega_{LO}^3(\omega_c/\omega_{LO})^{3/2}$, $b_0 = b\sqrt{\hbar/2m_b\omega_{LO}}$ and

$$D(t) = e^{-t^2} \int_0^t dx e^{x^2} \quad,$$

is the Dawson integral. The real and imaginary part of the memory function are connected by the Kramers-Kronig relations.

The cyclotron resonance absorption spectrum is shown in Fig. 4 around the resonant condition $\omega_c \sim \omega_{LO}$. The splitting of the cyclotron peak is apparent and there is a transfer of the oscillator strength from the low frequency peak (i.e. $\omega < \omega_{LO}$) to the higher frequency peak (i.e. $\omega > \omega_{LO}$) when ω_c increases above ω_{LO}.

If the cyclotron resonance absorption spectrum is investigated for frequencies ω above ω_{LO} another set of peaks is observed around $\omega \sim \omega_{LO} + n\omega_c$ which are the *phonon-assisted harmonics*. This is shown in Fig. 5 for the ideal 2D case and for different values of the magnetic field strength. The magnetic field dependence of the position of the first 4 peaks in the absorption spectrum are given in Fig. 6. The oscillator strength of these peaks are depicted in Fig. 7. Note that subsequent phonon-assisted harmonics have an oscillator strength which are an order of magnitude smaller.

3.2 Many-polaron theory

The one polaron theory is able to account for the basic *qualitative* behaviour of the cyclotron resonance peak. In a next step we will investigate how *many-particle* aspects of the electron system will influence the absorption spectrum and more specific the influence on the position of the cyclotron resonance frequency will be investigated in detail.

As a first step the *occupation* effect (i.e. the Pauli-blocking effect) and the influence of *static screening* will be described[24,25]. The occupation effect is taken into account if in Eq.(15) the

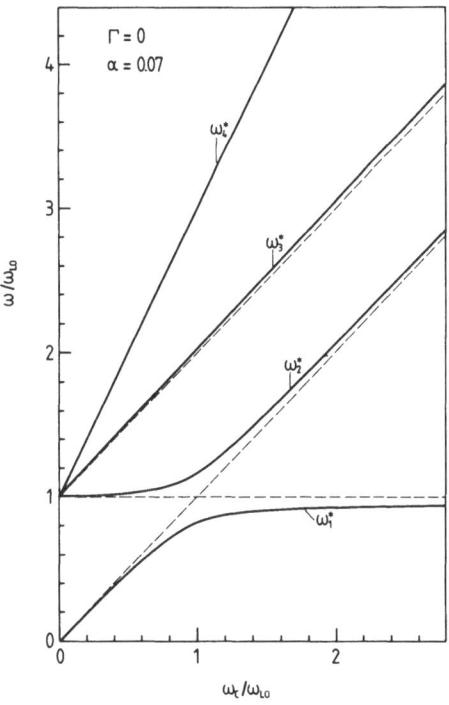

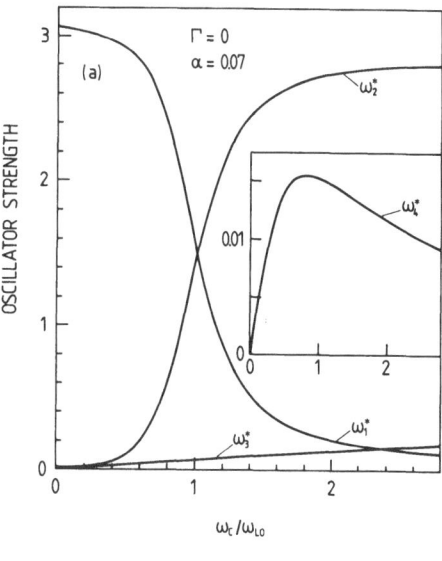

Fig. 6: The position of the first four peaks in the magneto-optical absorption spectrum as function of the magnetic field for an ideal 2D system.

Fig. 7: The oscillator strength of the first four peaks in the magneto-optical absorption spectrum as a function of the magnetic field for an ideal 2D system.

polarization function $\Pi = \hbar D$ is approximated by Π_0 which is the polarization of a non-interacting 2D electron gas[26,27]. In the following we will concentrate on the zero temperature case where we only need the imaginary part of the polarization function. The real part of the memory function becomes

$$Re\Sigma(\omega) = \sum_{n=0}^{N_0} \sum_{n'=N_0}^{\infty} \sum_{k} \frac{k_{\parallel}^2}{2\nu m_b \hbar \omega} |V_k|^2 C_{n,n'}(k_{\parallel})[\frac{2}{(n'-n)\omega_c + \omega_{LO}} \\ - \frac{1}{(n'-n)\omega_c + \omega_{LO} + \omega} - \frac{1}{(n'-n)\omega_c + \omega_{LO} - \omega}]f_n(1-f_{n'}) \quad , \tag{22}$$

with $N_0 = [\nu]$ the integer part of the filling factor $\nu = (n_e/2)(2\pi\hbar/m_b\omega_c)$ which may also be written as $\nu = \sum_{n=0}^{\infty} f(\epsilon_n)$. At zero temperature the Fermi-Dirac function takes the simple form $f_n = 1(n < N_0), \nu - N_0(n = N_0)$ and $0(n > N_0)$.

To include the effect of the *static screening* into the memory function calculation $|V_k|^2$ in Eq.(22) is substituted by $|V_k|^2/\epsilon^2(k_{\parallel})$ where $\epsilon(k)$ is the static dielectric function which we took within the RPA approximation[26,27] and which depends on the magnetic field strength.

We have extended the above results recently[28] and have included a full *dynamical screening* of the electron-phonon interaction. Within such a calculation the real part of the memory function can be obtained by substituting Eq.(16) into Eq.(15)

$$Re\Sigma(\omega) = \sum_k \frac{k_\parallel^2}{n_e m_b \omega} |V_\mathbf{k}|^2 \frac{\omega^2}{\pi} \int_{-\infty}^{+\infty} dx \frac{(1 + n(x)) Im\Pi^R(k_\parallel, x)}{((x + \omega_{LO})^2 - \omega^2)(x + \omega_{LO})}$$

$$+ n(\omega_{LO}) \sum_k \frac{k_\parallel^2}{n_e m_b \omega} |V_\mathbf{k}|^2 \frac{1}{2} \big[Re\Pi^R(k_\parallel, \omega + \omega_{LO})$$

$$+ Re\Pi^R(k_\parallel, \omega - \omega_{LO}) - 2Re\Pi^R(k_\parallel, \omega_{LO}) \big] \quad , \tag{23}$$

where $\Pi = \hbar D$ is the polarization function of the 2D electron gas which has been extensively studied[26,27]. Note that in the calculation of the memory function the *frequency* dependence of the polarization function of the 2D electron gas is retained.

In the zero temperature limit Eq.(23) can be simplified to

$$Re\Sigma(\omega) = \sum_k \frac{k^2 f(k, b)}{n_e m_b \omega} \frac{|V_k|^2}{v(k)} \frac{\omega^2}{\pi} \int_{-\infty}^{+\infty} dx \frac{1}{[(x + \omega_{LO})^2 - \omega^2](x + \omega_{LO})} Im \frac{1}{\epsilon(k, x)} \quad , \tag{24}$$

where we considered the zero temperature limit and $\mathbf{k}$ is a 2D wave vector. The sum over k_z has been performed and leads to the well-known form factor $f(k, b)$ where the width of the 2D electron layer is given by $\sqrt{< (z- < z >)^2 >} = \sqrt{3}/b$. In Eq.(24) $|V_k|^2$ gives the strength of the 2D electron-phonon coupling for a phonon with wave vector $\mathbf{k}$, $v(k)$ is the Fourier transform of the unscreened 2D Coulomb interaction and $\epsilon(k, \omega)$ is the dielectric function of the 2D electron gas which is considered here in the random-phase-approximation(RPA) with the inclusion of a magnetic field.

In deriving the above expressions for the cyclotron resonance spectrum the following assumptions are made: (1)the interaction of the electrons is with the bulk LO-phonons of GaAs; the LO-phonons are assumed to be dispersionless; (2)perpendicular to the electron layer the electron wave function is chosen to be of the Fang-Howard type. Only the occupation of the lowest electric subband is incorporated; (3)the calculation is done for the zero temperature limit which is reasonable in view of the low experimental temperature $(T \sim 4.2K)$; (4)the electron conduction band is taken to be parabolic.

3. Results

In order to make a detailed quantitative comparison between the experimental measured cyclotron resonance mass and the theoretical calculated values it is necessary to incorporate the electron band non-parabolicity which is quite important for the GaAs-heterostructures under investigation here. The band non-parabolicity will be taken into account via the assumption of a *local parabolic band* approximation. Larsen[29] has shown recently that this is valid for GaAs. In this approach the Landau levels in the absence of polaron coupling are determined from the two-level Kane theory as elaborated by Lassnig and Zawadzki[30]. They found

$$\epsilon_{np} = -\frac{E_g}{2} + < U > + \frac{E_g}{2} \sqrt{1 + \frac{4}{E_g} (< T > + \epsilon_\parallel)} \quad , \tag{25a}$$

with $< U >$ the average electron potential energy in the z-direction

$$< U >= \frac{12\pi e^2}{\epsilon_0 b}(n_d + \frac{11}{16}n_e) \quad , \tag{25b}$$

and

$$< T >= \frac{\hbar^2 b^{*2}}{8m_b} \quad , \tag{25c}$$

the average electron kinetic energy in the z-direction and $\epsilon_\parallel = \hbar\omega_c(n+1/2)$. $b^* = b\sqrt{x}$ is determined from the equation $x^3 - px - q = 0$ with $p = \hbar^2 b^2/2m_b E_g$ and $q = (1 + 4\epsilon_\parallel/E_g)$.

Due to the band non-parabolicity the cyclotron resonance frequency will be different from ω_c even in the absence of any polaron effects. The shifted cyclotron resonance frequency is given by $\hbar(\omega_c)_{np} = \epsilon_{np}(n = 1) - \epsilon_{np}(n = 0)$. In order to incorporate polaron effects together with the band non-parabolicity, we insert $(\omega_c)_{np}$ and b^*, as obtained from above, into the memory function. This results in the non linear equation

$$\omega - (\omega_c)_{np} - Re\Sigma(\alpha, (\omega_c)_{np}, b^*; \omega) = 0 \quad , \tag{26}$$

where the solution $\omega = \omega_c^*$ equals the theoretical cyclotron mass which is affected both by band non-parabolicity and by polaron effects.

Note that in the present approximation the correct behaviour for either vanishing electron-phonon coupling, or parabolic energy band limit, is obtained. Also in the limit of zero magnetic field the correct behaviour is obtained because we included *all* the intermediate states. Further the polaron effect and the band non-parabolicity are not considered to be additive.

A quantitative comparison of the polaron mass renormalization was made[28] for GaAs - heterostructures with a typical electron density of $n_e \sim 4 \times 10^{11} cm^{-2}$. Earlier we found[23] that the *non-zero width* of the 2D electron layer considerably reduces the 2D *one* polaron results for

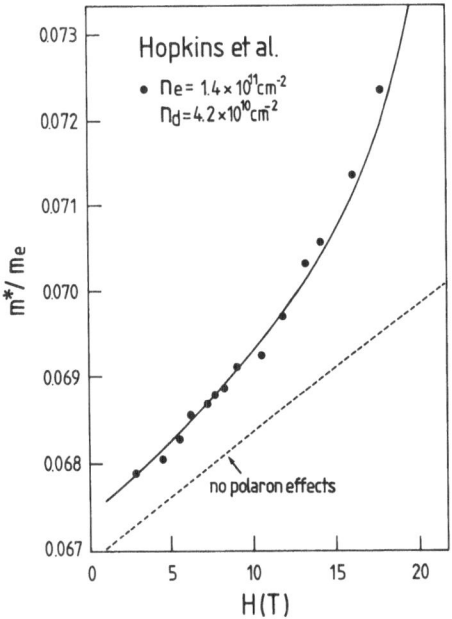

Fig. 8: Cyclotron mass as function of the magnetic field for a sample with electron density $n_e = 1.4 \times 10^{11} cm^{-2}$. The solid dots are the experimental results, the full curve is the present theoretical result and the dashed curve corresponds to the theoretical result without polaron effect.

the polaron mass(by a factor of 2 for $n_e \sim 4 \times 10^{11} cm^{-2}$ and $\omega_c/\omega_{LO} \sim 0.8$). Inclusion of the occupation probabilities for the Landau levels leads to a further reduction[25] (a factor of 1.7 for $n_e \sim 4 \times 10^{11} cm^{-2}$ and $\omega_c/\omega_{LO} \sim 0.8$). We found[28] that screening of the electron-phonon interaction further weakens the effect of the electron-phonon interaction but that treating the screening dynamically does not alter the results significantly as compared to the static screening results. Static screening overestimates the reduction of the mass renormalization. This is most apparent in the weak field case (i.e. $\omega_c/\omega_{LO} < 0.5$) where the polaron mass renormalization is small. In the magnetic field range where polaron effects become observable (i.e. $\omega_c/\omega_{LO} > 0.5$) a static screeening theory gives, within 5%, the same results as a full dynamical screening theory. Thus for all practical purposes the numerical much simpler static screening approach will be sufficiently reliable when we compare theory with experiment.

In Fig. 8 the experimental result for the cyclotron mass as obtained by Hopkins et al[16] (solid dots) is compared with the present theoretical results. A very close agreement between theory and experiment could be obtained over the whole magnetic field range if the band mass is taken to be $m_b/m_e = 0.0661$. Polaron effects are evident because band non-parabolicity alone would give only a linear(roughly) increase of the cyclotron mass with magnetic field strength(dashed curve).

If a higher density sample is investigated additional complications occur. When the filling factor $\nu > 1$ different Landau levels can be occupied and for $1 < \nu < 2$ not only the transition $n = 0 \rightarrow n = 1$ is present but also the transition $n = 1 \rightarrow n = 2$ can occur. This complicates the picture considerably especially for the calculation of the band non-parabolicity where, up to now, all existing calculations assume delta-function Landau levels which leads to a discontinuous effective mass at integer filling.

Earlier we found[23] that for $\nu > 1$ the polaron correction to the polaron mass is small and approximately independent of the magnetic field (in Refs. 24 and 25 it was found that the polaron contribution to the effective mass m^* slightly oscillates as function of H). In Fig. 9 the cyclotron mass is shown as function of the magnetic field for a sample with an electron density $n_e = 3.4 \times$

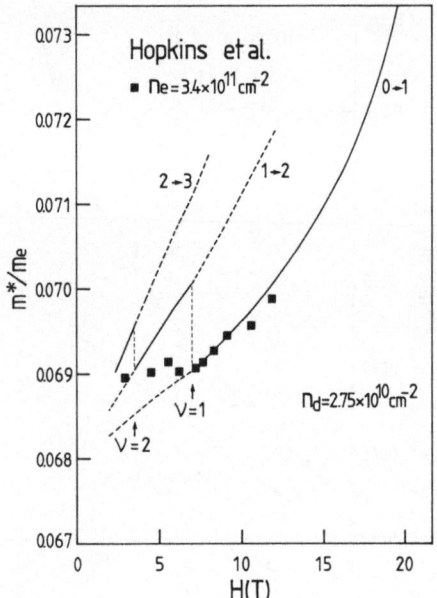

Fig. 9: The same as Fig. 8 but now for a higher electron density of $n_e = 3.4 \times 10^{11} cm^{-2}$. The transition between higher Landau levels are also included.

$10^{11} cm^{-2}$. We take the electron 3D band mass the same as obtained from the lower electron density sample and consequently no fitting parameter is present. For $\nu < 1$ good agreement is found with the experimental results of Hopkins $et\ al$[16]. When $1 < \nu < 2$ the second Landau level is partially filled and not only the transition $0 \to 1$ but also the transition $1 \to 2$ is possible and we expect that the experimental result lies between the theoretical curves for those two transitions. This is indeed also the behaviour found experimentally. At $\nu = 2$ the second Landau level is completely filled and only the transition $1 \to 2$ is possible. This agrees with the experimental data because around $H \sim 3.5T$ the results coincides with the theoretical result for the $1 \to 2$ transition. For higher filling factors the same reasoning can be applied.

The situation is less clear for the cyclotron resonance data of Horst et al[15] where the electron density is $n_e = 4.07 \times 10^{11} cm^{-2}$. These results are shown in Fig. 10. In comparing these results with our theory we took the same 3D band mass as before and detemine the depletion charge density such that a good fit was obtained with the $H > 15T$ data. We found $n_d = 1.4 \times 10^{11} cm^{-2}$. From Fig. 10 we notice that the experimental results for $H < 15T$ are larger than expected from theory. A possible explanation may be that in this sample the broadening of the Landau levels is at least a factor 3 larger than in the samples of Hopkins $et\ al$. Consequently the overlap between the Landau levels is large when $\nu > 1$. Even for ν just above 1 we expect that the Landau level $n = 1$ will already be partially occupied and thus already transitions $1 \to 2$ are able to contribute which results in a larger polaron cyclotron mass.

Recently[17] the polaron cyclotron resonance mass was measured in a very high density sample (i.e. $n_e = 5 \times 10^{11} cm^{-2}$). These results are shown in Fig. 11 and compared to our theoretical results. Reasonable good agreement is found with experiment. But in this case the polaron correction is very small as indicated in the lower part of Fig. 11. Almost the complete shift in the cyclotron resonance frequency can be explained by band non-parabolicity alone as was pointed out in Ref. 17.

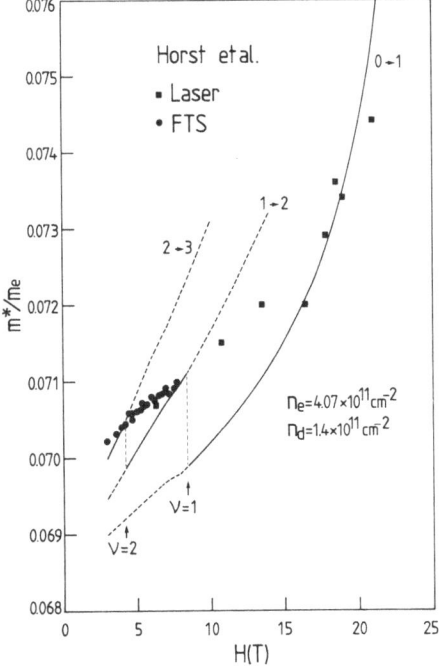

Fig. 10: The same as Fig. 9 but now for an electron density of $n_e = 4.07 \times 10^{11} cm^{-2}$. The experimental results are from Horst $et\ al$[15].

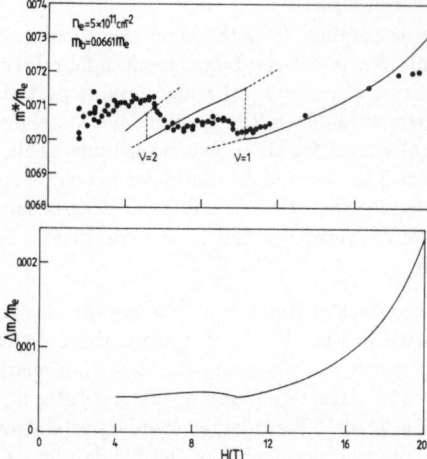

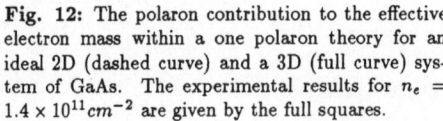

Fig. 11: The same as Fig. 9 but now for an electron density of $n_e = 5 \times 10^{11} cm^{-2}$. The experimental results are from Thiele *et al*[17].

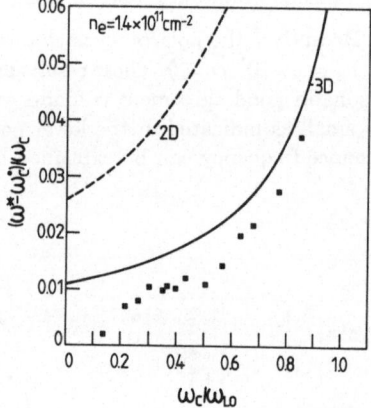

Fig. 12: The polaron contribution to the effective electron mass within a one polaron theory for an ideal 2D (dashed curve) and a 3D (full curve) system of GaAs. The experimental results for $n_e = 1.4 \times 10^{11} cm^{-2}$ are given by the full squares.

Next we want to address the question whether or not the polaron shift in the cyclotron resonance peak position is larger or smaller in real 2D systems than in 3D systems. We know that, due to the occupation effect and due to screening, the polaron effect decreases with increasing electron density. Therefore it is sufficient to analyse only the low electron density sample of Hopkins *et al*[16]. The electron-phonon contribution to the shift in the experimental cyclotron resonance frequency(ω^*) is extracted by substracting the shift due to band non-parabolicity $\omega_c^0 = (\omega_c)_{np}$. The result is shown in Fig. 12 where, for comparison, we also show the electron-phonon contribution to the polaron cyclotron mass for an ideal 2D system (one polaron) and for the 3D system. From Fig. 12 we may conclude that in GaAs/Al$_x$Ga$_{1-x}$As heterostructures polaron effects below the resonance condition are *smaller* than in bulk GaAs if $n_e \geq 1.4 \times 10^{11} cm^{-2}$.

Up to now all cyclotron resonance measurements in GaAs-heterostructures are for $\omega_c^* < \omega_{LO}$. Measurements are needed in the $\omega_c^* > \omega_{LO}$ region. The discontinuity at the resonance condition $\omega_c^* = \omega_{LO}$ is a direct consequence of the electron-phonon interaction and is not obscured by effects from band non-parabolicity. Splitting of the cyclotron resonance peak around $\omega_c^* \sim \omega_{LO}$ is a consequence of the removal of the degenarcy of the energy level $E_1 = \frac{3}{2}\hbar\omega_c$ and $E_0 + \hbar\omega_{LO} = \frac{1}{2}\hbar\omega_c +$

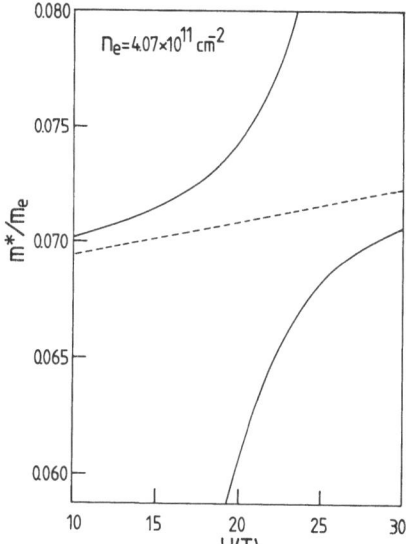

Fig. 13: The cyclotron mass around the polaron resonant condition as predicted from our theory for the same sample as in Fig. 8. Full curve is with inclusion of polaron effects and dashed curve is without the inclusion of polaron effects.

$\hbar\omega_{LO}$. This splitting has been observed in polar 3D semiconductors, in InSb-inversion layers[12,13] and in $Ga_xIn_{1-x}As$-heterostructures[31] but not in GaAs-heterostructures (because magnetic fields such that $H > 22T$ are needed). Fig. 13 shows what we expect for the cyclotron mass from our calculation for the low density sample of Hopkins et al[16].

Acknowledgements

One of us (FMP) is supported by the Belgian National Science Foundation. This work is sponsored by FKFO, project No. 2.0072.80. We would like to thank Wu Xiaoguang for a nice collaboration in a large part of the research presented here.

References

* Also at R.U.C.A., B-2020 Antwerp and University of Technology, NL-5600 MB Eindhoven.

0 Present address: Bell Communications Research, 331 Newman Springs Road, Box 7020, Redbank, New Jersey 07701-7020

1 T. Ando, A.B. Fowler and F. Stern, Rev. Mod. Phys. **54**, 437 (1982).

2 K. von Klitzing, G. Dorda and M. Pepper, Phys. Rev. Lett. **45**, 494 (1980).

3 H. Störmer, A. Chang, D.C. Tsui, J.C.M. Hwang, A.C. Gossard and W. Wiegmann, Phys. Rev. Lett. **50**, 1953 (1983).

4 H.L. Störmer, T. Haavasoja, V. Narayanamurti, A.C. Gossard and W. Wiegmann, J. Vac. Sci. Techn. **B1**, 423 (1981).

5 S. Das Sarma and A. Madhukar, Phys. Rev. **B22**, 2823 (1980).

6 S. Das Sarma and B.A. Mason, Ann. Phys. (N.Y.) **163**, 78 (1985).

7 Wu Xiaoguang, F.M. Peeters and J.T. Devreese, Phys. Rev. **B31**, 3420 (1985).

8 F.M. Peeters and J.T. Devreese, Phys. Rev. **B** (1987).

[9] S. Das Sarma, Phys. Rev. Lett. **52**, 859 (1984).

[10] D.M. Larsen, Phys. Rev. **B30**, 4807 (1984).

[11] F.M. Peeters and J.T. Devreese, Phys. Rev. **B31**, 3689 (1985).

[12] M. Horst, U. Merkt and J.P. Kotthaus, Phys. Rev. Lett. **50**, 754 (1983).

[13] U. Merkt, M. Horst and J.P. Kotthaus, Physica Scripta **T13**, 272 (1986).

[14] H. Sigg, P. Wyder and J.A.A.J. Perenboom, Phys. Rev. **B31**, 5253 (1985).

[15] M. Horst, U. Merkt, W. Zawadzki, J.C. Maan and K. Ploog, Solid State Commun. **53**, 403 (1985).

[16] M.A. Hopkins, R.J. Nicholas, M.A. Brummell, J.J. Harris and C.T. Foxon, Superlatt. and Microstr. **2**, 319 (1986).

[17] F. Thiele, U. Merkt, J.P. Kotthaus, G. Lommer, F. Malcher, U. Rössler and G. Weimann (to be published)

[18] M. Ziesmann, D. Heitmann and L.L Chang, Phys. Rev. **B35**, 4541 (1987).

[19] G. Lindemann, R. Lassnig, W. Seidenbusch and E. Gornik, Phys. Rev. **B28**, 4693 (1983).

[20] D.M. Larsen, Phys. Rev. **B30**, 4595 (1984).

[21] R. Lassnig, Surface Sci. **170**, 549 (1986).

[22] D.M. Larsen and E.J. Johnson, J. Phys. Soc. Jpn., Suppl. **21**, 443 (1966).

[23] Wu Xiaoguang, F.M. Peeters and J.T. Devreese, Phys. Rev. **B34**, 8800 (1986); F.M. Peeters, Wu Xiaoguang and J.T. Devreese, Physica Scripta **T13**, 282 (1986).

[24] Wu Xiaoguang, F.M. Peeters and J.T. Devreese, in *the Proc. of the 18th Int. Conf. on the Phys. of Semicond.*, Ed. O. Engström (World Scientific, Singapore, 1987) 589.

[25] Wu Xiaoguang, F. M. Peeters, and J. T. Devreese, (submitted for publication).

[26] N. J. Horing and M. M. Yildiz, Ann. Phys. (N.Y.) **97**, 216 (1976).

[27] M. L. Glasser, Phys. Rev. B **28**, 4387 (1983).

[28] Wu Xiaoguang, F. M. Peeters, and J. T. Devreese, (submitted for publication).

[29] D.M. Larsen (to be published).

[30] R. Lassnig and W. Zawadzki, Surface Sci. **142**, 388 (1984).

[31] R.J. Nicholas, L.C. Brunel, S. Huant, K. Karrai, J.C. Portal, M.A. Brummell, M. Razeghi, K.Y. Cheng and A.Y. Cho, Phys. Rev. Let. **55**, 883 (1985).

ON THE CONCENTRATION DEPENDENCE OF THE THERMAL ACTIVATION ENERGY OF IMPURITIES IN SEMICONDECTORS

B. Pödör

Research Institute for Technical Physics of the
Hungarian Academy of Sciences
H-1325 Budapest, P. O. Box 76, Hungary

Abstract

The thermal activation energies of impurities in semiconductors, when plotted against the cubic root of impurity concentration usually give linear plots. Based on an extensive analysis of the existing data for group IV, III-V, II-VI and II-IV-V$_2$ semiconductors in the literature it is demonstrated that the slopes of the linear plots are proportional to the static dielectric constant of the host material as suggested by the simple model based on Coulomb interaction.

Since the pioneering work of Pearson and Bardeen [1] it has been well known that the thermal activation energy of impurities decreases with increasing impurity concentrations. Pearson and Bardeen fitted their results obtained on Si to the half-empirical relation

$$E = E_o - \alpha N^{1/3} \tag{1}$$

where α is a constant, E_o is the activation energy at infinite dilution, and N is the concentration of majority impurities, i. e. of donors in n-type crystals. Later Debye and Conwell [2] argued that N should be replaced by N_i, where N_i is the average density of ionized majority impurities in the low-concentration or low-temperature region. Notwithstanding the further works in this field, neither the exact meaning of N in (1) nor the correlation of the slope α with any other parameter of the host semiconductor, kind of impurity, etc. have yet been cleared up to now, c. f. [3, 4] and references therein. However there are strong indications in the literature [3 to 9] that the concentration which has to be substituted into (1) is the concentration of ionized impurities, which e. g. in the case of n-type crystals is equal to $N_a + n \simeq N_a$, except for high temperatures or heavy doping.

Several different suggestions have been put forward to explain the observed concentration dependence of the impurity activation energy (see e. g. [3, 10] for a review of the different models). No attempt will be made here to discuss the various theoretical ideas which had been put forward so far. However it seems that the idea put forward by Pearson and Bardeen [1] and also by Castellan and Seitz [11], i. e. that the decrease of the thermal activation energy of impurities was due to Coulomb attraction between charge carriers and ionized impurities can give at least a qualitative explanation for the numerous experimental observations. According to Pearson and Bardeen [1] (c.f. also [11]) the slope α in (1) is calculated as

$$\alpha = 3.292 (4\pi/3)^{1/3} a_o E_o \qquad (2)$$

where a_o is the effective Bohr radius and E_o is the (hydrogenic) activation energy according to the effective mass theory. E_o can be identified with the thermal activation energy at infinite dilution (which is equal to the optical ionization energy). From (2) we obtain $\alpha = 3.82 \times 10^{-4}/\varepsilon_r$ (meVcm) where ε_r is the (static) relative dielectric constant of the material. The above model suggests that the α values obtained for different impurities (donors and acceptors as well) in the same host semiconductor should be equal, moreover the α values obtained in different semiconductors should scale as the reciprocals of the dielectric constants. In other words the product $\alpha\varepsilon_r$ should be constant, independently of the semiconductor.

To test the predictions of the above model relevant data were collected from the literature and also from our laboratory for a number of semiconductors with various impurities. Lack of space prevents to give a full discussion of the data analysis, but we note here, that in some cases the literature data were reevaluated and reanalyzed where it seemed to be necessary. The available data for α are compiled in Table 1 for 15 semiconductors, and also plotted against $1/\varepsilon_r$ in Fig. 1. In most of the cases the data for a given

Table 1

Values of α in different semiconductors

Material	ε_r	α	$\alpha\varepsilon_r$	Notes
		(10^{-3}meVcm)	(10^{-3}meVcm)	
C	5.7	10.9	62.1	using data from [12]
Ge	16.0	2.4	38.4	[2, 13]
Si	11.8	3.80	44.8	using data from many sources
SiC(α)	10.2	5.5	56.1	[5, 14]
SiC(β)	9.92	3.1	30.8	using data from [15]
GaP	11.1	3.75	41.6	[3, 8, 9, 16]
GaAs	12.5	3.55	44.4	[6] and using data from [17, 18, 19]
GaSb	15.7	3.2	50.2	[20]
InP	12.4	3.10	38.4	[4]
InSb	17.8	2.26	40.2	using data from [21]
CdS	9.13	4.0	36.5	[7, 22]
CdSe	9.44	4.5	42.5	[7]
CdTe	9.65	3.8	36.7	[7] and using data from [23]
ZnSe	8.66	4.1	35.5	using data from [24]
ZnTe	9.3	4.47	41.6	using data from [25, 26]
ZnSiP$_2$	11.2	3.1	34.7	[3]

semiconductor in Table 1 represent results for several impurities, donors and acceptors as well. It should also be noted that for several semiconductors the data are sparse and partially

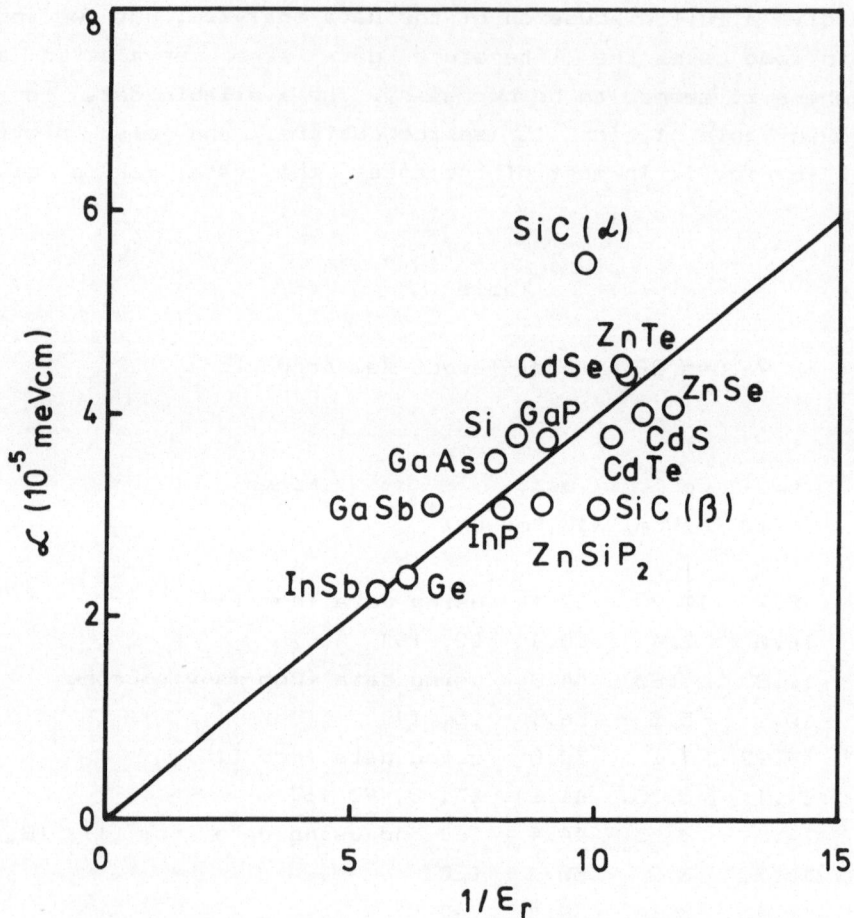

Fig.1. Plot of the values of α versus the reciporical of the dielectric constant for various semiconductors. The straight line corresponds to αε$_r$ = 40×10^{-5} meVcm. The datum point for C has been omitted from the figure.

contradictory. The data analysis has born out that the α values for a given host material are nearly independent of the impurity. It can also be seen from tha data in Table 1 that except for C and SiC(α) the product αε$_r$ is really roughly constant, with an average value of (40 ± 5)×10^{-5} (meVcm), which coincides with the value predicted by the model of Pearson and Bardeen [1]. However it should be kept in mind that the data compiled in Table 1 and also plotted in Fig. 1 are mostly based on primary data sometimes showing wild scatter. Therefore the seemingly good numerical agreement with the model of Pearson and Bardeen [1] should perhaps be viewed with some caution.

It can be concluded however, that the values of the slope α in (1) derived from experimental data for various semiconductors are actually proportional to the reciprocal of the static dielectric constant. This also indicates, that the basic mechanism responsible for the decrease of the thermal activation energy is connected with electrostatic interactions.

References

1. G. L. Pearson and J. Bardeen, Phys. Rev. 75 865 (1949).

2. P. P. Debye and E. N. Conwell, Phys. Rev. 93 693 (1954).

3. J. Monecke, W. Siegel, W. Ziegler and G. Kühnel, phys. stat. sol. (b) 103 269 (1981).

4. B. Pödör, Semiconductor Science and Technology 2 177 (1987).

5. J. van Daal, W. F. Knippenberg and J. D. Wasscher, J. Phys. Chem. Solids 24 109 (1963).

6. J. S. Blakemore, W. J. Brown, M. L. Stass and D. A. Woodbury, J. Appl. Phys. 44 3352 (1973).

7. H. H. Woodbury and M. Aven, Phys. Rev. B9 5195 (1974).

8. J. Pfeifer, B. Pödör, L. Csontos and N. Nádor, Revue de Physique Appliquée 13 741 (1978).

9. B. Pödör, J. Pfeifer, L. Csontos, N. Nádor and F. Deák, phys. stat. sol. (a) 76 695 (1983).

10. T. F. Lee and T. C. McGill, J. Appl. Phys. 46 373 (1975).

11. C. W. Castellan and F. Seitz, in: Semiconducting Materials, Ed. H. K. Henisch, London, Butterworths, 1951, p. 8.

12. J. C. Bourgoin, J. Krynicki, B. Blanchard, phys. stat. sol. (a) 52 293 (1979).

13. P. Norton and H. Levinstein, Phys. Rev. B6 470 (1972).

14. G. A. Lomakina, Fiz. Tverd. Tela 7 600 (1965).

15. L. S. Aivazova, S. N. Gorin, V. G. Sidjakin and I. M. Shvarts, Fiz. Tekh. Poluprovodnikov 11 1826 (1977).

16. W. Siegel and G. Kühnel, phys. stat. sol. (a) 84 251 (1984).

17. D. V. Eddols, phys. stat. sol. (a) 17 67 (1966).

18. C. M. Wolfe, G. E. Stillaman and W. T. Lindley, J. Appl. Phys. <u>49</u> 3088 (1970).

19. Q. H. Hua, Y. Z. Sun, S. R. Xue and G. P. Li, Progr. Crystal Growth and Charact. <u>11</u> 291 (1985).

20. I. Gyúró and B. Pödör, unpublished

21. K. I. Vinogradova, V. S. Ivleva, G. V. Ilmenkov, D. N. Nasledov and Ju. S. Smetannikova, Fiz. Tekh. Poluprovodnikov <u>6</u> 185 (1977).

22. K. Morimoto, M. Kitagawa and T. Yoshida, J. Cryst. Growth <u>59</u> 254 (1982).

23. J. Saraie, H. Shinohara, H. Edamatsu and T. Tanaka, phys. stat. sol. (a) <u>62</u> 665 (1980).

24. T. Yao, J. Cryst. Growth <u>72</u> 31 (1985).

25. B. C. Crowder and W. N. Hammer, Phys. Rev. <u>150</u> 541 (1966).

26. F. El-Akkad and A. Ghazali, Crystal Lattice Defects <u>6</u> 7 (1975).

THE USE OF HYDROSTATIC PRESSURE AND ALLOYING TO INTRODUCE
DEEP LEVELS IN THE FORBIDDEN GAP OF InSb AND GaAs

R.A. Stradling

Physics Department, Imperial College of Science and Technology,
London, SW7 2BZ, U.K.

Summary: In both GaAs and InSb common residual contaminants introduce
both shallow donor states and other levels normally resonant with the
conduction band but taking their predominant character from higher
conduction band minima. These states can be made to emerge into the
forbidden gap by applying hydrostatic pressure.   Judged by the
measured pressure coefficients two sets of levels are closely tied to
the L- and X-conduction band minima.   At low temperatures the
X-associated states show local relaxation effects and metastability
(i.e., D(X) behaviour). Following on these experiments with bulk
material, pressure experiments were extended to spike-doped GaAs both
in single layer and in superlattice form. These experiments were in
collaboration with A. Zrenner and F. Koch of the Technical University
at Munich. Shubnikov-de Haas peaks can be observed above 3T. Fourier
analysis of these shows that up to six sub-bands can be occupied. The
application of hydrostatic pressure of up to 20kbar dramatically
increases the mobility of the electrons in the lowest sub-band by as
much as an order of magnitude with smaller increases occurring the
higher sub-bands. The increases in mobility arise from the diffusion
of the silicon donors during growth. Localised donor states
associated with the L-conduction band drop below the Fermi energy in
the centre of diffused distribution of donors on the application of
pressure and become neutralised. This reduction of ionised scattering
centres affects the most confined electron states and thus the
mobility of the lowest sub-band increases very substantially.

1.   Introduction

The main effects of pressure are to increase the value of the direct

(Γ) energy gap (typically at ∼ + 15meV/kbar), to increase the indirect band gap between the valance band maximum and the L-conduction band minima at a lower rate (approximately + 5meV/kbar) and to decrease the indirect gap at the X-point (-5meV/kbar). It should be noted that alloying a direct-gap semiconductor with another of wider gap (e.g., GaAs with AlAs to form $Ga_{1-x}Al_xAs$) usually has a similar effect, the the alloy composition x playing an analogous role to pressure in changing the relative positions of the bands. In addition, the application of a magnetic field increases the energy gaps because of the shift in the band edge by the zero point cyclotron energy $(1/2)\hbar\omega_c$ where ω_c is the cyclotron frequency. The direct gap generally increases more rapidly than the indirect gaps because of the smaller effective masses associated with the Γ-minimum. In this case the magnetic field has an analogous role to pressure.

This paper reviews infrared experiments on the non-Γ donors in InSb and GaAs which can be introduced into the forbidden gap by pressures. A major advance in recent years is the ability to perform optical experiments at low temperatures under hydrostatic pressure. This can be done with pressures exceeding 50kbar using diamond clamp cells [1], where the working space is limited to a fraction of a mm in diameter and, consequently, the experiments are mainly restricted to those using near band-gap radiation, or with beryllium copper cells [2], there the diameter of the pressure chamber is of the order of 1 cm, and therefore experiments can easily be performed in the far infrared or even mm microwave region of the spectrum, but the maximum pressure is limited to less than about 20kbar.

2. Pressure Experiments with n-InSb

The first indication that donor states in InSb might not be associated completely with the Γ conduction band came from Hall and resistance measurements undertaken as a function of hydrostatic pressure and temperature by the group led by Porowski. With a high degree of doping it was possible to detect resonance states deep in the conduction band arising from the Te, Se, and S impurities, and by applying pressure it was possible to study deionisation of the resonance states associated with higher order minima [4]. By this means it was possible to determine the precise position of the resonance levels as a function of pressure. It was found that the depth of these states with respect to the higher order conduction band minima increases with decreasing atomic number (increasing electronegativity) and that the slopes for Se and S show distinct breaks with the resonance levels at low pressure following quite closely the L-minima and at higher pressures the X-minima. These breaks in slope observed suggested a crossing of two levels from the same impurity, one have predominantly L character and the other X-character.

With high purity material (i.e., donor concentrations below $10^{15} cm^{-3}$ two dominant levels are again observed in electrical measurements but these lie much closer to the conduction band edge. Again the pressure coefficients are close to those expected for the X- and L-minima: a surprising result in view of the more than 0.5 eV separation between the localized states and extrema providing their dominant characters. Nevertheless, the separations are far too great for the levels to be described by simple effective mass theory and contributions from many points of the zone must be substantial. The transfer of electrons from the Γ-band to the X-like impurities is inhibited by a large energy barrier (0.3 eV), which becomes apparent below 120K due to a strong lattice relaxation around the impurity below this temperature. The time constant for transfer of electrons into or out of this state can become extremely long at low temperatures (e.g., 270 days at 77K) and the state can then be considered to be metastable. On occupancy of the non-Γ associated states the mobility is found to increase substantially [6].

Fig 1 clearly shows the evidence for an anticrossing of Γ and L-associated levels obtained by far-infrared spectroscopy. In InSb the effects of 4kbar and a field of 10T are comparable, so that the strength of the maximum interaction as judged by the anticrossing can be followed over quite a range of magnetic field, simply by adjusting the pressure accordingly. The effect of increasing the magnetic field

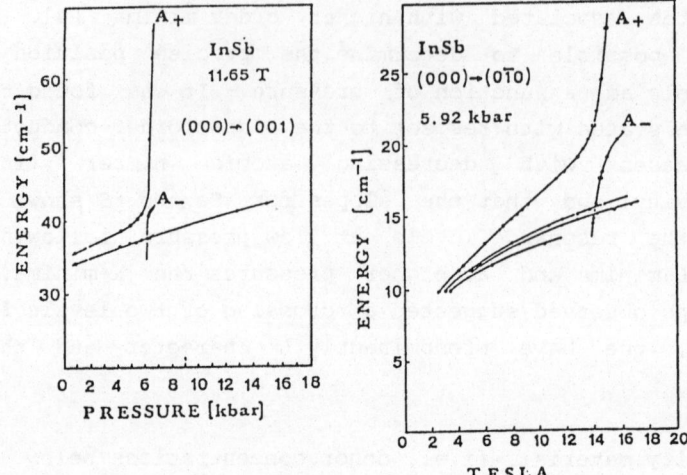

Fig. 1.   (a) shows the result of Fourier transform spectroscopy of the (000) $\longrightarrow$ (001) transition (1s-2p$_0$) of donor A and the unresolved group of other residual donors (B-D) as a function of pressure at a fixed magnetic field (11.65T). The anticrossing is seen at 6kbar at this field.

(b) shows the results of far infrared laser spectroscopy of the (000) $\longrightarrow$ (010) transition (1s-2p) at a pressure of 5.92kbar obtained by sweeping the field. This transition is much sharper than the (000) $\longrightarrow$ (001), so the other three residual donors are fully resolved.

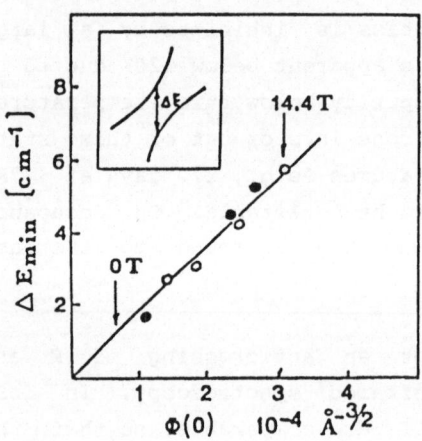

Fig. 2   shows the variation of the interaction energy Δ E defined as the minimum separation of the two branches A$^+$ and A$^-$ against the probability of finding the electron at the origin.

is to coil the electrons more tightly about the donor site and, therefore, to increase the contact term involving the local non-Coulombic potential, as is demonstrated in Fig. 2. It is not possible to observe a similar anticrossing of the X- and Γ-associated levels at 4K because of the lattice relaxation and associated effects discussed above. However, these D(X) like centres can be filled by slowly cooling the sample. At 77k or below it is then possible to remove the pressure in the gas cell and to maintain both the occupancy of the D(X) centres and the improvement in the mobility after returning to the original energy gap and effective mass. The removal of the Γ-like states of donor A from the spectrum because of their conversion to deep states after such a pressure treatment is illustrated in Fig. 3, thus demonstrating unambiguously that the same donor species (A) gives rise to Γ, L-, and X-like levels.

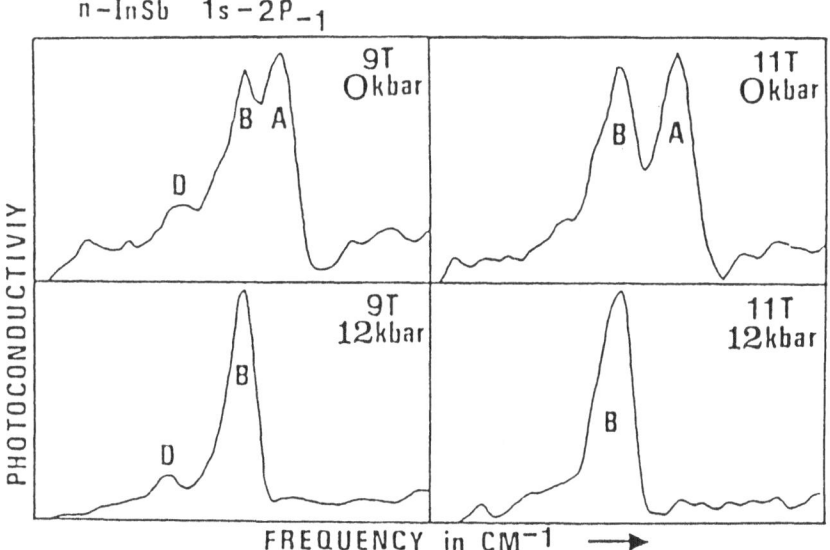

Fig. 3   A comparison of the central cell spectrum of a sample of n-InSb at two magnetic fields. The upper recordings are for the sample cooled normally and show three out of the four common donor species resolved (A, B, D). The lower recordings are taken after the sample had been slowly cooled to 77 K with 12kbar applied and the pressure released before being finally cooled to 4K. It is seen that this treatment has completely removed donor A from the spectrum. Thermal depopulation with increasing field would tend to enhance donor A, as is seen from the disappearance of the shallowest donor D on going to fields above 10T. In order to maintain a reasonable resistance for the sample after the pressure treatment, an uncompensated sample with a lower proportion of A centres was chosen for this experiment.

As can be seen from Fig. 4 a similar anticrossing interaction between L- and Γ-like states takes place for the deepest of the shallow donors in GaAs [9] (X_3 or Ge).

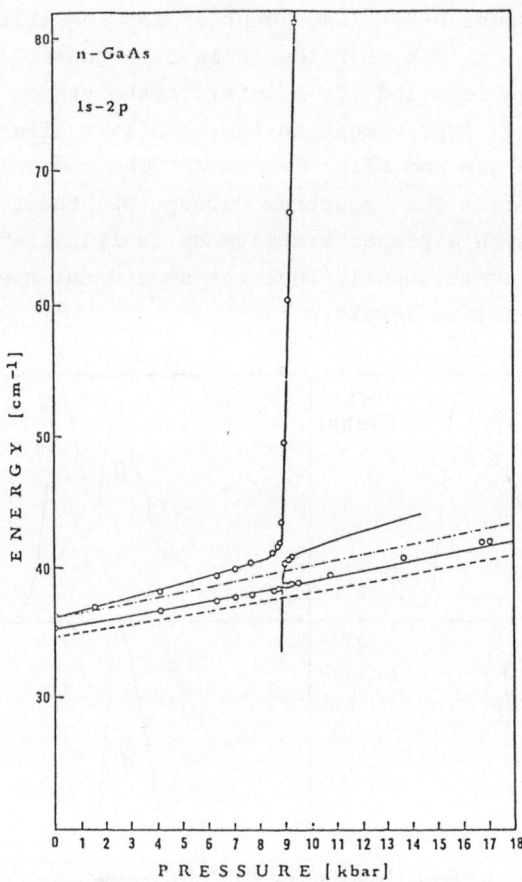

Fig. 4   Fourier-spectroscopy results for n-GaAs showing level crossing of Γ-band and L-levels of the deepest common residual 'effective mass' donor at 8.6 kbar.

With the high purity epitaxial samples of GaAs currently available, very sharp impurity and cylotron resonance lines are observable in the far-infrared. The cyclotron resonance can be used to derive an effective mass (m^*) with a precision approaching 0.1%. At the same time the magnetic field dependence of the excited states of the donor are fitted very accurately by effective mass theory if the effective Rydberg (= 13.6 (m^*/m_e)/ε^2eV) is used as a fitting parameter. Consequently the dielectric constant (ε) can be obtained to a precision [9] determined by the experimental line-widths and by small corrections to simple effective mass theory for polaronic and non-parabolicity effects [10]. These corrections are small and their pressure dependencies can be neglected to within the accuracy of the experiment. Consequently far infrared spectroscopy can be used to derive the pressure dependence of the dielectric constant with high precision. With n-GaAs a value of - 0.173 $\pm$ 0.002% $(kbar)^{-1}$ was obtained from fitting the magnetic field and pressure dependence of the $2p_-$ to $3d_{-1}$ impurity transition. This line was chosen because it is extremely sharp and, as the wavefunctions involved both have nodes at the origin, is unaffected by any non-effective mass chemical shift for the donor species involved which can be significant for the 1s ground state. With InSb [8] a value of -1.1% $(kbar)^{-1}$ was found from a similar procedure.

The pressure coefficient of the fast moving component (8.6 meV/Kbar) is within the range of values quoted for the difference in coefficients of the L and Γ minima but is somewhat different from the figure of 7.5 meV/kbar currently accepted. In diamond cell experiments crossings of the Se [11] and nitrogen [12] donors have been observed at approximately 30 kbar with rather similar pressure coefficients. However, Wolford and Bradley [13] failed to see any states emerge into the forbidden gap with high purity VPE GaAs before 39.5 kbar. At this pressure two nearly effective mass like-levels could be observed which tracked with the X-minima on increasing the pressure (the band structure of GaAs becomes indirect at 41.3 kbar) VPE GaAs normally contains Si as the domanant shallow donor [14]. The chemical shift for Si on the Γ-associated donor states is less than for other common residual donors and it would therefore be expected to give rise to effective-mass states close to the X-minima. It seems likely that the other residual donor species which were probably present at lower abundance were not detected because of the loss of luminescent intensity as the transitions involved switched from being direct to indirect in character.

3.   Spike or δ-doped GaAs

At high doping levels the Fermi energy can become greater than the levels resonant with the conduction band. If these states are sufficiently localised then carrier transfer can take place with a loss of carriers from the Γ-band. Very recently Maude et al. [14] reported a filling of metastable D(x) centres in both Si and Sn doped samples grown by MBE when the carrier concentration reached 2 x $10^{19} cm^{-3}$. At this carrier concentration the Fermi energy is 280 meV. By using pressures up to 12 kbar it was possible to populate this level at lower concentrations and to measure the position of the level with respect to the conduction band edge. This was found to quite concentration dependent dropping to a value of ~ 150 meV at a concentration of 3 x $10^{18} cm^{-3}$. This result may arise from band-gap renormalisation effects. The figure of 150 meV is consistent with the result of Tachikawa et al. [15] who observed D(x) centres in purer GaAs above ~ 15 kbar. There were only differences (10 meV) between the energies derived for Si and Sn. It should be noted that there is no inconsistency between this result which would indicate that the metastable level in high purity material would cross the Γ-band edge and become populated in the pressure range between 20 and 30 kbar and the results of Wolford and Bradley [13] which showed another level crossing close to band inversion at 40 kbar as the metastable level would not be observed directly in optical experiments, as was found with InSb. Any weak transistions are even less likely to be detected in photoluminescence as they would be swamped by the presence of intense direct transitions.

Extremely high local concentrations of silicon donors can be obtained with MBE by the spike of δ-doping technique where growth is interrupted by shutting off the Ga beam while at the same time opening a silicon shutter. The areal densities which can be achieved by this technique ($\sim 3 \times 10^{13} \text{cm}^{-2}$), exceed by between one and two order of magnitude than found in GaAs; GaAlAs heterostructures. The Munich group [16,17] have made extensive magneto-transport measurements of δ-doped layers and from Fourier analysis of the Shubnikov-de Haas effect have shown very pronounced two-dimensional behaviour with up to six sub-bands occupied.

The sum of the occupancies of the sub-bands was accurately equal to the doping level until a concentration of $2 \times 10^{13} \text{cm}^{-2}$ was reached. At this point the carrier concentration saturated with no further increase being found with increasing doping. A concentration of $2 \times 10^{13} \text{cm}^{-2}$ corresponds to a bulk doping level of $\sim 10^{20} \text{cm}^{-3}$ if the dopant is localised on a single atomic plane. At this concentration the Fermi energy is extremely high even when many sub-bands are filled and it seemed reasonable that the saturation in the free carrier concentration as a function of doping might arise from occupancy of localised states associated with higher order conduction-bands. However, it was also noted that the relative occupancies of the lowest (i = o) and first excited sub-band (i = 1) did not fit self-consistent calculations unless it was assumed that the silicon diffused significantly away from the initial dopant plane. Approximately 50 Å diffusion was required at a doping level of $5 \times 10^{12} \text{cm}^{-2}$ and 200 Å at $2 \times 10^{13} \text{cm}^{-2}$ to fit the experimental data. These areal densities would then both correspond to a bulk density of 10^{19}cm^{-3} which is close to the value at which silicon ceases to substitute solely on the gallium site and act a donor in bulk growth.

In order to determine the extent to which resonant impurities were playing a role in the δ-doped layers, high pressure measurements were undertaken as a collaborative project between Imperial College and the Technical University at Munich. The effect of the application of high pressure is unexpected in that the first order effect is an increase in amplitude rather than a very pronounced shift in position of the Shubnikov-de Haas oscillations. The increase in strength is associated with an improvement in the mobility of the carriers. This

effect is particularly pronounced in the case of lowest (i = o) sub-
band. In the case of the results shown in Fig. 5 for sample PD14

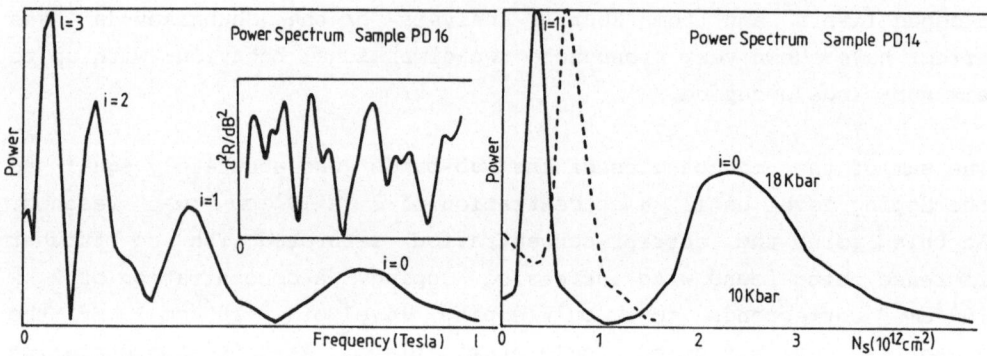

Figure 5 shows the Fourier analysis of the Shubnikov-de Haas data for two
-doped GaAs samples grown at Imperial College. The results for the lower
concentration sample are shown for two pressures. At 10 kbar the signal for
the i = 0 sub-band was below the noise in the spectrum and is not shown.

(grown in the Imperial College MBE Facility) the increase in mobility
approaches an order of magnitude for the i = o sub-band at 18 kbar.
In contrast, the decrease in carrier concentration is about 25% for
i = o sub-band with proportionally greater falls for the higher order
sub-band. An identical effect was seen with samples grown at
Stuttgart. The key to the understanding of this effect lies in the
diffusion of the dopants coupled with a consideration of the extent
and symmetry of the wavefunctions in the z-direction for the different
sub-bands. The i = o sub-band has an antinode at the original doping
plane and for $n = 5 \times 10^{12} cm^{-2}$ and no dopant diffusion,
self-consistent calculations show that $(z^2) \sim 25$ A. In contrast
$(z^2) \sim 75$ A for the i = 1 sub-band and, because there is a node in

the wavefunction at z = o the probability of finding the electron within 20 A of the doping plane is extremely low. Suppose that the dopants diffuse a distance of 50 A (a typical distance for n = 5 x $10^{12} cm^{-2}$) and that, for simplicity, the diffusion profile is slab of total extent 100 A. The extent of localised electronic states associated with higher order conduction band minima is likely to be < 10A as the binding energy to the minima which give them their character is > 100 meV. Consequently, these deep levels will have eigenvalues which are a function of distance z and will have a minimum value when z = o as this is where the potential energy is a minimum. Consequently, the impurities will become neutralised first at z= o as pressure is increased and then a neutralised zone will spread outwards from the z = o plane with increasing pressure. For the simple model chosen, when 50% of the donors are neutralised, this zone will extend 25 A from z = o. The carriers in the i = o sub-band will then move in a zone where there are no ionised impurity scatterers and a spatial separation of scatterers and carriers will have been achieved giving the mobility enhancement observed. In contrast, although the carriers in the i = 1 sub-band will sense the fall in the number of scatterers, the scattering sites which have disappeared are from a region which is remote from where the carriers are located. Although these ideas remain to be put on a quantitative footing, they can readily explain the features observed and provide corrobative evidence for diffusion of the dopant impurities which was first deduced from relatively small changes in relative occupancy of the lower sub-bands compared with predictions of self-consistent theory calculated with the assumption that the dopants were located on a single atomic plane.

Evidence that diffusion is of the correct order is provided by studies of a δ-doped superlattice of GaAs with 100 A period grown by MBE at the Philips Research Labs at Redhill. The amplitude of the Shubnikov-de Haas oscillations increased rapidly as a function of pressure and Fourier analysis revealed the occupancy of two sub-bands (see Fig. 6). The relative numbers are rather similar to a single δ-doped sample but the amplitude (mobility) of the contribution from the i = 1 sub-band is much smaller either because of superlattice effects which will be much greater for this sub-band because of the

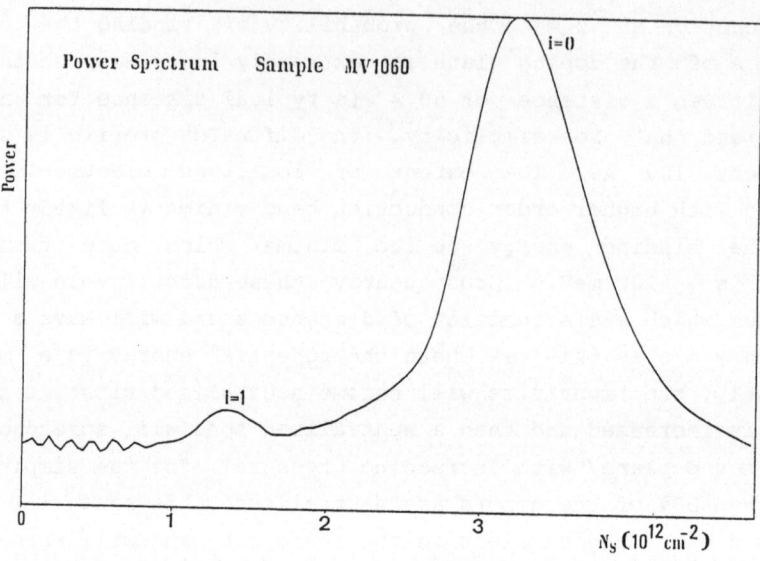

Figure 6   shows the Fourier analysis of the Shubnikov-de Haas effect from a
-doped GaAs superlattice at 15 kbar. The increase in amplitude of the signal
can be judged from the fact that Shubnikov-de Haas oscillations coud not be
observed at zero applied pressure.

greater z-extent of the wavefunctions or because scattering from
adjacent doping layers is significant in the superlattice. The
existence of a superlattice is confirmed by electrochemical profiling
which shows a modulation of at least a factor of two in carrier
concentration with the superlattice period. Both these results
demonstrate unambiguously that the diffusion of silicon is less than
50 A. Because of their sensitivity to diffusion, the Shubnikov-de
Haas measurements combined with high pressure techniques promise to
provide a measurement technique for determining diffusion profiles
with a precision unrivalled by SIMS or other profiling methods. The
contributions from Z. Wasilewski and R.L. Williams at Imperial College
and from F. Koch and A. Zrenner at Munich to the work presented at
this Summer School are gratefully acknowledged.

References

[1] J.D. Barnett, S. Block, and G.J. Piermarini. Rev. Sci. Inst. $\underline{44}$ 1 1973) and B. Welber, Rev. Sci. Inst. $\underline{47}$, 183 (1976).

[2] Z. Wasilewski, S. Porowski, and R.A. Stradling. J. Phys. E, $\underline{19}$, 480 (1986), also Solid State Comm. $\underline{57}$, 123 (1986).

[3] L. Konczewicz, E. Litwin-Staszewska, and S. Porowski, Proc. 3rd Conf. on Narrow Gap Semiconductros (Warsaw 1977), p. 211.

[4] S. Porowski, Proc. of the 4th Int. Conf. on Narrow Gap Semiconductors (Linz 1980), p. 420.

[5] S. Porowski, L. Konczewicz, M. Konczykowski, R. Aulombard and J.L. Robert, in: Proc. Int. Conf. on Phys. of Semiconductors (Kyoto 1980), p. 271.

[6] E. Litwin-Staszewska, W. Szymonska, and R. Piotrzkowski. Springer Lecture Notes in Physics (Narrow Gap Semiconductors) $\underline{152}$. [/ 397 (1981).

[7] Z. Wasilewski, A.M. Davidson, R.A. Stradling, and S. Porowski, Lecture Notes in Physics $\underline{177}$, p. 233. "Applications of High Magnetic Fields to Semiconductor Physics" (Grenoble 1982).

[8] M. Baj, L.C. Brunel, S. Huant, W. Trzeciakowski, Z.Wasilewski, and R.A. Stradling, Proc. Int. Conf. on Physics of Semiconductors (San Francisco 1984); C.J. Armistead, F. Kuchar, S.P. Najda, S. Porowski, C. Sotomayor-Torres, R.A. Stradling and Z. Wasilewski, Proc. Int. Conf. on Physics of Semiconductors (San Francisco 1984).

[9] Z.Wasilewski, and R.A. Stradling. Semicond. Sci. and Tech. $\underline{1}$, 264 (1986).

[10] G.E. Stillman, D.M. Larsen, C.M. Wolfe, and J.O. Dimmock. Solid State Comm. $\underline{9}$, 2245 (1971).

[11] M. Kobayashi, T. Yokoyama, and S. Narita, Jap. J. App. Phys. $\underline{22}$, 2612 (1983).

[12] M. Leroux, G. Pelous, F. Raymond, and C. Verie. App. Phys. Lett. $\underline{46}$, 288 (1985).

[13] D.J. Wolford, and J.A. Bradley, Solid State Comm. $\underline{53}$, 1069 (1985).

[14] D.K. Maude, J.C. Portal, L. Dmowski, L. Eaves, M. Nathan, M. Heiblum, J.J. Harris, and R.R. Beall. To be published.

[15] M. Tachikawa, T. Fujisawa, M. Kukimoto, A. Shibata, G. Oomi, and S. Minomura. Jap. J. Appl. Phys. 24, L893 (1985).

[16] F. Koch, A. Zrenner, and M. Zachau. Springer Series in Solid State Sciences Two Dimensional Systems 67, p175 (1986).

[17] A. Zrenner, H. Reisinger, F. Koch, K. Ploog, J.C. Maan. Phys. Rev. B33, 5607 (1986).

ELECTRONIC STRUCTURE OF COMPLEX DEFECTS IN SILICON

José R. Leite, Lucy V.C. Assali, and Vivili M.S. Gomes
Instituto de Física, Universidade de São Paulo
CP 20516, São Paulo, CEP 01498 SP, Brazil

Ab initio self-consistent-field electronic state calculations are carried out for transition metal-related and hydrogen-related complexes in silicon. The first theoretical investigation of the chemical trends in electronic properties of transition metal impurity pair complexes in a semiconductor is reported. Chemical trends in the electronic structure of pair complexes involving a boron atom and a transition metal impurity in silicon are also investigated. Results are presented for the hydrogen-boron and hydrogen-carbon complexes in silicon and attempts are made to explain the electrical properties of these systems. The calculations do not provide support for the currently accepted ionic model where the pair is described as two electrostatically bound point charges.

1. Introduction

Isolated point defects and/or impurities in a semiconductor may interact in the lattice giving raise to complexes. Defect complexes can be formed in heavily doped semiconductors, during irradiation and ion implantation processes, by doping the samples with two or more species and may also appear as native defects during crystal growth or can be formed by contamination. Complex of point defects and/or impurities have been studied for many years now by using several experimental techniques [1-8]. In the limit of strongly interacting impurities the complexes induce deep-levels, resonances and hyper-deep-levels in the electronic structure of the otherwise perfect crystal. Despite the large amount of experimental work that has been conducted for the class of deep-level complexes in semiconductors, only recently considerable efforts have been made to interpret the results in terms of rigorous theoretical calculations. Of course, we have to bear in mind that under a theoretical point of view, these systems represent a more complicated problem than the isolated defects, which themselves require a considerable computational effort to be described.

Starting with the early works conducted in the sixties and seventies [9-11] and concluding with recent publications [12-19], we found about sixty papers in the literature reporting on elaborated electronic state calculations for complex defects in semiconductors. As for the case of isolated impurities, the theoretical methods applied to the study of complexes are separated in two general groups. In one group are the cluster-type methods [10-14] and the other contains the band-based methods [15-19], depending on one starts with the complex surrounded by a finite cluster of host atoms or with the perfect-crystal electronic structure and a defect complex potential, respectively. In both cases semi-empirical and ab-initio approachs are used to solve the one-electron Schroedinger equation. As for the case of isolated impurities, these theoretical works have been contributing for the enhancement of our knowledge of the physical properties displayed by several defect complexes in semiconductors.

In this work we review our recent achievements in the understanding of the electronic properties of transition metal (TM) related- and hydrogen related-pair complexes in silicon. We report the first theoretical investigation of the chemical trends in the electronic properties of TM-related pair complexes in a semiconductor. We show that the chemical trends in the pairs of Au with V, Cr, Mn, Fe, Co, Ni and Cu in silicon arise mainly from covalency. Self-consistent one-electron state calculations are also carried out for the first time for the pairs of B with Fe, Mn and Cr. Finally, results are reported for the B-H and C-H complexes in silicon. The calculations yield microscopic models for the structure of these complexes in clear contrast with the currently used ionic model. We conclude that the covalent effects play a basic role in determining the physical properties of these defect complexes.

2. Theoretical Model

The calculations carried out in the present investigation were performed within the framework of the molecular cluster model. All the complexes analysed were considered in an atomic configuration where one impurity replaces a silicon host atom and the other sits at a nearest-neighbor interstitial site. The pair complex is then surrounded by 25 silicon atoms. The cluster model utilized in our studies of the complexes is schematically shown in Fig. 1. As a tetrahedral (T) site centered cluster in a C_{3v} symmetry, it comprises one central interstitial atom, one nearest-neighbor substitutional impurity and eight shells of silicon atoms.

The one-electron Schroedinger equation is solved for the molecular cluster by using the multiple-scattering theory developed by Slater and Johnson [20,21]. The molecular self-consistent potential is ob-

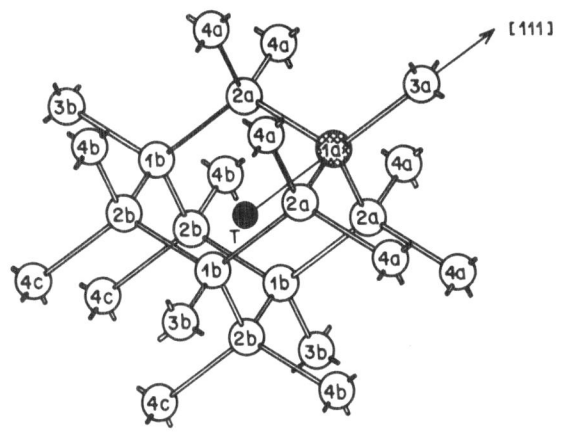

Fig. 1. Schematic representation of the cluster of 27 atoms simulating a pair complex in silicon formed by a substitutional atom nearby an interstitial impurity. As a T-centered cluster it comprises one central interstitial impurity and four shells of silicon atoms, numbered from 1 to 4, in T_d symmetry. By replacing a nearest neighbor silicon atom by another impurity the cluster symmetry is lowered to C_{3v}. The atoms in this symmetry belong to classes of equivalence which are indicated by the labels a, b, and c.

● INTERSTITIAL   ⊛ SUBSTITUTIONAL   ○ Si

tained within the framework of the local density functional theory according to the Xα approximation to the exchange-correlation energy [21]. The undesirable dangling-bond effects at the cluster surface are avoided by adopting the Fazzio, Leite and De Siqueira solution for the boundary condition problem [22]. According to this model the effects of the dangling bonds are neutralized by transferring the electrons filling these dangling bonds to a sphere surrounding the cluster (Watson sphere).

The so called Watson-sphere-terminated cluster model has been used by us to investigate several complex defects in silicon [23-35]. The particular cluster adopted here has already been used to obtain the electronic states of the Au-Fe, Fe-B and B-H pair complexes in silicon [30-35].

3. Pair Complexes Involving 3d Transition Metals in Silicon

3-1. Experimental aspects and the ionic model

It is well known since more than twenty years ago that the isolated interstitial 3d transition metals are very mobile in silicon, even at room temperature, forming complex pairs with both, shallow and deep level impurities [1]. Since the pioneering work [1] by Ludwig and Woodbury (LW) the electron paramagnetic resonance (EPR) technique has been used to determine the effective spin and the structure of these pairs [36-43]. They are associated to electrically active gap levels which have been characterized by several techniques such as deep level transient spectroscopy (DLTS) [44-51], thermally stimulated capacitance [44], photocapacitance [48], photoconductivity [48], photoluminescence (PL) [7,52], Hall effect [47,48], resistivity [53] and correlated EPR and Hall effect [54] measurements. Pair formation and dissociation reactions around room temperature have been extensively investigated by using DLTS [45-47]. Recently, stress and Zeeman PL measurements have contributed for the identification of the pair structure [7].

Since the early LW experiments [1] it has been confirmed by EPR [36-43] and PL [7] techniques that the pairs are usually aligned along the $\langle 111 \rangle$ direction, indicating that they may consist of a substitutional impurity with a TM occupying a nearby interstitial site. The neutral Fe-In pair was though to be the only exception to this rule, since the Fe impurity was found to be aligned along a $\langle 100 \rangle$ direction, probably placed at the next-nearest neighbor interstitial site [1]. However, recent EPR [41] and DLTS [49,50] investigations show that the bistable $\langle 111 \rangle$ and $\langle 100 \rangle$ configuration is also ascribed to the Fe-Al and Fe-Ga pairs.

Despite the great deal of experimental work that has been placed on the characterization of the TM-related pair complexes in silicon, the only theoretical tool which has been applied hitherto to deal with such systems is the so called ionic model [1]. No attempts to apply the recently developed band-based theoretical methods to the study of TM-related complexes have been made yet [56]. According to the ionic model, the pair stable configuration corresponds to a classical system consisting of an interstitial positive TM ion electrostatically bound to a nearest or next-nearest substitutional negative ion, embedded in a dielectric medium.

Although it seems to be clear from DLTS experiments that the electrostatic interactions between Fe^+ and B^- ions play a fundamental role in the Fe-B pair association reaction, these experiments do not demonstrate that the complex stabilizes as an ionic pair [45-47]. We point out that it would be difficult to explain by using the ionic model the formation mechanism of a pair like Au-Fe. As far as the Fe^+-Au^- structure is concerned [36], we remind that Fe has only one well established donor level [54] at $E_v + 0.38$ eV, therefore below the acceptor level of Au, at $E_c - 0.55$ eV [31]. On the other hand, the fact that isolated Fe has no acceptor levels in the gap [55] avoids the possibility of the pair formation reaction to be driven by the Coulomb interaction between the Fe^- and Au^+ ions in a compensate sample.

Recently the ionic model was applied to show that the Fe-B acceptor transition, (-/0), which occurs at 0.2 eV below the bottom of the conduction band, can be obtained from the Fe_i donor transition, (0/+), which occurs at 0.4 eV above the top of the valence band, by considering the pairing interaction energy [57]. The observed transitions suggest an intra-defect Coulomb energy of 0.55 eV, which corresponds to the nearest-neighbor Fe^+-B^- Coulomb interaction screened by the silicon dielectric constant. The same argument is used to obtain the Fe second donor transition, (+/++), from the measured value of the Fe-B donor transition, (0/+), which occurs at 0.1 eV above the top of the valence band. According to the model the intra-defect Coulomb interaction would position the isolated Fe second donor transition at 0.4 eV below the top of the valence band, in good agreement with theoretical calculations. Although this description seems to be fairly consistent as far as the Fe-B pair is concerned, it fails when applied to the Cr-B, Cr-Aℓ and Cr-Ga pairs.

The bistable properties of Fe-A pairs in silicon, A being a group-III A acceptor, have been recently described by using the ionic model [49,50]. By assuming that the two observed stable configurations correspond to the Fe-A pair in the ⟨111⟩ and ⟨100⟩ symmetries, the DLTS data taken for the pair were described in terms of a double-site configuration-coordinate diagram. By using the ionic model for the pairs, it was possible to correlate directly the experimental results obtained for the two configurations. A very good and a reasonable good agreement between the model and the data for Fe-Aℓ and Fe-Ga, respectively, was found. In contrast, the simple ionic model fails completely to describe the properties of the Fe-In pair [50].

Except for Fe-Aℓ and Fe-Ga, all the EPR spectra of the M-A complexes (where M represents a TM atom) have been described with g ≅ 2. Within the framework of the ionic model for Fe-B this fact implies in a strong quenching of the angular momentum L = 1 observed for the isolated Fe^+ ion [1]. Recently, this reduction in the orbital magnetism of the pair has been attributed to the covalent effects which bind the Fe impurity [42]. The pair binding energy determined by applying the ionic model is close to 0.5 eV [49]. The comparison between this result and the measured binding energy of the Au-V pair [44], 1.6 eV, provides another evidence that the effects of covalency are important for a realistic description of the TM-related pair complexes in silicon.

Therefore, we conclude from the analysis made above, that any effort to go beyond the ionic model in order to address the problem of modelling the microscopic structure of TM-related pair complexes in silicon, is highly desirable.

3-2. Gold-Transition Metal Complexes

In this section we report the results of our self-consistent one-electron state calculations for the neutral gold-transition metal pairs in silicon where Au replaces a Si atom (Au_s) and the TM impurity is placed at the nearest-neighbor interstitial site (M_i). The pair is simulated by using the cluster $25Si + Au_s M_i$, schematically shown in Fig. 1, where the TM atom is at the center of the cluster (T-site) and the Au impurity replaces one of the four nearest-neighbor Si atoms.

Fig. 2 summarizes the main features of our calculations. The Au 5d- and the TM 3d-derived energy levels, relative to the band edges, identified in the energy spectra of the clusters $25Si + Au_s M_i$, with M = Ni, Co, Fe, Mn and Cr, are shown. The energy spectra of these clusters simulate the electronic structure of the neutral Au-Ni, Au-Co, Au-Fe, Au-Mn and Au-Cr pairs, respectively. All the gap levels are shown. The band edges were defined according to the energy spectrum of the 26Si atom cluster [27]. The calculations were carried out within the framework of the spin restricted approximation and the one-electron levels were filled according to the ordering of increasing energy.

The two highest gap levels, which play a fundamental role in determining the physical properties of the complexes, are labelled as 1e and 2e in the spectra. We are indicating by triangles, for comparison, the approximate location of the 3d-derived $e(d)$ and $t_2(d)$ resonances induced by the T-site TM impurity in Si according to previous cluster model calculations [58]. The results depicted in Fig. 2 are complemented by the entries displayed in Table 1. There we show the difference between the 1e and 2e one-electron energies (ΔE), the charges, normalized to one electron, inside the TM spheres, for the 1e and 2e states ($Q(1e)$ and $Q(2e)$, respectively), and the total charge, $Q(M)$, in number of electrons, inside the TM spheres.

The analysis of the results leads to the conclusion that the pair impurity levels arise from the covalent interactions between the molecular orbitals of the isolated impurities. Neutral Au_s in Si gives rise to a dangling-bond-like t_2 level within the band gap, occupied by three electrons [59,60]. The 1e gap level and the a_1 resonance in the conduction band shown in Fig. 2, result from the crystal field splitting of the t_2 gap level when the symmetry is lowered from T_d for the Au_s to C_{3v} for the Au-M complex. On the other hand, an isolated T-site TM impurity gives rise to 3d-derived gap states (or resonances), with t_2 and e symmetries, which split in the crystal field symmetry C_{3v} of the pair into a non degenerate a_1 level and a pair of twofold degenerate e levels [56]. The highest of those

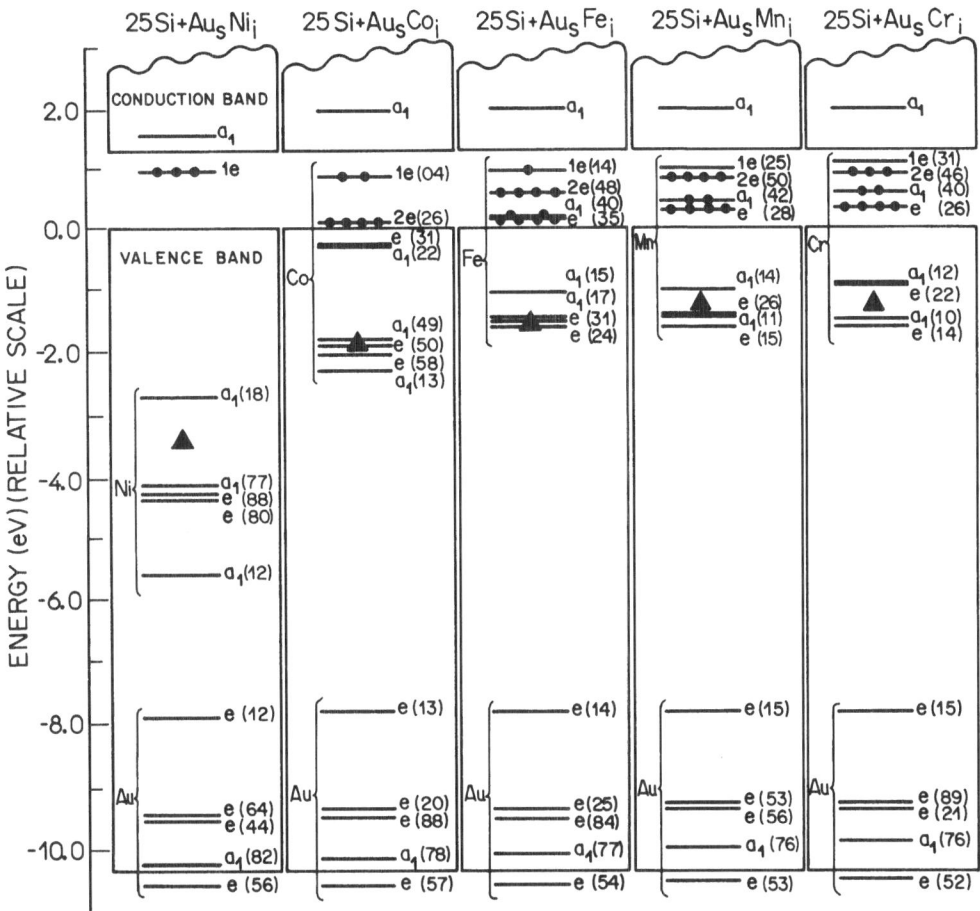

Fig. 2. Gold 5d- and TM 3d-derived energy levels, relative to the band edges, identified in the energy spectra of the clusters 25Si + Au$_S$M$_i$, with M = Ni, Co, Fe, Mn and Cr. All the obtained gap levels are shown. The numbers between parenthesis give the percentage of charge, normalized to one electron, within the Au and the corresponding TM spheres (radii = 1.18 Å). The solid circles indicate the occupancy of the gap levels and the triangles denote the approximate positions of the 3d-derived resonances e(d) and t$_2$(d) , induced by the TM impurities in silicon, according to Ref. 58.

levels is labelled as 2e in Fig. 2.

Fig. 3 shows schematically the model that we are going to use to describe the electronic properties of the Au-M complexes in silicon as it applies to the Au-Fe pair. The impurity gap levels of the pair, as originated from the interaction between molecular orbitals of the isolated impurities, are shown. The model has been previously applied to describe the physical properties of the Au-Fe and Au-Mn complexes in silicon [31,32].

The EPR spectra obtained for the pair of Au with Fe and Mn have

Table 1. Ground state properties of the neutral Au-M complexes: ΔE, in eV, is the difference between the 1e and 2e one-electron energies (see Fig. 2). Q(1e) and Q(2e) are the charges, normalized to one electron, inside the TM sphere for the 1e and 2e states, respectively. Q(M) is the total charge, in number of electrons, inside the TM sphere.

Complex	ΔE	Q(1e)	Q(2e)	Q(M)
Au–Ni	——	0.01	——	28.22
Au–Co	0.78	0.04	0.26	27.36
Au–Fe	0.37	0.14	0.48	26.36
Au–Mn	0.20	0.25	0.50	25.30
Au–Cr	0.18	0.31	0.46	24.31

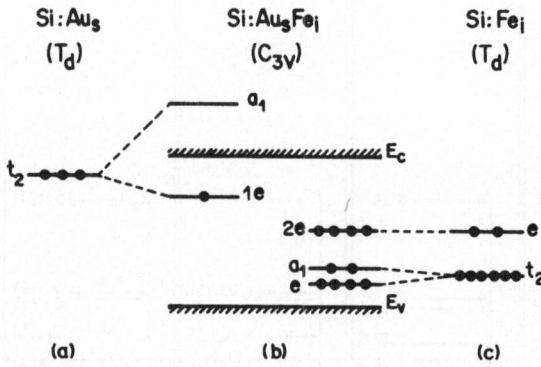

Fig. 3. Schematic representation of the impurity levels of the Au-Fe complex in silicon as originated from the interaction between molecular orbitals of the isolated impurities. The band edges are indicated and the full circles represent the occupancy of the levels. (a), (b), and (c) indicate the one-electron impurity levels for substitutional gold, for the Au-Fe complex, and for the T-site iron, respectively.

been interpreted previously by using the ionic model [36,43]. According to this description, the observed signals originate from the coupling between the angular momenta of two magnetic centers, described by the LW model, one centered on the gold and the other on the TM impurity. It has been shown, however, that rigorous calculations do not provide support for this description for the Au-Fe and Au-Mn pairs [31,32].

The first interesting feature that emerges from the present calculations is that the conclusion mentioned above can be extended to the pairs of gold with the remaining elements of the TM series. Firstly, we remark the strong localization of the Au 5d-derived resonances observed along the series. These compact and fully occupied levels result from the splitting by the C_{3v} crystal field of the e(d) and t_2(d) levels induced close to the bottom of the valence band by the Au_s impurity [59]. The perturbation caused by the TM impurity on the gold t_2(d) and e(d) levels is small, thus the 5d-derived levels remain regularly close to the bottom of the valence band along the series. Since

there is no gap levels with Au 5d-contribution, this impurity can not
be described by the LW model. Secondly, the values of Q(M) displayed
in Table 1 show clearly that there is no transfer of one electron from
the TM atom to the Au impurity. The values are systematically larger
than the corresponding atomic numbers of the TM atoms. It is worth mentioning that the TM spheres have radii equal to 1.18 Å, therefore much
smaller than the atomic radii of the TM atoms or of the same order of
the covalent radii of these impurities.

The overall analysis of the charge distribution associated to the
one-electron spectra shown in Fig. 2 allows us to conclude that the
pairs are formed by a covalent mechanism which includes, besides Au
and the TM impurities, also the Si neighbors. Therefore, the EPR parameters of the pair are related to molecular orbitals spread out over
the entire cluster rather than being derived from the interactions between two localized magnetic centers as has been assumed in the ionic
model [31,32].

We analyse now the trends displayed by the TM impurity - induced
levels which originate from the 3d atomic states. For the Au-Ni pair
the 3d-derived states are highly localized in the middle of the valence
band, displaying a small crystal field splitting. As we proceed to
lighter impurities the 3d states interact with the host states and
move up into the valence band.   Due to the hybridization with the a_1
and e valence states the 3d-derived resonances become progressively
more delocalized when we move to lower values of the TM impurity atomic
number. For the Au-Ni pair there are no gap levels with 3d character.
As we proceed to lighter impurities the 3d orbitals of the TM interact
with the host states and push the a_1 and e levels towards the band
gap. We observe the striking similarity between the chemical trends of
the 3d-derived impurity levels for the pair and for the TM themselves
(see the triangles in Fig. 1) [58].

The 1e is the only level in the gap for the Au-Ni pair. As we proceed to lighter TM impurities two simultaneous effects occur. The 2e
level moves up, becoming closer to the 1e level and the 3d composition
of the 1e level increases. The values of ΔE and Q(1e) displayed in
Table 1 show these effects quantitatively. The 1e gap state behaves as
a typical dangling-bond-like state for the Au-Ni and Au-Co pairs and
as a typical TM 3d-state for the Au-Mn and Au-Cr. Although there is a
non negligible 3d contribution for the 1e state of the Au-Fe pair,
the value of ΔE is large enough to assure that the ground state properties of this neutral pair are dictated by the 1e state occupied by
one electron [31].

We observe that the properties of the Au-Ni , Au-Co and Au-Fe pairs are defined by the 1e gap state since the 2e level is expected to remain fully occupied and below the 1e level. It is interesting to point out that the 1e gap state for the pairs and the t_2 gap state for Au_s have quite analogous dangling-bond-like nature [59,60]. Since the electrons filling the 1e level are occupying a delocalized state, the angular momentum is expected to be quenched and an effective low spin configuration is ascribed to the ground state of the complex. We observe that one and two electrons are transferred from the Au-derived 1e level to the TM 3d-derived 2e level for Au-Co and Au-Fe , respectively. Here the Haldane and Anderson mechanism [56], which is inherent in our covalent model of the complexes, prevents the crossing between the 1e and 2e levels and keeps the net charge inside the TM sphere approximately neutral.

For the Au-Mn and Au-Cr pairs the 1e gap state has a significative contribution from the 3d states, behaving as a typical gap level of a TM impurity. On the other hand, this level is so close to the 3d-derived 2e level that the exchange interaction may drive the system to a high spin configuration involving also the 2e level.

In Table 2 our microscopic model of the pairs is used to predict their ground state effective spins. The electronic configurations which determine the electrical, optical and magnetic properties of the complexes are indicated by $(1e\uparrow)^x(1e\downarrow)^y(2e\uparrow)^z(2e\downarrow)^w$, where x , y , z and w are parameters which give the occupancy of the levels and $\uparrow(\downarrow)$ denotes up(down) spin. The complexes which have been observed are indicated. We are defining the ground state of the complexes according to the criterium that for Cu, Ni, Co and Fe they correspond to low effective spin and for Mn, Cr and V they correspond to high effective spin configurations. The results for the Au-V and Au-Cu pairs were inferred by extrapolating the results of the calculations to the low and high atomic number limits of the TM impurities, respectively. The model, schematically shown in Fig. 3, was used to define the electronic configurations of the pair complexes.

Our predictions for the spins of the pairs are in agreement with the EPR data, except for the (Au-Mn)⁻ complex [1]. We argue, however, that in this case the observed spin S = 5/2 may correspond to the pair formed by Au and Mn, both at nearest-neighbor substitutional sites. This assumption is supported by the fact that Mn has been detected as an isolated impurity replacing a Si atom in the lattice, and the pair in this configuration would also display a trigonal symmetry [55]. Recent EPR experiments ascribe the value S = 3/2 for the spin of both, the (Cu-Mn)⁺ and (Cu-Mn)⁻ pairs, which are assumed to be stable according to the Cu_s-Mn_i structure [43]. The fact that Cu_s and Au_s are

Table 2. Electronic structure of Au-M complexes in silicon. The entries x , y , z and w are the occupancies of the highest gap levels 1e and 2e indicated in Fig. 3. The electronic configurations which determine the spin S of the complexes are $(1e\uparrow)^x(1e\downarrow)^y(2e\uparrow)^z(2e\downarrow)^w$, where $\uparrow(\downarrow)$ denotes up(down) spin. The complexes which have been observed are indicated. For the DLTS observations the indicated pair refers to the final state of the transition.

Complex	x	y	z	w	S	Observed
Au-Cu	2	2	-	-	0	No
(Au-Cu)+	2	1	-	-	1/2	No
Au-Ni	2	1	-	-	1/2	No
(Au-Ni)+	1	1	-	-	0	No
(Au-Ni)-	2	2	-	-	0	No
Au-Co	1	1	2	2	0	No
(Au-Co)+	1	0	2	2	1/2	No
(Au-Co)-	2	1	2	2	1/2	No
Au-Fe	1	0	2	2	1/2	EPR[a,b]
(Au-Fe)+	0	0	2	2	0	DLTS[b,c]
(Au-Fe)-	1	1	2	2	0	DLTS[b,c]
Au-Mn	2	0	2	0	2	No
(Au-Mn)+	1	0	2	0	3/2	EPR[d], DLTS[c]
(Au-Mn)-	2	0	2	1	3/2	EPR[d], DLTS[c]
Au-Cr	1	0	2	0	3/2	EPR[d]
(Au-Cr)+	0	0	2	0	1	DLTS[c]
(Au-Cr)-	2	0	2	0	2	No
Au-V	0	0	2	0	1	No
(Au-V)+	0	0	1	0	1/2	DLTS[c]
(Au-V)-	1	0	2	0	3/2	DLTS[c]

[a]Reference 36   [c]Reference 44

[b]Reference 37   [d]Reference 1

quite similar impurities in Si , indicates that the value S = 3/2 obtained by us for the spin of the (Au-Mn)⁻ pair may be correct [61].

3-3. Transition Metal - Boron Complexes

In this section we report the results obtained by us for the self-consistent electronic states of the Fe-B , Mn-B and Cr-B complexes in silicon. The pairs are considered in the $\langle 111 \rangle$ configuration with the TM impurity at the interstitial site (M_i) nearest to a substitutional boron atom (B_s). The pairs were simulated by considering the cluster schematically show in Fig. 1, where the TM impurity is at the center of the cluster and boron replaces a nearest-neighbor silicon atom.

The energy spectra of the $25Si + B_sM_i$ clusters, with M = Fe, Mn and Cr, simulating the electronic structure of the Fe-B, Mn-B and Cr-B pair complexes, respectively, are shown in Fig. 4. In order to better characterize the impurity levels induced by the pair complex, only the TM 3d- and B 2s- and 2p-derived levels are shown in the spectra. The band edges were defined according to the energy spectrum of the 26Si atom cluster and the calculations were not carried out to the spin po-larized limit.

According to a recent Watson-sphere-terminated cluster model calcu-lation, the relevant effect of introducing a B_s impurity into the si-licon lattice is the appearance of a t_2 acceptor level slightly above the top of the valence band [62]. Our calculations for the pairs show that there are strong covalent interactions between the molecular or-bitals of the isolated impurities. As a result the B_s acceptor level is pulled down into the valence band and splits into a non degenerate σ -like and a doubly degenerate π-like pair states. The t_2 acceptor state of B_s interacts primarily with the $t_2(d)$ resonance of the Fe_i im-purity which also splits into an a_1 and e levels. The Fe_i-derived e(d) resonance is only shifted by the crystal field of the complex. All these levels are placed near the crystal band edges and are label-led as $1a_1$, $2a_1$, 1e, 2e and 3e in Fig. 4.

The t_2 and e gap states induced by the isolated interstitial Fe, Mn and Cr impurities in silicon [56] give rise to the levels labelled as 4e, 5e and $3a_1$ in the spectra depicted for the pairs in Fig. 4. The interactions with the B_s molecular orbitals split the t_2 state into the 4e and $3a_1$ levels, with the latter crossing the 5e level and be-coming the highest gap level for the pair. This splitting is mostly due to the upwards displacement of the σ-like $3a_1$ level. One im-portant feature of our calculations is finding that the gap states of the M-B pairs are TM-like states. The strong localization of the pair gap states within the cluster central TM spheres emphasizes the TM 3d character of these impurity states.

Based on the analysis made above we are now able to propose a model which describes the formation of the impurity levels induced by the M-B pairs from the interaction between the molecular orbitals of the isolated impurities. The model is schematically shown in Fig. 5 as it applies to the Fe-B pair. The 3d-derived resonances induced by the isolated interstitial Mn and Cr impurities in the silicon valence band are slightly higher than the resonances induced by the isolated Fe atom. This is shown by the triangles in Fig. 4. As a consequence,

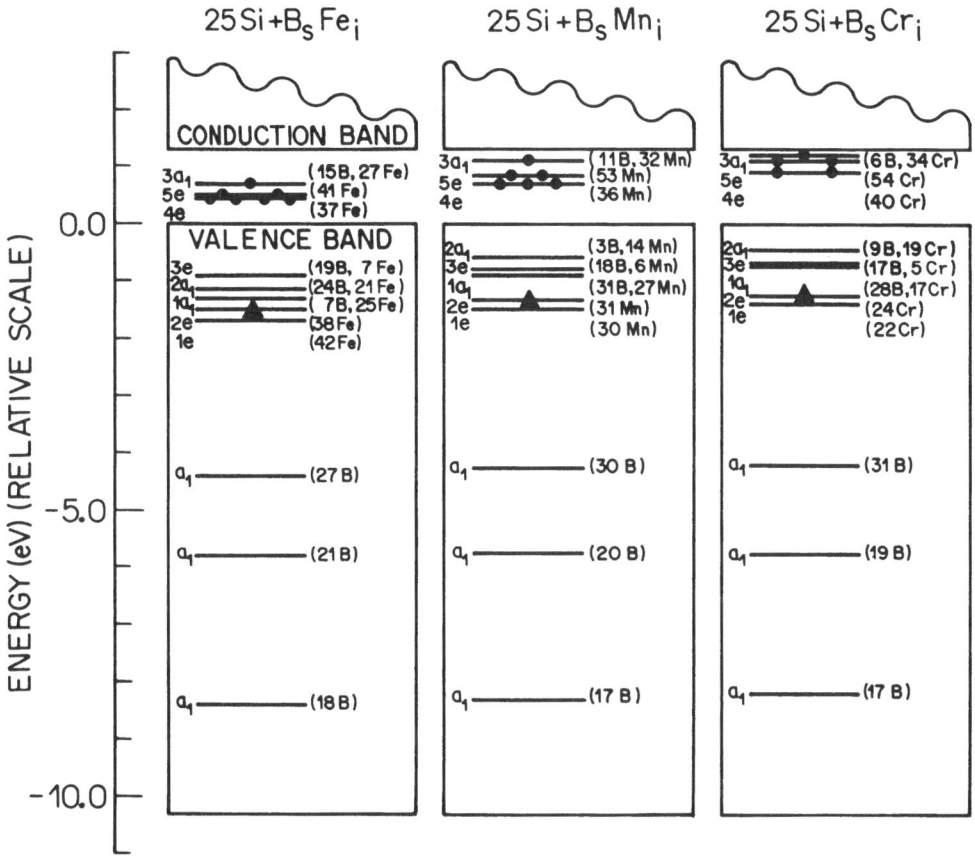

Fig. 4. TM 3d- and B 2s- and 2p- derived levels, relative to the band edges, identified in the energy spectra of the clusters $25Si + B_sM_i$, with M = Fe, Mn and Cr. All the obtained gap levels are shown. The number between parenthesis give the percentage of charge, normalized to one electron, within the B_s and the corresponding TM spheres (radii = = 1.18 Å). The solid circles indicate the occupancy of the gap levels and the triangles denote the approximate positions of the 3d-derived resonances e(d) and t_2(d), induced by the TM impurities in silicon, according to Ref. 58.

the interactions between the acceptor t_2 state derived from B_s and the t_2(d) and e(d) resonances induced by the TM impurities are stronger for the Mn-B and Cr-B than for the Fe-B pair. The results displayed in Fig. 4 show that the Mn-B and Cr-B induced $1a_1$, $2a_1$, 1e, 2e and 3e states mix together and a non degenerate level becomes the uppermost resonance within the valence band.

The analysis of the charge distribution within the TM spheres for the isolated TM and M-B gap states leads to an interesting conclusion related to the electronic structure of the pairs. Taking the Fe-B pair as an example, we observe from Fig. 4 that the seven electrons occupying

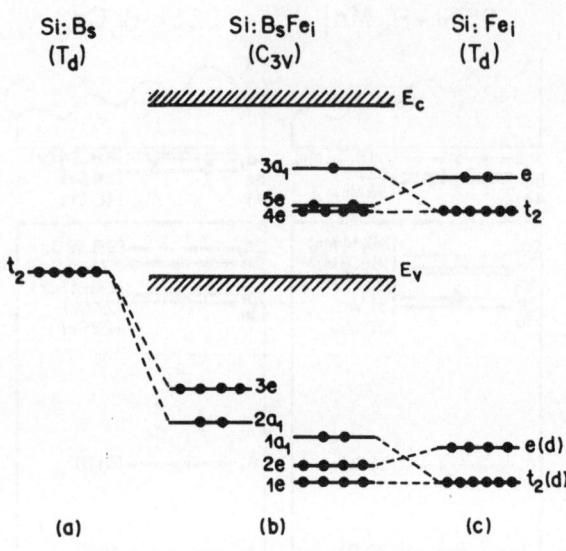

Fig. 5. Schematic representation of the Fe–B impurity levels, relative to the silicon band edges, as derived from the interactions between the impurity states of the isolated impurities. The filled circles indicate the occupancy of the levels. (a) Indicates the B_s shallow acceptor level. (b) Indicates the Fe–B resonances and gap levels, labelled according to Fig. 4. (c) Shows the Fe_i-derived $t_2(d)$ and $e(d)$ resonances and t_2 and e gap levels.

the Fe–B gap levels give the value 2.57e for the charge within the Fe_i sphere. On the other hand, the eight electrons occupying the t_2 and e gap levels lead to the value 2.56e for the charge associated with these states within the Fe_i sphere of the isolated impurity, according to the results estimated from Ref. 58. Therefore, the same amount of charge within the Fe_i atomic sphere is obtained in both cases. This result is in clear contrast with the assumption established in the ionic model that a Fe_i^+ ion is created by transferring one electron from the donor level of the isolated iron to the B_s impurity. Although there is a decrease in one unit of the gap level occupancy of the iron impurity when the pair is formed, the covalent effects, inherent in the Haldane and Anderson mechanism, prevent charge transfer out of the Fe_i impurity [56]. This phenomenon has been observed for the three pairs analysed here.

Although the calculations carried out for these pairs are not taking into account the spin polarization effects, we can infer from the significative localization of the pair gap states on the TM impurity spheres that the exchange interactions drive the complexes to high-spin configurations. The results shown in Fig. 4 lead to the conclusion that the C_{3v} crystal-field effects are strong enough to push the $3a_1$ level above the $5e$, keeping them close enough to favor spin alignment.

Our calculations yield a ^4A orbital singlet for the ground state of the Fe–B pair, implying that the values $L = 0$ and $S = 3/2$ are ascribed to the angular momentum and spin eigenvalues of the complex, respectively. The calculations indicate that the $(Mn-B)^+$ pair is a high spin complex with a ^6A ground state ($L = 0$, $S = 5/2$). Finally, the values

$L = 0$ and $S = 5/2$ are ascribed to the Cr-B pair, implying in a ^6A orbital singlet for the ground state of the complex. These results are in perfect agreement with the data obtained from EPR measurements [1]. We conclude that the pair complexes do not undergo Jahn-Teller distortions, therefore they are stable in a C_{3v} symmetry as has been observed.

As a last comment in this section we add that there is no clear indication from the calculations that there is a transfer of one electron from the TM to B_s, as is assumed in the ionic model. For the Fe-B pair, for example, the Fe_i atomic sphere contains 26.13e when the neutral impurity is isolated [58], whereas this value changes to 26.07 when the pair is formed.

4. Hydrogen-Related Complexes in Silicon

The passivation effects of group-III A acceptors in silicon by hydrogen have been deserving a great deal of attention in the last few years [63]. It has been demonstrated recently that the major part of the passivated acceptors results in acceptor-H pairs [64]. However, the microscopic structure of the acceptor-H complex formed have not unambiguously established so far [64,65].

Two microscopic models for the complex structure have been proposed. Pankove and collaborators have suggested that H is directly attached to a Si dangling bond with the acceptor being displaced towards the plane of its three Si neighbors, remaining effectively threefold coordinated [66]. This model has been assumed by DeLeo and Fowler to obtain the electronic structure and vibrational frequencies of the B-H and Aℓ-H pairs [67]. A rather different model has been proposed by Assali and Leite who suggested that H is placed at an interstitial site nearest to the acceptor impurity, in a $\langle 111 \rangle$ direction [33]. It has been demonstrated that both microscopic models account for the passivation of the acceptor activity and yield the infrared vibrational frequency 1870 cm^{-1} (1360 cm^{-1} for deuterium) that is observed [68] after passivation of the B acceptor [33,67]. The Pankove and collaborators' model accounts also for the frequency 2201 cm^{-1} recently measured for the Aℓ-H pair by Stavola and co-workers [64,67].

An alternative model to explain the passivation mechanism has been proposed by Sah and co-workers and recently by Pantelides [63]. These authors suggest that the passivation is a direct consequence of electronic compensation between substitutional B and interstitial H, which is assumed to have a donor level in the crystal band gap. However, the

electrical activity of H in a perfect Si crystal has not been established so far [63]. Some authors assume that H induces no levels in
the crystal band gap, others suggest that H has a donor level and
others ascribe an acceptor activity to H in silicon. As far as the
theoretical calculations are concerned the description of H in silicon
is also rather incomplete and ambiguous [63]. Moreover, there are other
pair complexes involving H in silicon whose structures are unknown,
despite the fact that several experiments have been conducted to investigate their properties [8]. It has been found, for example, that
the pair of H with group-IV impurity in germanium produces a shallow
acceptor. The source of acceptor activities found in these complexes
is also unknown. As has been pointed out previously [63], it would be
very informative to determine whether a C-H or Ge-H pair in silicon is
a shallow acceptor or shallow donor.

Now we address ourselves to the problems raised above, starting
with the questions whether isolated H is an electrically active or
inactive center in silicon. Self-consistent one-electron state calculations have been carried out by us for H at a tetrahedral interstitial
site in silicon. The cluster model adopted to perform the calculations
is schematically shown in Fig. 1. The H impurity, which is placed at
the center of the cluster, is surrounded by 26 Si atoms.

The one-electron energy spectra of the $26Si + H_i$ cluster are shown
in Fig. 6, along with the results obtained for a 26Si cluster used to
simulate the bulk band edges. The self-consistency in the calculations
was achieved firstly by filling the one-electron energy levels in order
of increasing energy. It is found that H induces a hyperdeep 1s-de-
rived a_1 level below the bottom of the valence band which is filled
with two electrons. Moreover, the t_2 level at the top of the valence
band is shifted down and appears as a resonance occupied with five
electrons. These results are depicted in Fig. 6(a). One electron is
them promoted from the H 1s-derived a_1 state to the highest occupied
t_2 state. The obtained spectrum is depicted in Fig. 6(b).

The overall analysis of the energy spectra shown in Fig. 6 allows
us to conclude that T-site H in silicon introduces a hyperdeep 1s
atomic like level below the bottom of the valence band, which suggests
a small interaction with its four silicon neighbors. The hyperdeep a_1
level shown in Fig. 6(b) is associated to a strongly localized state,
with 82% of charge inside the H sphere. The comparison between the
spectrum obtained for the 26Si atom cluster and that depicted in Fig.
6(b) shows that the perturbation caused by H in the bulk valence band
states is small. These conclusions are consistent with the observations that H is very mobile in silicon, which explains why no direct

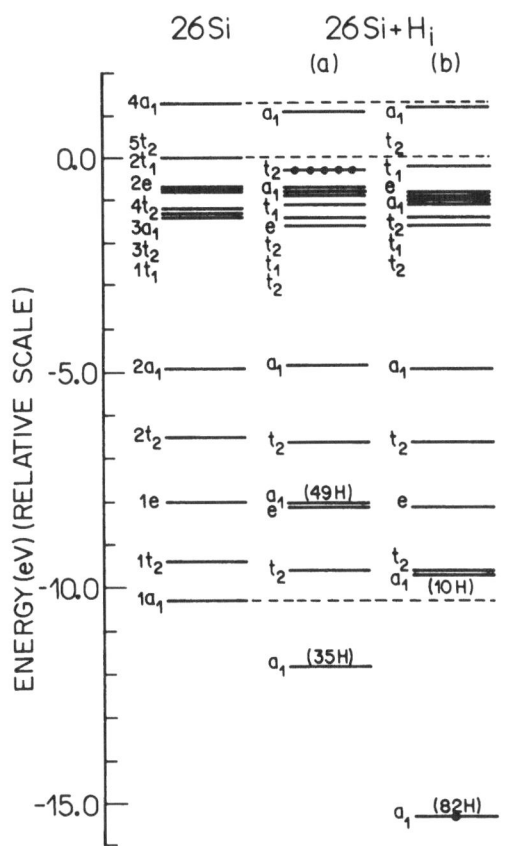

26Si   26Si+H_i

(a)   (b)

Fig. 6. Energy spectra of the 26Si and 26Si + H_i clusters simulating the electronic structure of Si and T-site H impurity, respectively. The dashed lines indicate the crystal band edges and the solid circles indicate the occupancy of the a_1 and t_2 levels in two different configurations indicated by (a) and (b). All the remaining levels below the zero of energy are fully occupied. The numbers between parenthesis give the percentage of charge within the H sphere (radius = 1.18 Å).

experimental information about its electrical activity is available.

The results of our calculations seem to indicate that T-site H in silicon does not induce impurity levels in the band gap. It is worth mentioning that this conclusion remains valid even with the inclusion of the spin polarization effects in the calculations, contrary to what has been pointed out recently [69]. We have verified that the results do not change significatively if the calculations are carried out to the spin polarized limit.

Nevertheless, it has been verified from theoretical work that the T-site is a position of minimum energy in the possible diffusion paths of H in silicon [70]. Thus, we are assuming that when H reaches a T-site near a substitutional impurity it gets trapped and a complex may be formed. This assumption, which was put forward by Assali and Leite to explain the passivation mechanism of boron [33], is here applied to investigate the electronic properties of the C-H pair. The cluster model depicted in Fig. 1 is used to simulate the electronic structure of the B-H and C-H pairs, where H is placed at the center of the cluster and B or C is placed at the nearest-neighbor substitutional site.

The energy spectra of the clusters $25Si + B_sH_i$ and $25Si + C_sH_i$ are shown in Fig. 7. The results obtained for the $26Si + H_i$ cluster, simulating the electronic structure of the T-site H impurity is also shown, for comparison. The first interesting feature that emerges from the spectrum obtained for the B-H pair is the strong interactions observed

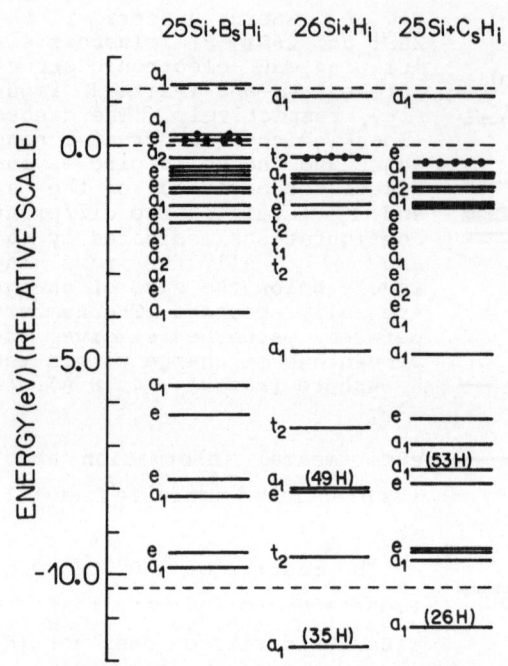

Fig. 7. Energy spectra of the 25Si + B_SH_i , 26Si + H_i and 25Si + C_SH_i clusters simulating the electronic structure of B-H, T-site H and C-H systems in silicon, respectively. The solid circles indicate the occupancy of the levels close to the top of the valence band. The meaning of the dashed lines and number between parenthesis are the same as in Fig. 6.

between the 1s state of H with the B and Si states. These interactions are such that the hyperdeep level induced by H disappears and no well identified H levels can be found in the spectrum. The acceptor t_2 level induced by the isolated B splits into a_1 and e levels which are fully occupied within the band gap. The results indicate that the pair is formed and the acceptor activity of B is neutralized.

However, a rather different picture emerges from the calculations carried out for the C-H complex. The interesting feature is the striking similarity observed between the spectra of the 26Si + H_i and 25Si + C_SH_i clusters. According to our calculations isolated substitutional C in Si is an electrically inactive center, in agreement with experiment [71]. Thus, the H impurity feels the presence of C as it was a host Si atom. The results seem to indicate that the pair is not formed in this configuration. They also seem to indicate that the C-H pair in Si, if exists, has a smaller binding energy than the B-H complex.

The comparison between the results obtained for the B-H and C-H pairs in Si may allow us to reach an interesting conclusion. The electronic structure of the C-H complex shows that the H 1s-derived state induces an a_1 level in the spectrum. This fact does not occur for the B-H complex. If we assume that H moves towards the C atom, allowing for a small binding, we expect that the 1s-derived level moves up penetrating the valence band. As a consequence the top of the valence band, a t_2 state, splits in the crystal field of the complex and a non-degenerate hole state may be pushed within the band gap. The small binding energy expected for the pair explains the tunnelling of the H impurity and the hole state in the gap explains the shallow acceptor activity observed for the C-H pair in germanium, for example [8].

References

1. G.W. Ludwig and H.H. Woodbury, in "Solid State Physics", Vol. 13 (Academic, New York, 1962) eds. F. Seitz and D. Turnbull, pp. 223-304
2. G.D. Watkins, in "Lattice Defects in Semiconductors, 1974" (Institute of Physics, London, 1975) ed. F.A. Huntley, pp. 1-22
3. For a general Review see "Defect Complexes in Semiconductor Structures", Lecture Notes in Physics, Vol. 39 (Springer-Verlag, Berlin, 1983) eds. J. Giber, F. Beleznay, I.C. Szép, and J. László
4. P. Wagner, C. Holm, E. Sirtl, R. Oeder, and W. Zulehner, in "Festkörperprobleme: Advances in Solid State Physics" Vol. XXIV (Vieweg, Braunschweig, 1984) ed. P. Grosse, pp. 191-228
5. C.A.J. Ammerlaan, M. Sprenger, R. van Kemp, and D.A. van Wezep, Mat. Res. Soc. Symp. Proc. 46, 227 (1985)
6. J.W. Corbett, J.C. Corelli, U. Desnica, and L.C. Snyder, Mat. Res. Soc. Symp. Proc. 46, 243 (1985)
7. R. Sauer and J. Weber, Physica 116B, 195 (1983)
8. E.E. Haller, in "Festkörperprobleme: Advances in Solid State Physics" Vol. XXVI (Vieweg, Braunschweig, 1986) ed. P. Grosse, pp. 203-229
9. J. Callaway, Phys. Rev. 164, 1043 (1967)
10. V.A. Singh, C. Weigel, J.W. Corbett, and L.M. Roth, Phys. Stat. Sol. (b) 81, 637 (1977)
11. C.A.J. Ammerlaan and J.C. Wolfrat, Phys. Stat. Sol. (b) 89, 85 (1978)
12. A. Oshiyama and S. Ohnishi, Phys. Rev. B 33, 4320 (1986)
13. G.S. Khoo and C.K. Ong, J. Phys. C 20, 1385 (1987)
14. A.A. Bonapasta, A. Lapiccirella, N. Tomassini, and M. Capizzi, Phys. Rev. B (1987), in press
15. G.G. DeLeo, W.B. Fowler, G.W. Barry, and M. Besson, Mater. Sci. Forum 10-12, 31 (1986)
16. N.M. Johnson, C. Herring, and D.J. Chadi, Phys. Rev. Lett. 56, 769 (1986)
17. R.W. Jansen and O.F. Sankey, Phys. Rev. B 33, 3994 (1986)
18. C.M. Weinert and M. Scheffler, Phys. Rev. Lett. 58, 1456 (1987)
19. G.A. Baraff and M. Schlüter, Phys. Rev. Lett. 56, 1214 (1986)
20. K.H. Johnson, in "Advances in Quantum Chemistry" Vol. 7 (Academic, New York, 1973) ed. P.O. Löwdin, pp. 143-185
21. J.C. Slater, "The Self-Consistent Field for Molecules and Solids" (McGraw-Hill, New York, 1974)
22. A. Fazzio, J.R. Leite, and M.L. De Siqueira, J. Phys. C 12, 513 (1979); 14, 3469 (1979)
23. M.J. Caldas, J.R. Leite, and A. Fazzio, Phys. Rev. B 25, 2603 (1982)
24. A. Fazzio, J.R. Leite, and M.J. Caldas, Physica 116B, 90 (1983)
25. J.R. Leite, V.M.S. Gomes, L.V.C. Assali, and L.M.R. Scolfaro, J. Electron. Mater. 14a, 885 (1985)
26. V.M.S. Gomes, L.V.C. Assali, J.R. Leite, A. Fazzio, and M.J. Caldas, Solid State Commun. 53, 841 (1985)
27. V.M.S. Gomes and J.R. Leite, Appl. Phys. Lett. 47, 824 (1985)
28. V.M.S. Gomes, L.V.C. Assali, and J.R. Leite, Int. J. Quant. Chem. S20, 749 (1986)
29. V.M.S. Gomes and J.R. Leite, Mater. Sci. Forum 10-12, 905 (1986)
30. L.V.C. Assali and J.R. Leite, Mater. Sci. Forum 10-12, 55 (1986)
31. L.V.C. Assali, J.R. Leite, and A. Fazzio, Phys. Rev. B 32, 8085 (1985)
32. L.V.C. Assali and J.R. Leite, Solid State Commun. 58, 577 (1986)
33. L.V.C. Assali and J.R. Leite, Phys. Rev. Lett. 55, 980 (1985); 56, 403 (1986)

34. L.V.C. Assali and J.R. Leite, in "18th International Conference on the Physics of Semiconductors" Vol. 2 (World Scientific, Singapore, 1987) ed. O. Engstrom, pp. 999-1002
35. L.V.C. Assali and J.R. Leite, Phys. Rev. B 36, 1296 (1987)
36. E.G. Sieverts, S.H. Muller, C.A.J. Ammerlaan, R.L. Kleinhenz, and J.W. Corbett, Phys. Status Solidi B 109, 83 (1982)
37. S.D. Brotherton, P. Bradley, A. Gill, and E.R. Weber, J. Appl. Phys. 55, 952 (1984)
38. J. Kreissl and W. Gehlhoff, Phys. Status Solidi (b) 112, 695 (1982)
39. W. Gehlhoff and K.H. Segsa, Phys. Status Solidi (b) 115, 443 (1983)
40. H. Conzelmann, K. Graff, and E.R. Weber, Appl. Phys. A 30, 169 (1983)
41. J.J. van Kooten, G.A. Weller, and C.A.J. Ammerlaan, Phys. Rev. B 30, 4564 (1984)
42. C.A.J. Ammerlaan and J.J. van Kooten, Mat. Res. Soc. Symp. Proc. 46, 525 (1985)
43. H. Dietrich, H. Vollmer, and R. Labush, Solid State Commun. 58, 811 (1986)
44. H. Lemke, Phys. Stat. Sol. (a) 64, 215 (1981); 75, 473 (1983)
45. K. Graff and H. Pieper, J. Electrochem. Soc. 128, 669 (1981)
46. L.C. Kimerling and J.L. Benton, Physica 116B, 297 (1983)
47. K. Wunstel and P. Wagner, Appl. Phys. A 27, 207 (1982)
48. K. Wunstel, K.-H. Froehner, and P. Wagner, Physica 116B, 301 (1983)
49. A. Chantre and D. Bois, Phys. Rev. B 31, 7979 (1985)
50. A. Chantre and L.C. Kimerling, Mater. Sci. Forum 10-12, 387 (1986)
51. S.D. Brotherton, P. Bradley, and A. Gill, J. Appl. Phys. 57, 1783 (1985); 1941 (1985)
52. T.E. Schlesinger and T.C. McGill, Phys. Rev. B 28, 3643 (1983)
53. W.H. Shepherd and J.A. Turner, J. Phys. Chem. Solids 23, 1697 (1962)
54. H. Feichtinger, J. Oswald, R. Czaputa, P. Vogl, and K. Wunstel, J. Electron. Mater. 14a, 855 (1985)
55. E.R. Weber, Appl. Phys. A 30, 1 (1983)
56. A. Zunger, in "Solid State Physics", Vol. 39 (Academic, New York, 1986) eds. H. Ehrenreich and D. Turnbull, pp. 275-464
57. L. Samuelson and P. Omling, in "18th International Conference on the Physics of Semiconductors", Vol. 2 (World Scientific, Singapore, 1987) ed. O. Engström, pp. 931-934
58. G.G. DeLeo, G.D. Watkins, and W.B. Fowler, Phys. Rev. B 23, 1851 (1981)
59. J.L.A. Alves and J.R. Leite, Phys. Rev. B 30, 7284 (1984)
60. A. Fazzio, M.J. Caldas, and A. Zunger, Phys. Rev. B 32, 934 (1985)
61. J.G. Angelo, L.V.C. Assali, R. Pintanel, V.M.S. Gomes, and J.R. Leite, Solid State Commun. (submitted)
62. M.A. Bunin, Yu.A. Matveev, N.A. Petrov, and Yu.V. Sukhetskii, Sov. Phys. Semicond. 17, 1291 (1983)
63. For a recent discussion related to the subject, see S.T. Pantelides, Mater. Sci. Forum 10-12, 573 (1986); Appl. Phys. Lett. 50, 995 (1987)
64. M. Stavola, S.J. Pearton, J. Lopata, and W.C. Dautremont-Smith, Appl. Phys. Lett. 50, 1086 (1987)
65. M. Stutzmann, Phys. Rev. B 35, 5921 (1987)
66. J.I. Pankove, P.J. Zanzucchi, C.W. Magee, and G. Lucovsky, Appl. Phys. Lett. 46, 421 (1985)
67. G.G. DeLeo and W.B. Fowler, Phys. Rev. B 31, 6861 (1985); J. Electron. Mater. 14a, 745 (1985)
68. N.M. Johnson, Phys. Rev. B 31, 5525 (1985)
69. H. Katayama-Yoshida and K. Shindo, Phys. Rev. Lett. 51, 207 (1983)
70. S.J. Pearton, J.W. Corbett, and T.S. Chi, Appl. Phys. A 43, 153 (1987)
71. R.C. Newman, Mat. Res. Soc. Symp. Proc. 46, 459 (1985)

ELECTRON MICROSCOPY IN SEMICONDUCTOR PHYSICS

J. Heydenreich
Institut für Festkörperphysik und Elektronenmikroskopie
der Akademie der Wissenschaften der DDR, Weinberg 2,
DDR-4050 Halle/Saale, German Democratic Republic

1. Introduction

Electron microscopy (EM) with its high spatial resolution is especially suitable for the investigation of defects in semiconductors.
Here, particularly the investigation of crystal defects and of phase
boundaries is concerned including the detection of the interaction
between point defects and extended defects (point defect clustering,
precipitation, decoration of extended defects). With respect to the
defects to be investigated – as-grown defects, process-induced defects – based on their reliable geometrical characterization the
estimation of their electrical and/or optical activity is of importance. Within the materials to be investigated elemental semiconductors (Si, Ge) and compound semiconductors – A_3B_5 compounds, A_2B_6
compounds, also ternary compounds – are of equal interest. Furthermore, interfaces between semiconducting materials and isolating
layers (SiO_2, Si_3N_4) and conducting layers (e.g. silicides) are topical subjects of investigation.

From the wide field of applying electron microscopic methods to
semiconductor physics and semiconductor technology – transmission EM
(TEM), surface EM, especially scanning EM (SEM), in-situ EM- the
following topics have been selected and will be treated briefly:
– Geometrical characterization of defects by TEM
– Estimation of the electrical activity of defects by special SEM
  techniques

2. Geometrical characterization of defects by TEM

Because of the strong dependence of the properties of semiconducting materials and semiconductor devices on the presence of defects in the used materials the detection of defects and the exact
characterization of their geometrical structure are very important.
TEM (also scanning transmission electron microscopy /STEM/) is especially suitable for this. – In the characterization of defects in
semiconductors (see e.g./1,2/) apart from the investigation of extremely pure materials /3/ for basic research in semiconductor phys-

ics, the interest in detecting process-induced defects /4-8/ is increasingly growing in semiconductor technology. This is connected with the ever-growing quality of as-grown semiconducting materials - for a review on microdefects (swirls) see /9/ - so that defects produced during treatments necessary for the production of microelectronic and optoelectronic devices (diffusion, oxidation, ion implantation, annealing, epitaxial processes, including graphoepitaxy) are becoming more and more influential. Special interest in the geometrical structure of defects (see e.g. /10/) is given to the structure of dislocation cores /11/, to impurities in the dislocation core /12/, to moving dislocations /13/, to defects in interfaces /14-21/, to defect arrangements in surfaces /22/, and to defect complexes, which arise during irradiation (e.g. ion implantation and subsequent Laser annealing) of materials (see e.g. /23/). Furthermore, the formation of defect complexes is of great importance.

To the imaging of defects in semiconductors (crystalline specimens) TEM is applied at different levels of imaging. On the one hand, diffraction contrast imaging is used (see Fig. 1a), where the image is formed by using only the undiffracted beam (bright-field image) or alternatively also only one diffracted beam (dark-field image). In this way solely disturbed regions of the crystal, i.e. crystal defects, are detected, which differ from the ideal lattice arrangement, with no lattice structure being imaged. As an example of this kind of imaging a CuSi precipitate is shown, the stress field around which has created punching dislocations in a three-dimensional arrangement. The defects were formed in long-time annealed FZ-Si containing a large amount of Cu. In the case of high-resolution imaging (ray path in Fig. 1b) the image is formed by the interference of electron waves diffracted at the specimen in defined directions and by the undiffracted electron wave, thus allowing the resolution of the lattice structure itself under optimum conditions. As an example the $\langle 110 \rangle$ lattice image of a stacking fault tetrahedron (intrinsic stacking fault) in silicon (deformed at $680^{\circ}C$) is given.

2.1 Diffraction contrast imaging

The EM diffraction contrast, first applied in 1956 by Hirsch and co-workers /24/ and by Bollmann /25/, and theoretically founded by Howie and Whelan /26,27/ is now used as a routine method of crystal defect imaging. The mathematical formalism of describing the interaction process between the imaging electrons and the specimen is based either on plane waves /26/, or on Bloch waves /28/, or on

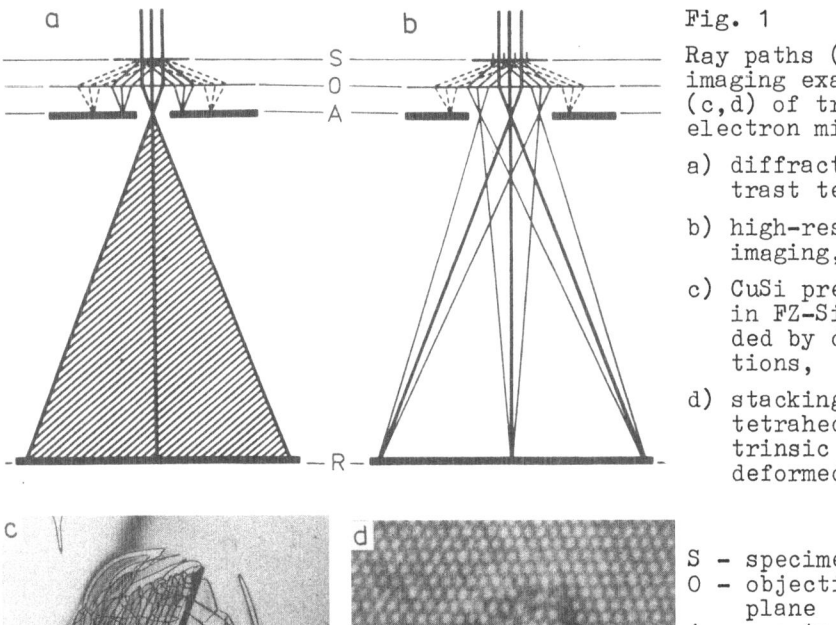

Fig. 1

Ray paths (a,b) and imaging examples (c,d) of transmission electron microscopy:

a) diffraction contrast technique,

b) high-resolution imaging,

c) CuSi precipitate in FZ-Si surrounded by dislocations,

d) stacking fault tetrahedron (intrinsic type) in deformed Si.

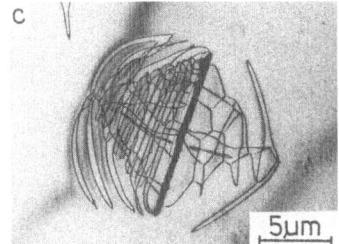

5µm

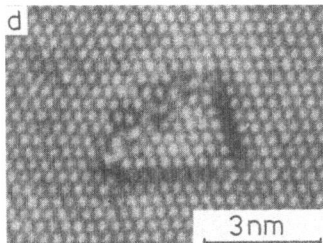

3nm

S – specimen
O – objective lens plane
A – aperture
R – registration plane

modified Bloch waves /29/. The interaction between the electron beam and the specimen necessary to produce the image contrast of a thin transmissible crystal foil is generally regarded in terms of the so-called column approximation, in which only those interactions between the electron beam and the specimen are regarded which occur within a column of atomic dimensions. The result, which is of interest, is the intensity of the transmitted and the diffracted electron beam – a two-beam case is assumed – at the exit surface of the crystal foil. For the transmitted intensity the distribution at the exit surface of the crystal corresponds to the bright-field image, and for the diffracted intensity it corresponds to the dark-field image. Important parameters for the imaging process are: ξ_g –the extinction distance (which implies the specifics of the diffracting material and of the applied Bragg reflection), ξ_o' and ξ_g' – the parameters of the normal and the anomalous absorption, s_g – a quantity defining the deviation from the Bragg orientation, $\vec{g}$ – the diffraction vector, and $\vec{u}$ – the displacement vector for the atoms or molecules in the neighbourhood of the crystal defect. The last-mentioned parameter $\vec{u}$ depends on the strain field of the defect. The elasticity theory offers analytical expressions for the displacement vector, e.g. for different types of dislocations or for spherical inclusions. The types of defects investigated in simple cases can successfully be analysed by applying standard rules of interpretation /30,31/, which in the majority of cases are based on the use of special specimen orientations in such a way that the defect contrast vanishes. In more complicated cases a computer simulation of the imaging process is necessary /32,33/. Using the classical diffraction contrast imaging technique it can be shown that the image width of dislocation profiles is about 1/3 to 1/5 of the extinction distance, which usually ranges from some 10 nm to a few 100 nm. In the so-called weak-beam technique first applied by Cockayne et al. /34/ image widths of dislocation profiles of some nm can be reached, thus allowing e.g. the observation of dissociated dislocations. In the weak-beam mode the orientation of the crystal relative to the incident beam is turned away from the exact Bragg orientation by realizing the deviation parameter of $s_g > 2 \quad 10^{-2} \, \text{Å}^{-1}$ and by taking a dark-field image with a weak reflection. Furthermore, in high-voltage electron microscopy (HVEM) the so-called high-order bright-field technique /35/ yields also defect micrographs with a resolution comparable to that of the weak-beam technique. Finally, as further higher-resolving diffraction contrast techniques, methods applying forbidden reflections in the imaging process shall be mentioned /36-38/.

Fig. 2, a weak-beam electron micrograph, shows the transformation of rod-like defects in boron-implanted Si to Frank partial loops (see arrows). The micrograph is related to a silicon sample implanted with an ion dose of 10^{15} cm^{-2} (implantation energy: 60 keV) and annealed at about 1000 K (for 30'). The transformation process itself can be explained by the agglomeration of self-interstitials to the initially rod-like defects, which is supported by the high supersaturation of interstitials due to the relatively high ion dose.

As an example of the investigation of compound semiconductors, a growth-induced stacking fault pyramid in a VPE epitaxial layer of InP on ⟨100⟩ InP is shown in Fig. 3. Defects of this type act as efficient obstacles to the movement of glide dislocations. The micrograph clearly reveals the capture of glide dislocations, which – depending on the Burgers vector relations – results in the formation of regions of multiple stacking faults bound by partial dislocations. Dislocations with screw components pinned in such a way can give rise to unwanted hillocks /39/ due to spiral growth.

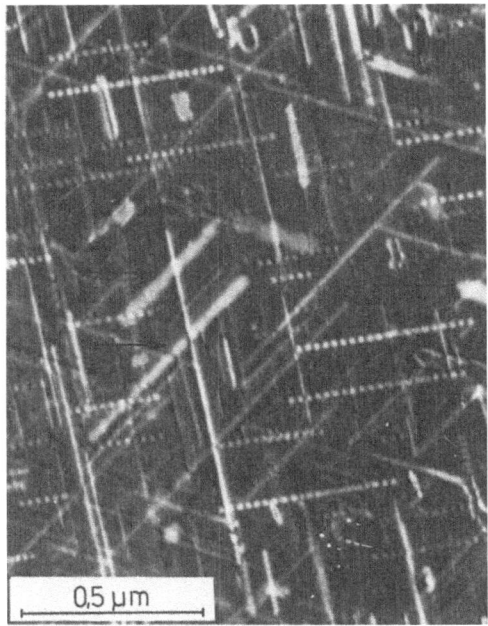

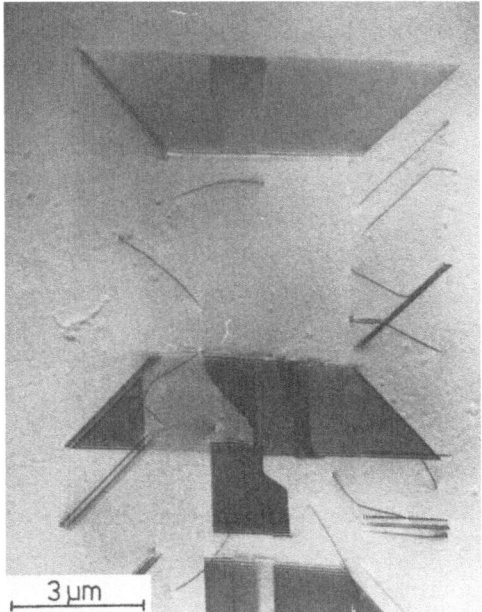

Fig. 2
Frank partial loops in boron-
implanted Si (see arrows), formed
by transformation of rod-like defects

Fig. 3
Stacking fault pyramid in a VPE
epitaxial layer of InP on ⟨100⟩
InP acting as obstacle to the
movement of glide dislocations.

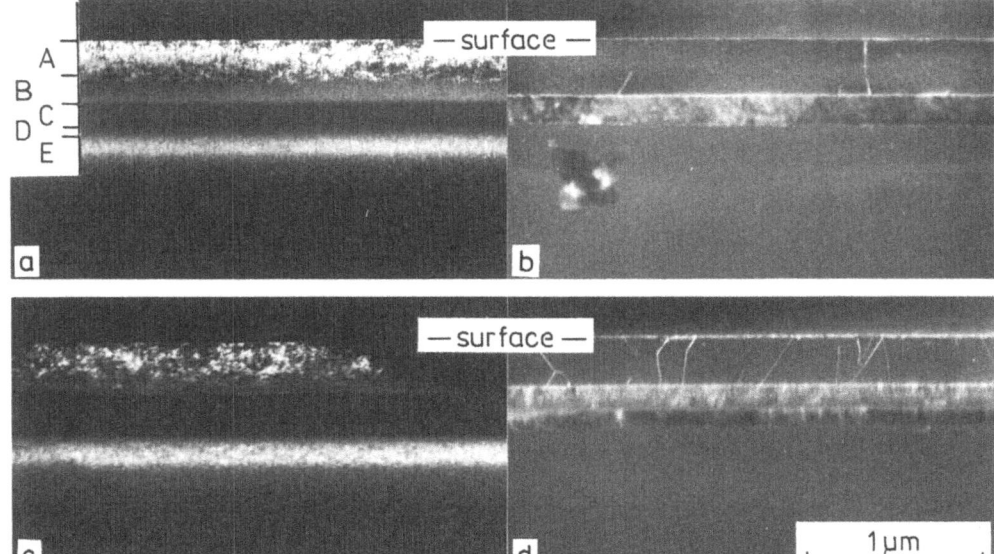

Fig. 4   Vertical views of N^+-implanted SOI structures (TEM dark-
field images): a) as-implanted (D = 6 10^{17} cm^{-2}, E = 165 keV),
b) annealed (T_a = 1470 K, t_a = 5 h), c) as-implanted (D = 10^{18} cm^{-2},
E = 165 keV), d) annealed (T_a = 1470 K, t_a = 5 h).

The investigation of interfaces and layer structures is successfully realized by the observation of specimen cross sections (for review see /40/). As an example Fig. 4 shows cross section electron-micrographs of N^+-implanted SOI structures (implantation energy: 165 keV) /41,42/. The dark-field images refer to the following treatments: a) as-implanted ($D = 6 \cdot 10^{17}$ cm^{-2}), b) annealed ($T_a = 1470$ K, $t_a = 5$ h), c) as-implanted ($D = 10^{18}$ cm^{-2}), d) annealed ($T_a = 1470$ K, $t_a = 5$ h). Beneath the surface the as-implanted specimens (a,c) show (from top to bottom) a monocrystalline Si layer (A) containing defects, a polycrystalline Si_xN_y layer (B), an amorphous Si_3N_4 layer (C), again a polycrystalline Si_xN_y layer (D), and again a monocrystalline Si layer (E) containing defects. Beneath the monocrystalline Si layer (with some dislocations) the annealed specimens (b,d) show a monocrystalline Si_3N_4 layer containing defects. In the higher-dose implanted specimen (d) the monocrystalline Si_3N_4 layer is divided into an upper and a lower part by an amorphous superstoichiometric intermediate layer (having a high quantity of N).

For the routine diffraction contrast work the tolerable specimen thicknesses are in the range of some 0.1 /um. In situ investigations demand higher specimen thicknesses (some /um), since solely sufficiently thick specimens allow conclusions to be drawn on the defect behaviour in bulk materials. These investigation modes require the application of high-voltage electron microscopy (HVEM) in which the dynamic behaviour of materials under the action of defined influences can be observed. Investigations of this kind are related to the formation, the transformation and the dynamic properties (e.g. mobility /43,44/) of extended crystal defects, phases and phase boundaries. According to the technological processes applied to microelectronics and optoelectronics in situ investigations during defined heat and radiation treatments (see e.g. /45-49/) are of special interest to semiconductor research.

2.2 High-resolution imaging

Since the first lattice fringe observations carried out by Menter /50/ in 1965 and the lattice images (of complex oxides) taken by Allpress et al. /51/ in 1969, high-resolution EM has been widely applied, especially to semiconductor research. With respect to the aspired lattice resolution (a few 0.1 nm) - unlike in diffraction contrast imaging - not only the interaction process between specimen and electron beam should be considered but the whole imaging process, which has to be regarded as a double Fourier transform. The short-comings of the imaging system (especially the spherical aberration of the objective lens, defocusing problems and the limited aperture) have to be taken into account. The real electron optical imaging process can be analytically treated by means of the Fresnel approximation (for detailed considerations, see e.g. /52,53/). Of the numerous conditions to be fulfilled for a successful high-resolution work in practice (for a review see e.g. /54/), sufficiently thin specimens (thickness: about 10 nm) is an important requirement.

The various types of high-resolution images possible are determined by the place and the size of the aperture diaphragm used in the diffraction plane (see Fig. 5, top). Examples of the feasible imaging types are given in Fig. 5 (bottom); they are related to ⟨110⟩ oriented silicon. If the only interference allowed is that between the non-diffracted beam and one of the diffracted beams (type I), a lattice fringe image is obtained. Using a larger aperture (type II) leads to a many-beam image, which facilitates a rough detection of the crystal lattice. If a large number of electron beams (type III) are allowed to contribute to the imaging process, a true image is obtained, which is called structure image under favourable conditions. A dark-field lattice image is obtained if only the diffracted beams are allowed to interfere with each other (type IV) /55/. The application of the type V aperture is particularly important. Here, the undiffracted beam as well as the diffracted ones are prevented from contributing to the imaging process, which means that the image is obtained by using only the scattering background, which is formed by the action of disordered structures. This technique has so far

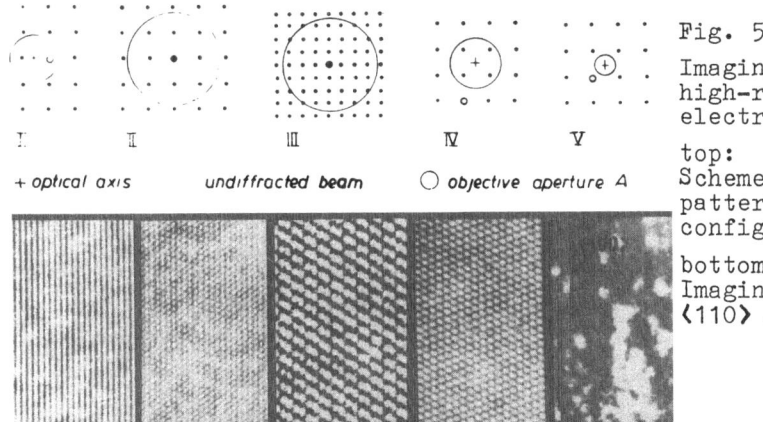

I   II   III   IV   V

+ optical axis   undiffracted beam   ○ objective aperture A

Fig. 5

Imaging modes in high-resolution electron microscopy:

top:
Scheme of diffraction patterns and aperture configurations;

bottom:
Imaging examples of ⟨110⟩ silicon.

been successfully employed for the imaging of individual atoms /56/.

For the investigation of ⟨110⟩ oriented semiconducting materials the defects are imaged by using either the imaging mode type II (one contrast dot is related to two closely neighboured atom columns in the beam direction) or - as was shown for the first time by Izui et al. /57/ - type III (each contrast dot corresponds to one column of atoms (see e.g. Hutchison /58/, Bourret /59/). With the advent of the new generation of high-voltage high-resolution electron microscopes in the 300/400 kV range, for the semiconductor research it is decisive that now structure images can be obtained not only from ⟨110⟩ oriented materials but also from ⟨111⟩ and ⟨100⟩ oriented specimens of elemental and compound semiconductors /60/, and that sublattices in compound semiconductors can be clearly distinguished /61/. - High-resolution EM with its possibility of a detailed defect characterization on an atomic scale is applied to a wide field of semiconductor research, yielding information, for instance, on the core structure of dislocations, on the structure of planar defects, and on the structure of interfaces. Special attention has to be paid to the atomic or molecular processes during the interaction between defects with impurities /62/ or the formation of point defect clusters.

Fig. 6 reveals the core structure of dislocations. The micrograph shows the typical structure of a dissociated 60° dislocation in silicon. Characteristic of this is, on the one hand, the stacking fault of intrinsic type, which - in this ⟨110⟩ orientation - is visible in the cross section. On the other hand, the partial dislocations bordering the stacking fault can be clearly recognized: left - 30° partial dislocation, right - 90° partial. The overlapping diffraction contrast pointing to an appreciable lattice strain is characteristic of the 90° partial. A more detailed analysis of micrographs of this type reveals that in this special dislocation configuration two additional opposite stair rod dislocations are present. This means, strictly speaking, the dislocation configuration is a dipole, which might be due to the interaction of two neighboured dissociated dislocations.

Fig. 7 illustrates the formation of point defect agglomerates during inhomogeneous plastic deformation of Si /63/. For a rough survey HVEM studies were carried out for ⟨111⟩ , ⟨110⟩ and ⟨100⟩ oriented specimens. During the operation of an indenter (at about 1000 K), on the one hand, 60° dislocations (line direction: mainly

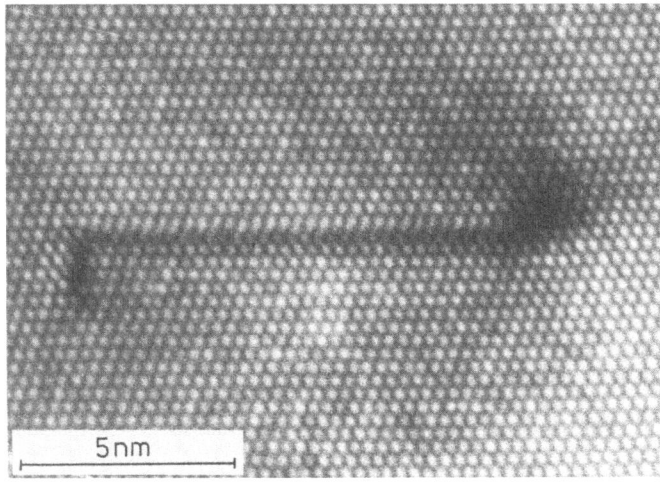

Fig. 6

7-beam bright-field lattice image of partial dislocations in ⟨110⟩ oriented silicon.

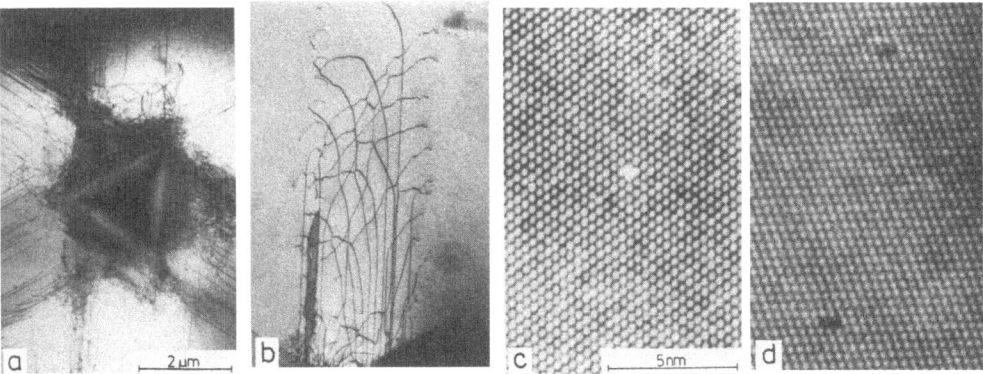

Fig. 7   Indenter region of a silicon sample (for details: see text) a) survey micrograph, b) 60° dislocations, c) HREM micrograph of interstitial agglomerate, d) HREM micrograph of vacancy agglomerate.

⟨110⟩) are formed on (111) slip bands, as Figs. 7a,b show for ⟨111⟩-oriented material (a: complete indenter region, b: one of the dislocation bands). On the other hand, self-interstitials are created, which diffuse during a suitable annealing process, entailing the formation of stable self-interstitial agglomerates. These self-interstitial agglomerates were found preferentially in the dislocation-free regions near the indenter groove. Figs. 7c,d show dark-field lattice images, taken with a conventional 100 kV EM, of the above-mentioned specimen regions of ⟨110⟩ material (where the dislocation rosettes show a 4-fold symmetry). Fig. 7c represents the usual situation of a self-interstitial agglomerate. But seldom also vacancy agglomerates are observable (Fig. 7d). Under the imaging conditions used - according to computer simulations - each white dot in the micrographs represents two atom columns. The visible **agglomerates** should contain at least 10 point defects.

3. Estimation of the electrical activity of defects by special SEM techniques

Besides the geometrical characterization of crystal defects the detection of their electrical activity is of special importance. It is of great interest, for instance, which of the crystal defects present in integrated circuits, influence the function of the latter, e.g. by causing leakage currents. As far as unipolar (MOS) or bipolar microelectronic devices are concerned it is proven that crystal defects transversing pn junctions are especially critical. In MOS structures these are defects in the channel region, in bipolar structures these are defects causing short circuits between regions that have been distinctly separated before.

As an example Fig. 8 shows defects in the boundary region between emitter and base in a bipolar Si integrated circuit. Micrograph a) presents a dislocation network in the emitter region and a stacking fault (SF) in the base region. The dislocations lie parallel to the pn junction. As no defects are crossing the pn junction the electrical activity of this defect arrangement is expected to be low. On the other hand, micrograph b) shows clearly defect-free emitter and base regions. But there is one defect, a stacking fault, crossing the pn junction and thus being expected to be electrically harmful. The action is similar to that of an emitter-collector pipe, which is usually formed by a crossing dislocation and which degrades the device function. Of course, for defects of this type the state of being decorated with impurities is an important feature.

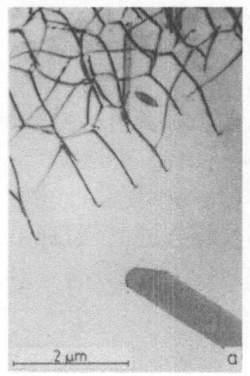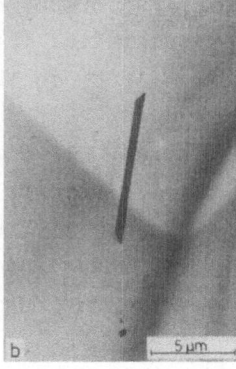

Fig. 8
Boundary between emitter and base in a Si IC:

a) dislocation network in the emitter region, stacking fault in the base region

b) stacking fault crossing the boundary

Apart from the above-mentioned rough considerations on the position of defects with respect to pn junctions the direct detection of
the electrical and/or optical activity of crystal defects is of great
interest. For this purpose special SEM techniques, applying the following signals are used: electron beam induced currents (EBIC), emitted cathodoluminescence radiation (CL), and capacitance transients
created by the impact of a pulsed electron probe (SDLTS: scanning
deep level transient spectroscopy).

3.1 SEM(EBIC)

The SEM(EBIC) technique was proposed by Lander and Czaja /64,65/
at the beginning of the sixties. Its main principle consists in the
following: During the impact of electrons on a semiconductor surface,
which is comparable to the action of the electron probe in a SEM,
electron-hole pairs are created, which contribute to the conduction
events in the region chosen by the electron probe; the current addi-
tionally created can be used for the imaging process. In general, the
created electron-hole pairs recombine quickly, and only in the region
of a present electrical field, as e.g. the inner field of a pn junction (or also a Schottky barrier), the charge carriers are sufficiently separated. Crystal defects lying in the neighbourhood of pn
junctions cause local inhomogeneities in the recombination behaviour
of the charge carriers so that these crystal defects can be imaged
by the action of local disturbances in the recombination process.
Thus, the crystal defects can be estimated with respect to their
electronic activity, of course, taking into account the position of
the defect relative to the pn junction /66-68/. Especially advantageous is the combination of TEM, used for the reliable characterization of the geometrical structure of individual defects with a high
spatial resolution, with the SEM(EBIC) technique, which allows the
detection of the recombination activity of the defect, unfortunately
only in the resolution range of micrometres. The combined application
of both methods was proposed by Ravi and Varker /69,70/ in the seventies.

Fig. 9 is to show the practice of combining TEM and SEM(EBIC) investigations. The micrographs present diffusion-induced dislocations
in the emitter region of a Si-npn-transistor /71/.   The pn junction
was located 3.5 /um in depth. First the specimen was investigated by
EBIC in a conventional SEM at a primary beam voltage of 10 kV. After
the preparation of the interesting region by chemical thinning trans-

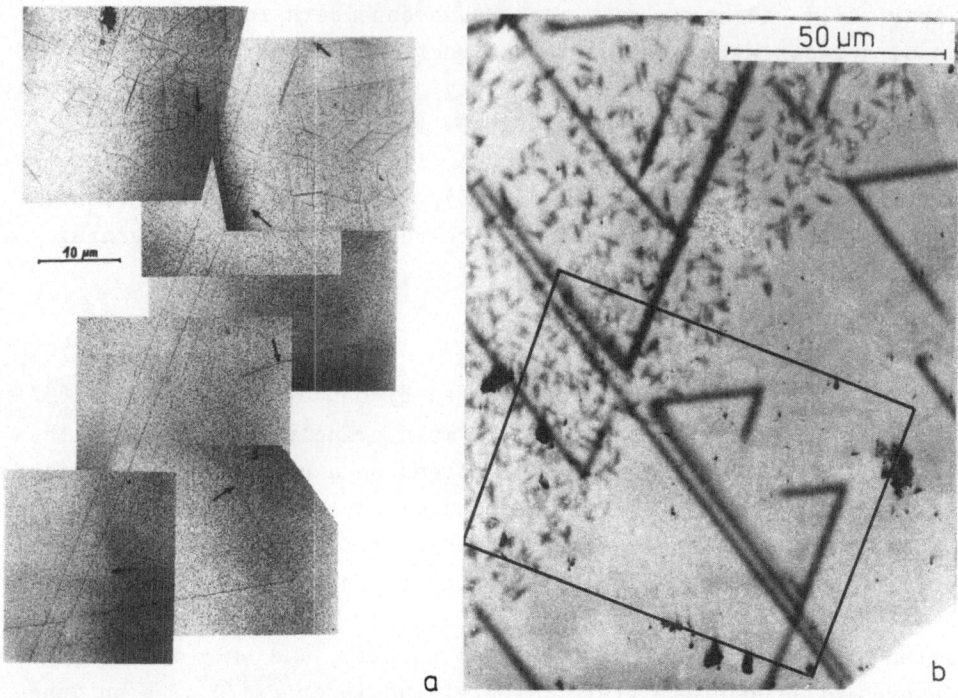

Fig. 9   Diffusion-induced dislocations in the emitter region of a
Si npn transistor: a) HVEM micrograph, b) SEM(EBIC) micrograph

mission images were taken in a 1000 kV HVEM. According to the TEM investigations (Fig. 9a) a dislocation network (mostly dislocation
loops) covers the upper part of the imaged region; its mean depth
position is about 0.5 /um - 1 /um. Furthermore, there are straight
dislocations elongated over the whole image region (directions:
⟨110⟩) lying near the pn junction. Diffraction contrast observations
yielded type 60° dislocations (Burgers vector: a/2 ⟨110⟩). In interpreting the EBIC micrograph (Fig. 9b) one can estimate the electrical
activity of the dislocations from the strength of the contrast. In
the positions of the arrows the dislocations penetrate the pn junction, and thus their corresponding EBIC contrast lowers rapidly and
vanishes, resp..

Investigations of this type, including weak-beam observations,
have shown that mainly the structure of the dislocation core determines the electrical activity of a dislocation. For 90° dislocations
Ourmazd and Booker /72/ convincingly showed that segments of dissociated dislocations are electrically more active than non-dissociated
ones; the same was proven for 60° dislocations /67/. Many investigations /73-76/ point to the fact that the electrical activity of a

dislocation (or another crystal defect) strongly depends on the agglomeration of impurity atoms at these defects ("decorated" dislocations). Obviously is the dissociated dislocation more attractive to impurities than the non-dissociated one. Thus, the topical problems in estimating the electrical activity of crystal defects first of all lie in the detection of reconstruction processes at the dislocation core and – associated with this – in the detection of the stage of agglomeration of impurities. In this connection observations of Bourret /77/ on Ge are of special interest. HREM investigations of partial dislocations lying end-on did show only low contrast, on the one hand, and distinctly visible cloud-like structures, on the other. These cloud-like structures are supposed to be connected with the agglomeration of impurities at the dislocation core. By applying the electron energy loss analysis in the nano-metre region Bourret and Colliex /78/ convincingly showed the presence of oxygen at increased concentration in the positions of the imaged cloud-like structures. – Furthermore, combined SEM(EBIC), TEM and EPR measurements were used to study the electrical activity of screw and 60° dislocations in Si /79/. Thus, it was possible to point out the influence of point defects on the electrical activity of these dislocations by investigating samples with and without dangling bond EPR centres.

3.2 SEM(CL)

The cathodoluminescence (CL) mode in the SEM, which allows the detection of radiative and non-radiative regions in semiconductors, esp. in III/V semiconductors – see e.g. /80-85/ – is used also in combination with TEM or STEM defect structure investigations and often also for the comparison with EBIC results /86/. Progress in this field refers to sensitive detection and amplification systems, with special respect to spectral and time-resolved CL. In consideration of this the study of CL decay time distributions around dislocations /87/ is mentioned here. The technique of transmission CL (TCL) /88/ will be advantageous in this field.

Of the combined application of HVEM and SEM(CL) an example is given, which is related to the optical action of misfit dislocation networks in diodes within graded $GaAs_{1-x}P_x$ layers /89/. The direct correlation (Fig. 10) between dislocation bundles, imaged by HVEM (on the left) and dark lines in the corresponding SEM(CL) image (on the right) shows that not all the dislocation bundles give rise to a decrease in the CL intensity, but only a few of them (labelled A,B,...).

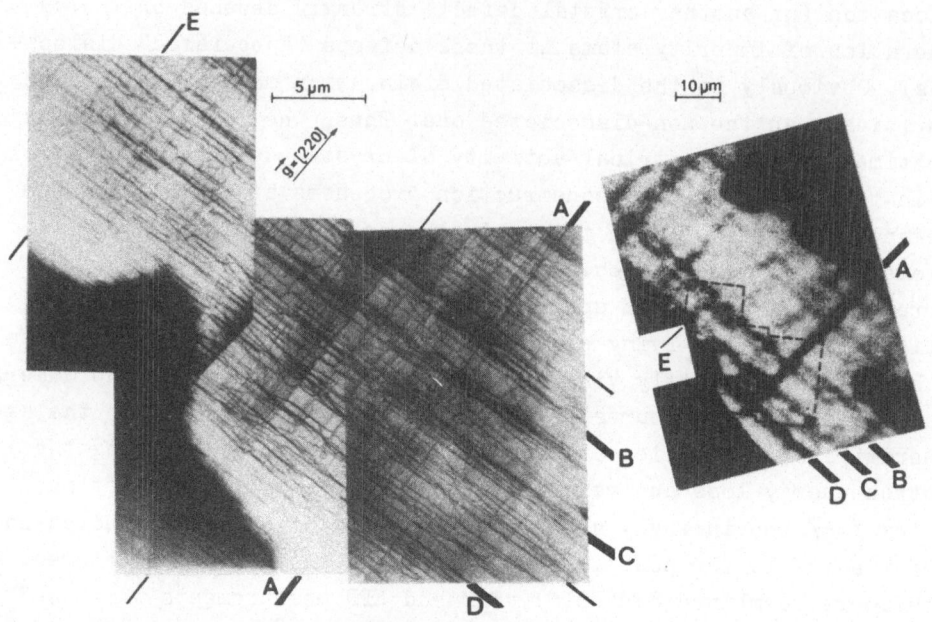

Fig. 10   Misfit dislocation network in a graded GaAs$_{1-x}$P$_x$ layer:
a) HVEM micrograph, b) SEM(CL) micrograph.

To explain this a diffraction contrast analysis of the imaged dislocations in bundle D is carried out. The dislocations in this bundle
are shown to be mixed and edge dislocations, of course, having the
Burgers vector type a/2 ⟨110⟩. Some of the dislocations present in
the bundle were proved to be Lomer dislocations, i.e. non-glissile
edge dislocations. Thus, the decrease in the CL intensity in line D
seems to be caused by the existence of Lomer dislocations in this
bundle. The same should also hold for lines A,B, and C. Of course,
the state of decoration of the Lomer dislocations with impurity
atoms should be taken into account. As it was already shown for silicon, the Lomer dislocations are decorated with impurity atoms more
strongly than other types of dislocations, thus leading to a lower
CL yield.

3.3 SDLTS

Scanning deep level transient spectroscopy (SDLTS) allows the detection of the energy levels of defects acting as recombination centres as well as of their relative concentrations in specimen regions
excited by an electron probe. Here, point defects are particularly
concerned (vacancies, interstitials, impurity atoms, antisite de-

fects), which are inhomogeneously distributed (including the special case of decorating extended defects, like dislocations). The spatially resolved investigation of these defects - they are called deep level defects if their binding energy exceeds that of the shallow doping atoms - seems to be promising.

The SDLTS technique is based on the DLTS method proposed by Lang as early as in 1974 /90/. It implies the measurement of capacitance changes (or current changes) in a space charge structure after an excitation pulse as a function of the temperature. The aspired statements on the energy levels and on the concentration of the interesting defects can be gained from the analysis of the temperature - dependence of the signal, which delivers the DLTS spectrum. Employing the principle of scanning DLTS - which means the excitation of the levels by an electron probe - enables spatially resolved investigations of deep level states to be carried out. Scanning the pulsed electron beam (at a selected temperature and rate window, both corresponding to a chosen energy level) yields an SDLTS image that shows the local variation of the density of the selected deep level defects. The experimental arrangement for SDLTS investigations is schematically drawn in Fig. 11. The SDLTS technique was introduced by Petroff and Lang /91/ in 1977. In their original version a resolution of 2 /um and a sensitivity of $10^{14} - 10^{15}$ deep levels /cm^3 were reached. In the meantime, using the correlator principle - which means the analogue multiplication of a capacitance signal with a transient function - at a measuring frequency of 2 MHz or 28 MHz under optimum conditions (integration time: 1 s) has yielded a sensitivity of about 2 10^{12} deep levels /cm^3 /92-94/, which means about 100 excited atoms in the excited volume of the space charge region (say 10 /um^3).

As an example of the great number of SDLTS investigations (see e.g. /95-98/), mainly carried out on A_3B_5 semiconductors, results

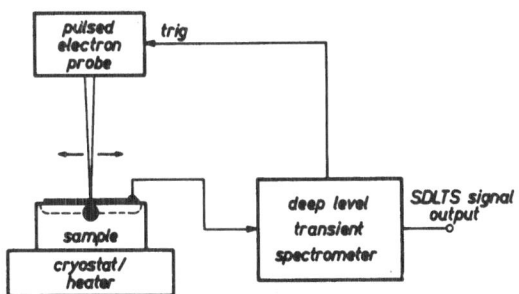

Fig. 11

Experimental arrangement for SDLTS investigations

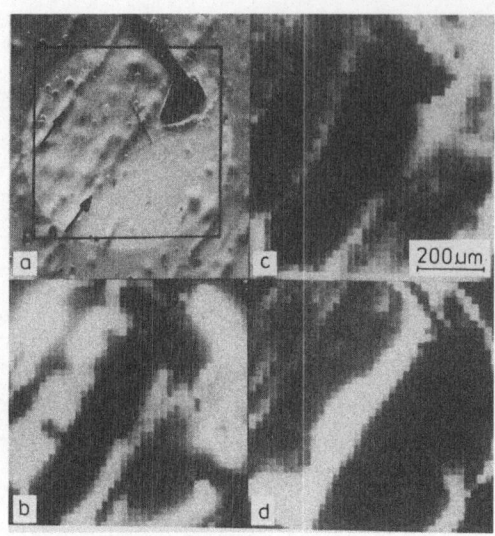

Fig. 12
GaAs:Cr,In crystal containing "streamer" defects:

a) etched surface,

b-d) SDLTS micrographs, taken at
    b) $T = 20^{\circ}C$,
    c) $T = 70^{\circ}C$,
    (both negative bias) and at
    d) $T = 50^{\circ}C$
    (positive bias)
rate window: $100\ s^{-1}$

will be quoted of an SDLTS investigation on semi-insulating GaAs /99/ (in the current detection mode under an applied bias). The observations are related to "streamer" defects /100/ in GaAs:Cr,In. Fig. 12 shows the surface etch pattern (a) together with SDLTS micrographs (b-d) taken at different temperatures and bias polarities in order to detect different species of levels. In all cases the correlation of the SDLTS contrasts with the streamers is obvious. Although the nature of the levels is not entirely clear, it should be possible to relate the identification to the EL2-level or to the chromium-induced midgap level.

4. Conclusions

The trend of applying electron microscopy to semiconductor physics coincides - at least as far as TEM and STEM investigations are concerned - with the general trend in solid state research. First of all, besides diffraction contrast imaging, high-resolution electron microscopy is increasingly applied. Of the specimens to be investigated, besides defects in the bulk material more and more interfaces, surfaces and defects in these regions become the objects of interest. Furthermore, more detailed knowledge of the dynamical behaviour of semiconducting materials is highly desirable so that in situ investigations (at well-defined heat- and beam treatments) are imperative, requiring the use of high-voltage electron microscopes. Additional information from the use of special SEM techniques for detecting the electrical and/or optical activity of crystal defects is expected by interpreting the time- and temperature-dependence of the signals.

Surface imaging methods and electron beam testing techniques are rapidly developing; they are of great importance for the investigation of semiconducting specimens. As they cover wide fields of their own, they were not treated within the scope of this paper.

Finally, the topical method of analytical electron microscopy (for review see /101-103/) - imperative to elucidate the chemical composition of materials (and/or defects) in micro- and nanometre regions, not treated in this paper - should be mentioned. The combined application of imaging (high-resolution EM), diffraction (microdiffraction, esp. convergent beam diffraction), and spectroscopic (energy-dispersive X-ray spectroscopy/EDX/and electron energy loss spectroscopy /EELS/) methods in a TEM or in a STEM enables the complex characterization of inhomogeneities and defects in nanoregions (nanophase analysis).

5. Acknowledgements

The electron micrographs shown originate from the research work in the IFE Halle. The author is indebted to the following colleagues engaged in this work: Drs. H. Bartsch, H. Blumtritt, O. Breitenstein, W. Erfurth, R. Gleichmann, P. Werner, U. Richter.

References

/1/ B.O. Kolbesen, H.P. Strunk, VLSI Electronics Microstructure Science 12, 143, 1985

/2/ J. Heydenreich, Electron Microscopical Characterization of Electronic Materials, in: Crystal Growth of Electronic Materials (Ed. E. Kaldis), Elsevier, Amsterdam 1985, p. 325

/3/ H. Alexander, J. Phys. (Paris) 40, C6-1, 1979

/4/ W. Coene, H. Bender, S. Amelinckx, Philos. Mag. A52, 369, 1985

/5/ W. Bergholz, J.L. Hutchison, P. Pirouz, J. Micr. 141, 143, 1986

/6/ A.K. Petford-Long, D.J. Smith, Philos. Mag. A54, 837, 1986

/7/ L.I. Fedina, A.L. Aseev, Phys. Stat. Sol. a 95, 517, 1986

/8/ A. Armigliato et al., J. Electrochem. Soc. 133, 2560, 1986

/9/ J. Chikawa, S. Shirai, J. Cryst. Growth 39, 328, 1979

/10/ H. Bethge, J. Heydenreich, Proc. 8th Europ. Reg. Congr. Electron Micr., Budapest 1984, Vol. 1, p. 663

/11/ J.L. Hutchison, J. Phys. (Paris) 44, C4-3, 1983

/12/ A. Bourret, C. Colliex, Ultramicroscopy 9, 183, 1982

/13/ D.J.H. Cockayne, A. Hons, J.C.H. Spence, Philos. Mag. A42, 773 1980

/14/ O.L. Krivanek et al., Proc. 41st EMSA Meeting, Phönix, Arizona 1983, p.118

/15/ J.L. Hutchison et al., J. Micr. _142_, 153, 1986

/16/ N. Otsuka et al., J. Vac. Sci. Technol. _B4_, 896, 1986

/17/ H. Kawarada et al., Philos. Mag. _A54_, 729, 1986

/18/ F.A. Ponce, G.B. Anderson, J.M. Ballingall, Surf. Sci. _168_, 564, 1986

/19/ A.H. Carim, R. Sinclair, Mater. Lett. _5_, 94, 1987

/20/ C.J. Kiely, D. Cherns, D.J. Eaglesham, Philos. Mag. _A55_, 237, 1987

/21/ C. D'Anterroches et al., J. Cryst. Growth _81_, 121, 1987

/22/ R. Sinclair et al., Nature _290_, 386, 1981

/23/ A.G. Cullis, Phys. Rev. Lett. _49_, 219, 1982

/24/ P.B. Hirsch, R.W. Horne, M.J. Whelan, Philos. Mag. _1_, 667, 1956

/25/ W. Bollmann, Phys. Rev. _103_, 1588, 1956

/26/ A. Howie, M.J. Whelan, Proc. Roy. Soc. _A263_, 217, 1961

/27/ A. Howie, M.J. Whelan, Proc. Roy. Soc. _A267_, 206, 1962

/28/ M. Wilkens, Phys. Stat. Sol. _6_, 939, 1964

/29/ M. Wilkens, Phys. Stat. Sol. _13_, 529, 1966

/30/ P.B. Hirsch et al., Electron Microscopy of Thin Crystals, Butterworths, London 1965

/31/ G. Thomas, M. Goringe, Transmission Electron Microscopy of Materials, Wiley & Sons, New York 1978

/32/ A.K. Head, Aust. J. Phys. _20_, 557, 1967

/33/ P. Humble, Aust. J. Phys. _21_, 325, 1968

/34/ D.J.H. Cockayne, I.L.F. Ray, M.J. Whelan, Philos. Mag. _20_, 1265, 1969

/35/ W.L. Bell, G. Thomas, Applications and Recent Developments in Transmission Electron Microscopy, in: Electron Microscopy and Structure of Materials, University of California Press, Berkeley 1972, p. 23

/36/ D. Cherns, Philos. Mag. _30_, 549, 1974

/37/ A. Ourmazd, D. Phil. thesis, University of Oxford, 1979

/38/ H. Alexander et al., Philos. Mag. _A53_, 627, 1986

/39/ C. Frigeri et al., Proc. 16th Conf. Electron Micr. Italy, Bologna 1987, to be published

/40/ G.W. Cullen et al., J. Cryst. Growth _56_, 281, 1982

/41/ H. Bartsch, Proc. GADEST '87, to be published

/42/ W. Skorupa et al., Proc. 5th Int. Conf. Ion Beam Modification of Materials, Catania 1986, p. 285

/43/ F. Louchet, Philos. Mag. _A43_, 1289, 1981

/44/ M. Sato, M. Takebe, K. Sumino, Proc. Yamada Conf. IX Dislocations in Solids, Tokyo 1985, p. 429

/45/ H. Bartsch et al., Phys. Stat. Sol. a _83_, 543, 1984

/46/ F. Reynaud et al., J. Micr. Spectrosc. Electron. _9_, 199, 1984

/47/ K. Hiraga et al., Cryst. Res. Technol. 17, 189, 1982

/48/ M. Hirata, M. Kiritani, Physica 116B/C, 616, 1983

/49/ A.L. Aseev, D. Hoehl, Proc. 8th Europ. Reg. Congr. Electron Micr., Budapest 1984, p. 477

/50/ J.W. Menter, Proc. Roy. Soc. A236, 119, 1956

/51/ J.G. Allpress, J.V. Sanders, D.W. Wadsley, Acta Cryst. B25, 1156, 1969

/52/ J.M. Cowley, Diffraction Physics, North-Holland, Amsterdam 1978

/53/ R. Hillebrand, W. Neumann, J. Heydenreich, Ultramicroscopy 4, 305, 1979

/54/ J.C.H. Spence, Experimental High-Resolution Electron Microscopy, Clarendon Press, Oxford 1981

/55/ N.D. Zakharov, M. Pasemann, V.N. Rozhanski, Phys. Stat. Sol. a 71, 275, 1982

/56/ H. Hashimoto et al., Jap. J. Appl. Phys. 10, 1115, 1971

/57/ K. Izui, S. Foruno, H. Otsu, J. Electron Microsc. 26, 129, 1977

/58/ J.L. Hutchison et al., Proc. 7th Europ. Reg. Conf. Electron Micr., The Hague 1980, Vol. I, p. 304

/59/ A. Bourret, Proc. 7th Europ. Reg. Conf. Electron Micr., The Hague 1980, Vol. I, p. 306

/60/ A. Ourmazd et al., Appl. Phys. Lett. 47, 685, 1985

/61/ A. Ourmazd, J.R. Rentschler, D.W. Taylor, Phys. Rev. Lett. 57, 3073, 1986

/62/ J.H. Rose, R. Gronsky, Mater. Res. Soc. Symp. Proc. 62, 57, 1986

/63/ P. Werner, H. Puschkasch, J. Heydenreich, unpublished

/64/ J.J. Lander et al., Appl. Phys. Lett. 3, 206, 1963

/65/ W. Czaja, G.H. Wheatley, J. Appl. Phys. 35, 2782, 1964

/66/ C. Donolato, Optik 52, 19, 1978

/67/ J. Heydenreich et al., SEM 1981, Vol. I, p. 351

/68/ L. Pasemann, H. Blumtritt, R. Gleichmann, Phys. Stat. Sol. a 70, 197, 1982

/69/ K.V. Ravi, C.J. Varker, C.E. Volk, J. Electrochem. Soc. 120, 533, 1973

/70/ C.J. Varker, K.V. Ravi, J. Appl. Phys. 45, 272, 1974

/71/ H. Blumtritt et al., unpublished

/72/ A. Ourmazd, G.R. Booker, Phys. Stat. Sol. a 55, 771, 1979

/73/ J. Heydenreich et al., Cryst. Res. Technol. 16, 133, 1981

/74/ M. Kittler, E. Bugiel, Cryst. Res. Technol. 17, 79, 1982

/75/ R. Gleichmann, H. Blumtritt, J. Heydenreich, Phys. Stat. Sol. a 78, 527, 1983

/76/ W. Seifert, M. Kittler, Phys. Stat. Sol. a 99, K11, 1987

/77/ A. Bourret, J. Desseaux, J. Phys. (Paris) 40, C6-7, 1979

/78/ A. Bourret, C. Colliex, Ultramicroscopy 9, 183, 1982

/79/ A. Ourmazd et al., Inst. Phys. Conf. Ser. No. 60, 63, 1981

/80/ L.J. Balk, E. Kubalek, E. Menzel, SEM 1975, p. 447

/81/ Y.M. Chu et al., Inst. Phys. Conf. Ser. No. 60, 331, 1981

/82/ B.G. Yacobi, D. Holt, J. Appl. Phys. 59, R1, 1986

/83/ A.K. Chin et al., J. Appl. Phys. 57, 2203, 1985

/84/ H. Ono et al., Jap. J. Appl. Phys. 25, L130, 1986

/85/ W. Hergert et al., Phys. Stat. Sol. a 101, 611, 1987

/86/ G. Attolini, Appl. Phys. Lett. 45, 167, 1986

/87/ A. Steckenborn, H. Münzel, D. Bimberg, J. Luminescence 24/25, 351, 1981

/88/ A.K. Chin, H. Temkin, R.J. Roedel, Appl. Phys. Lett. 34, 476, 1979

/89/ W. Erfurth, Thesis, Univ. Halle 1987

/90/ D.V. Lang, J. Appl. Phys. 45, 3021, 1974

/91/ P.M. Petroff, D.V. Lang, Appl. Phys. Lett. 31, 60, 1977

/92/ O. Breitenstein, Phys. Stat. Sol. a 71, 159, 1982

/93/ O. Breitenstein, J. Heydenreich, J. Phys. (Paris) 44, Suppl. 9, C4-207, 1983

/94/ O. Breitenstein, J. Heydenreich, Scanning 7, 273, 1985

/95/ P.M. Petroff et al., Proc. 9th Int. Congr. Electron Micr., Toronto 1978, Vol. I, p. 130

/96/ O. Breitenstein, T. Wosinski, Phys. Stat. Sol. a 77, K107, 1983

/97/ J. Heydenreich, O. Breitenstein, J. Micr. 141, 129, 1986

/98/ O. Breitenstein, J. Heydenreich, Proc. XIth Int. Congr. Electron Micr., Kyoto 1986, Vol. II, p. 1515

/99/ O. Breitenstein, Inst. Phys. Conf. Ser. 1987, to be published

/100/ O. Breitenstein, L.J. Giling, Phys. Stat. Sol. a 99, 215, 1987

/101/ D.B. Williams, Practical Analytical Electron Microscopy in Materials Science, Verlag Chemie, Weinheim 1984

/102/ J.J. Hren, J.I. Goldstein, D.C. Joy (Eds.), Introduction to Analytical Electron Microscopy, Plenum Press, New York 1979

/103/ Y. Bando et al., Ultramicroscopy 18, 117, 1985

DETERMINATION OF THE LATERAL DEFECT DISTRIBUTION BY SDLTS IN GaAs

László Dózsa and Attila L. Tóth
Research Institute for Technical Physics , Hungarian Academy of
Sciences, H-1325 Budapest Ujpest 1, P.O. Box 76, HUNGARY

ABSTRACT

In the SDLTS invesigations it is usually supposed that the SDLTS signal
is proportional to the lateral defect concentration. It will be shown
that this approximation cannot be used without restrictions and
investigation of the SDLTS signal as a function of generation
parameters is necessary, as will be illustrated by measuring the EL2
distribution in GaAs Schottky barrier.

INTRODUCTION

The homogeneity of semiconductor materials is a critical parameter in a
number of applications, such as in LSI circuits and optoelectronic
devices. A number of techniques, based on the local excitation of a
scanning electron microscope, can be used for the measuring of this
property (CL, EBIC, SDLTS). Among them SDLTS is a promising technique
as in principle it can map the spatial distribution of specific deep
centers. In this paper it will be shown that an appropriate
determination of the local deep level concentration by SDLTS requires a
detailed investigation of the trap filling mechanism. As an example the
EL2 distribution is determined in a LEC GaAs Schottky barrier.

EXPERIMENTAL

Scanning Deep Level Transient Spectroscopy (SDLTS) means the local
excitation of an electron microscope combined with the sophisticated
signal detection and analysis of DLTS, proposed by Petroff and Lang in
1977 (1). To understand the signal generation process SEI, EBIC, DLTS
and EBICap (DC beam induced capacitance) imaging were carried out. The
EBICap is proportional to the EBIC in the first approximation, with the
practical advantage that it can be measured without disconnecting the
sample and in the very circumstances of the SDLTS excitation. To
produce a two dimensional SDLTS distribution a special control of the
beam scanning and an appropriate data acquisition system is required in
addition to the one of the SEM. An Apple II computer was used for this
purpose and a program was developed to reach this goal (2). The
measuring system is shown in Fig.1.

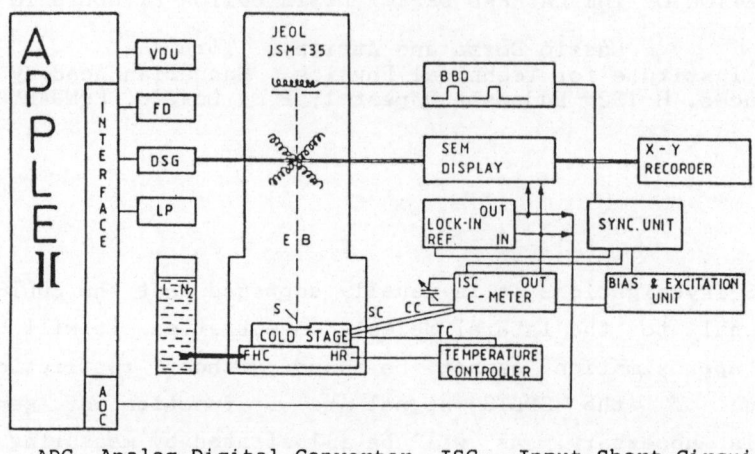

ADC	Analog Digital Converter	ISC	Input Short Circuit Unit
BBD	Beam Blanking Device	LN_2	Liquid Nitrogen
CC	Capacitance Compensator	S	Sample
DSG	Digital Scan Generator	LP	Line Printer
EB	Electron Beam	SC	Sample Connections
FD	Floppy Disc Drive	TC	Thermo Couple
FHC	Flexible Heat Conductor	VDU	Video Display
HR	Heating Resistor		

Fig.1   The bock diagram of the measuring system

CONSIDERATIONS ON SIGNAL GENERATION AND DETECTION

On the base of a theory given in (3), where complete filling is supposed in a uniform excitation volume and no filling outside, it is generally assumed that the SDLTS signal is proportional to the local defect concentration (4,5). Investigating GaAs samples it was found that besides the usual problems (surface irregularities, doping inhomogeneity, sample geometry etc) some other effects can cause spurious signal which can be avoided only by appropriate measuring setup and can be recognized by measuring the SDLTS signal as a function of the EBIC current and filling pulse length. Two of these effects are described below.

One of them is connected with the filling of minority traps during excitation. The process is sketched in Fig.2. The primary excited volume is a good approximation of the excited area only in that case, when the minority carrier capture process is negligible. When the minority carrier capture process is important the excited area of DLTS signal can extend to several diffusion length from the excitation spot and depends on the filling pulse length and EBIC current during excitation. In this case the concentration of deep centers (N_t) is proportional to the SDLTS signal only when the local minority carrier lifetime (τ) and mobility (μ) of the material do not vary in the material, as the SDLTS signal is proportional to:

SDLTS $\sim \mu \cdot \tau \cdot N_t$

The second effect is due to a nonlocal excitation. The sample irradiated by the electron beam acts as current generator with the EBIC current I_e This current, flowing through the resistive elements shown in Fig.3, generates a bias pulse on the sample, which fills the deep centers nonlocally. This spurious SDLTS is negligible compared to the useful SDLTS component when the relationship

dA/A $>>$ $I_e*(R_c+R_s+R_g)/U_r$

is satisfied, where dA is the area excited by the electron beam locally, A is the whole area of the junction, U_r is the reverse bias applied during the DLTS measurement. This nonlocal excitation depends also on the EBIC current and filling pulse length, and can be neglected only when:

   a. the conductivity of the measured material is good,

   b. appropriate bias unit is used,

   c. the capacitance (current) meter input is short circuited during excitation.

When the SDLTS signal saturates as a function of EBIC current and filling pulse length the usual approximation can be considered as valid.

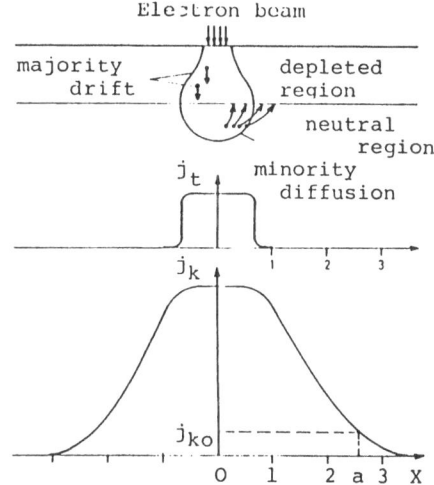

Fig.2. The sketch of the excitation of deep centers by majority and minority carriers in a Schottky barrier.
j_t: majority carrier current density
j_k: minority carrier current
X : distance from the excited spot in diffusion length units

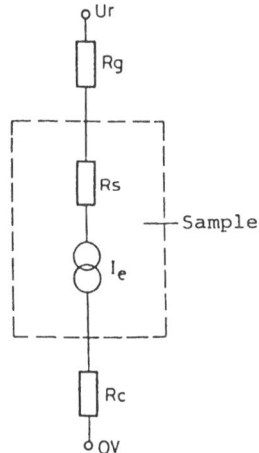

Fig.3 The equivalent circuit of the sample and its surroundings during excitation
R_g: bias unit output impedance
R_s: Series resistance of the sample
I_e: Induced EBIC current
R_c: Capacitance meter input impedance
U_r: Applied reverse bias.

RESULTS

In Pt-GaAs Schottky barrier the EBIC
and SDLTS line scans correlate and the
signal cannot be saturated as a
function of the EBIC current and
filling pulse width. In this case the
capacitance meter was not short
circuited during the excitation by the
electron beam. The detected center
(310 K, 1 kc/s) is a surface related
one which was created by the electron
beam irradiation during the
measurement. In spite of the larger
SDLTS than EBIC contrast the SDLTS
intensity has no direct relation to
the concentration of the deep centers
(Fig.4).

The EBICap, EBIC and SDLTS line scans
of a Au-GaAs Schottky barriers are
shown in Fig.5. The SDLTS signal does
not collerate with the other scans.
The SDLTS signal is reduced at
increasing or decreasing frequencies,
so the SDLTS map relates to a well
defined peak. The signal saturates as
a function of filling pulse length
with 0.2 ms time constant. The signal
saturates as a function of the EBIC
current at low current level. These
results prove that the measured SDLTS
signal reflects truely the EL2
distribution.

In Fig.6 and Fig.7 the EBICap and
SDLTS signals are shown. The regions
marked by A and B show the silver
pasted region and free GaAs surface.
The concentration of the EL2 increases
from the 10^{15}/cm^3 avarage value to
10^{16}/cm^3 in a few times 10 μm area,
(spots marked by C and D).

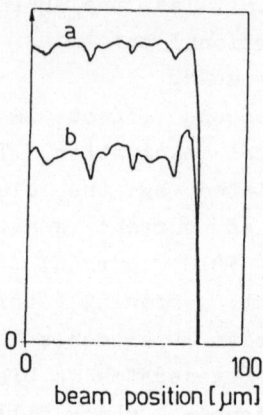

signal intensity [a.u.]

Fig.4 Line scan on the
GaAs-Pt Schottky barrier
a. EBIC signal
b. SDLTS signal of a surface
related defect at 310 K.

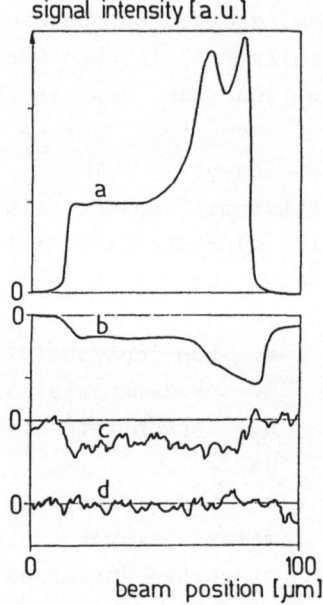

signal intensity [a.u.]

Fig.5 Line scan on a
GaAs-Au Schottky barrier
a. EBIC
b. EBICap
c. SDLTS signal of EL2 373 K
d. SDLTS noise level

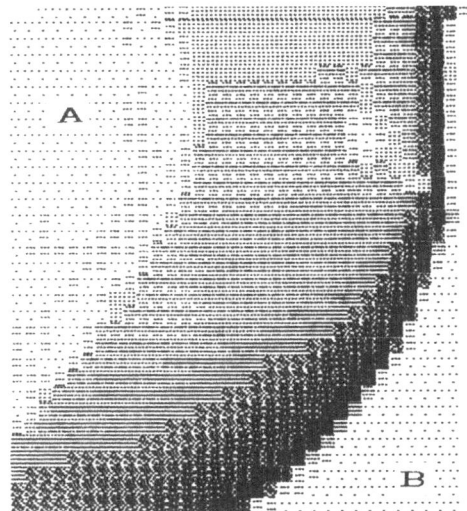

FILENAME: OL1.C (NEGATIVE)

Fig.6 EBICap image of a region
of the GaAs-Au Schottky barrier.
The maximal capacitance change
is 1 pF.
Resolution: 12 μm

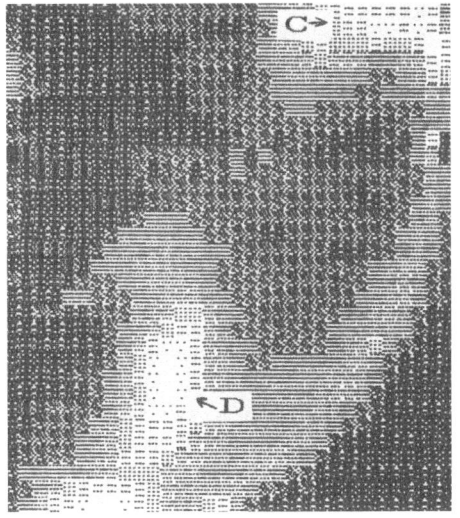

FILENAME: OL1.D2 (NEGATIVE)

Fig.7 SDLTS image of the region
shown in Fig.6. The concentration
of the EL2 level in the white
regions /spots marked by C and D/
is about 10^{16}/cm^3.

This fact suggests a precipitate formation of EL2 and this large, order
of magnitude contrast is a result of the careful investigation of the
excitation mechanism and was not found earlier, for example by Ikuta et
al(5).

CONCLUSION

It was shown that the SDLTS can give the distribution of deep centers
only when the SDLTS signal saturates as a function of EBIC current and
filling pulse length. It was also proved that using appropriate
measuring system the concentration distribution can be determined. The
concentration of the EL2 center in LEC GaAs has been found to have a
local increase of about one order of magnitude in a small (20x80 μm)
area, which suggests a precipitate formation.

REFERENCES

1. P. M. Petroff and D. V. Lang Appl. Phys. Lett. 31,60 (1977)
2. A. L. Toth and A. Konkol to be published
3.P.M. Petroff, D. V. Lang,R. A. Logan and W. D. Johnston Inst. Phys.
Conf. Ser. No.46,Chapter 7, page 427
4. O. Breitenstein and J. Heydenreich Scanning 7,273 (1985)
5. K. Ikuta, N. Inoue and K. Wada Semi-Insulating III-V Materials,
Ohmska, Ltd. 1986.

FORMATION OF RIBBON-LIKE DEFECTS DURING LOW-TEMPERATURE ANNEALING OF CZOCHRALSKI-GROWN SILICON

Manfred Reiche
Akademie der Wissenschaften der DDR
Institut für Festkörperphysik und Elektronenmikroskopie
Weinberg 2, Halle/S., 4050, G.D.R.

1. Introduction

Annealing Czochralski-grown silicon at temperatures from 450 to 500°C implies the formation of ribbon-like or rod-like defects (RLD). Based on high resolution electron microscope (HREM) investigations RLD's have been interpreted to be crystalline SiO_2 precipitates (coesite), analogous to the image characteristics detected in these defects after annealing at higher temperatures (for review see /1/). However, it was pointed out that the interpretation of RLD's as coesite precipitates cannot explain all experimental facts. Recently Bourret /2/ has shown by reinterpretation of the diffraction patterns as well as by comparison with analogous defects induced by ion-beam implantation that RLD's may consist of silicon self-interstitials (Si_I). The formation of the hexagonal (wurtzite-type) silicon phase inside the RLD's was expected /2/. Furthermore, the interpretation of RLD's as condensates of Si_I was supported by investigations into the kinetics of oxygen clustering at low temperatures /3,4,5/ as well as by further electron microscope analyses /6,7/. These analyses yielded also additional results as to the structure of RLD's, such as that besides the known {113} defects analogous ones exist parallel to {111} habit planes, too, or they are composed of {113} and {111} segments. In the present paper further results are dealt with. The significance of {111} defects for the RLD formation is discussed.

2. Experimental

Experiments were carried out in a commercial furnace in nitrogen ambiance. Before annealing, the temperature was calibrated by measurements via thermocouples; the maximum deviation during an annealing process was lower than 10K. Isothermal annealing procedures were applied between 440 and 500°C and up to 1500 hours.

Phosphorus- as well as boron-doped wafers were used (diameter 4in., $\langle 100 \rangle$ and $\langle 110 \rangle$ orientations, resp.). The doping concentrations of the material ranged from $1.1 \cdot 10^{15}$ to $4.5 \cdot 10^{14} cm^{-3}$ (P-doped) and from $1.4 \cdot 10^{15}$ to $7 \cdot 10^{14} cm^{-3}$ (B-doped), respectively.

The oxygen and carbon concentrations were measured by infrared spectroscopy following the DIN standard /8/. The defect structures produced by annealing were observed by both high voltage and high resolution electron microscopy.

3. Oxygen Reduction

Kinetic investigations of the oxygen reduction confirm the results of Newman /4/ as well as Messoloras et al. /5/. For annealing up to $500^{\circ}C$ the oxygen reduction can be described by second-order kinetics. Moreover, significantly low activation energies for oxygen diffusion were estimated from this model ($E_D \simeq 0.3 eV$) pointing to a characteristic dependence of the oxygen diffusion on the transport of Si_I produced by the molecule-like clustering of oxygen /7/.

Furthermore, the oxygen reduction was proved to depend on different factors: the initial oxygen and carbon concentrations, resp., and the doping type. These factors were previously reported to influence the oxygen reduction at higher temperatures ($T > 600^{\circ}C$). Here, particularly the effects of carbon and dopants on the oxygen reduction were discussed in terms of deviations of the self-interstitial flux from the growing oxide precipitates /9/. These precipitates, however, are not produced by annealing below $500^{\circ}C$ which suggests that the oxygen reduction at low temperatures should be caused by a mechanism resulting in the formation of self-interstitials, too.

4. Electron Microscope Investigations

Annealing at temperatures close to the maximum thermal donor formation temperature ($440^{\circ}C$) for up to 1500 hours does not cause any defects provable by electron microscopy. This suggests that irrespective of annealing time near the highest thermal donor activity after long annealing times /10/ only clusters of dimensions of about 1nm or lower exist.

First annealing-induced defects (RLD's) were observed at $465^{\circ}C$ (Fig. 1). The RLD's are elongated defects parallel to the $\langle 110 \rangle$

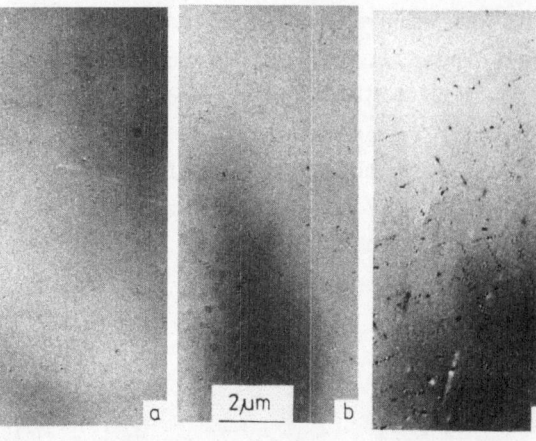

<u>Fig. 1:</u> Defect structure after an-
nealing at 465°C for 400 (a),
700 (b), and 1000 hours (c).

directions. Diffraction
contrast analyses enabled
Burgers vectors of the type
$\vec{b} = a/c\langle 100\rangle$ to be determined.

The density of RLD's in-
creases with increasing annealing time (Fig. 1). Increasing also the temperature
results in a higher density
of RLD's. However, trans-
formations into dislocation
dipoles occur with increasing
temperature (Fig. 2a). On the
other hand, with the an-
nealing time increasing at
lower temperatures (T~465°C) point-like, strong contrast effects
occur on the RLD's (Fig. 2b) which Bergholz et al. /11/ discussed to
be very small loop-like defects ("loopites"). Besides the formation
of RLD's, the generation of loopites, having {111} habit planes, can
refer to a second or alternative formation process of annealing-
induced defects at low temperatures.

The existence of RLD's depends on the same factors as those in-
fluencing the oxygen reduction and thermal donor formation. This
should indicate direct correlations between these processes. It was
found that the concentration of RLD's increases as the interstitial
oxygen concentration increases. On the other hand, an increasing
concentration of carbon reduces the RLD density. Furthermore,
retardations of the RLD formation are observed in samples preannealed
at high temperatures (T =
1250°C) or in phosphorus-
doped material. Also the

<u>Fig. 2:</u> Transformation of
RLD's.

a - Transformation into
dislocation dipoles
after annealing at
500°C.

b - Formation of loopites
after extended an-
nealing at 465°C.

c - Untransformed RLD's.

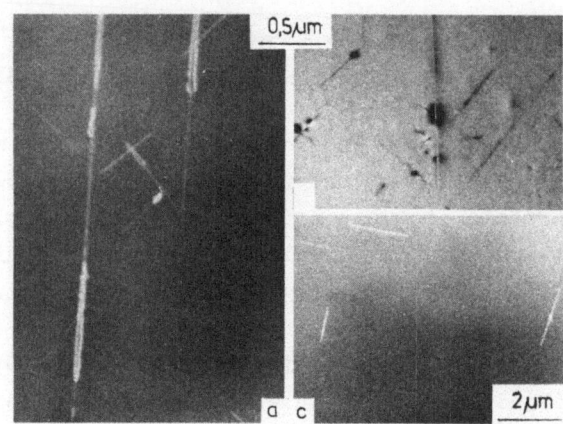

crystal growth conditions (thermal history) markedly affect the RLD formation. An explanation of all these interactions appears to be ingenious only if they are assumed to affect the concentration of silicon self-interstitials. This interpretation is also confirmed by the observed transformations of RLD's with increasing annealing time and temperature, respectively.

High resolution electron microscopy reveals an analogous structure of RLD's after annealing at 500°C as previously shown by Bergholz et al. /11/ at 485°C. Cross sections of the defects are on {113} habit planes (Fig. 3). A crystal structure within the defects as observed at higher temperatures /1/ cannot clearly be resolved; this may be caused by the large surface-to-volume ratio /11/. Furthermore, extended defects on {111} habit planes are observed having dimensions of more than 100nm. In part, they correspond to the loopites described by Bergholz et al. /11/. As shown in Fig. 4, the loopites can

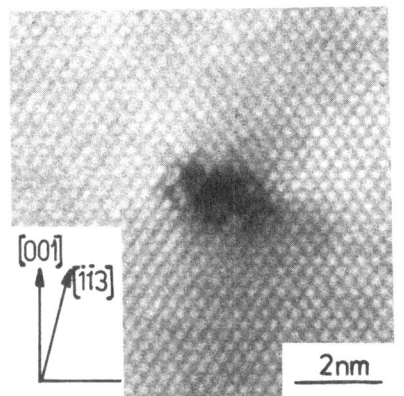

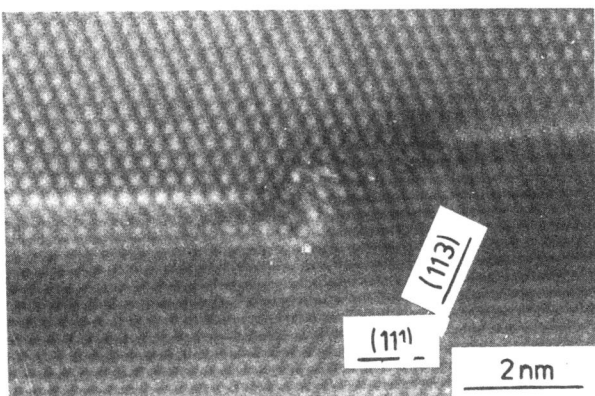

Fig. 3: Cross section of a RLD. Annealing at 500°C for 200 hours.

Fig. 4: Defect on {111} habit plane (loopite) contains {113} steps. Annealing at 500°C for 600 hours.

contain {113} steps between {111} segments. In addition, {113} steps at the ends of {111} segments are also found.

Besides the loopites further defects on {111} habit planes occur, but they do not contain {113} steps or end at these. Moreover, as shown in Fig. 5, contrasts inside the defects are proved by HREM structural imaging pointing to an atomic-ordered structure, analogous to that supposed for the RLD's lying parallel to {113} /1,2,11/. The defects are identified to be of extrinsic nature; the insertion between the $(111)_{Si}$ lattice planes is indicated in Fig. 5.

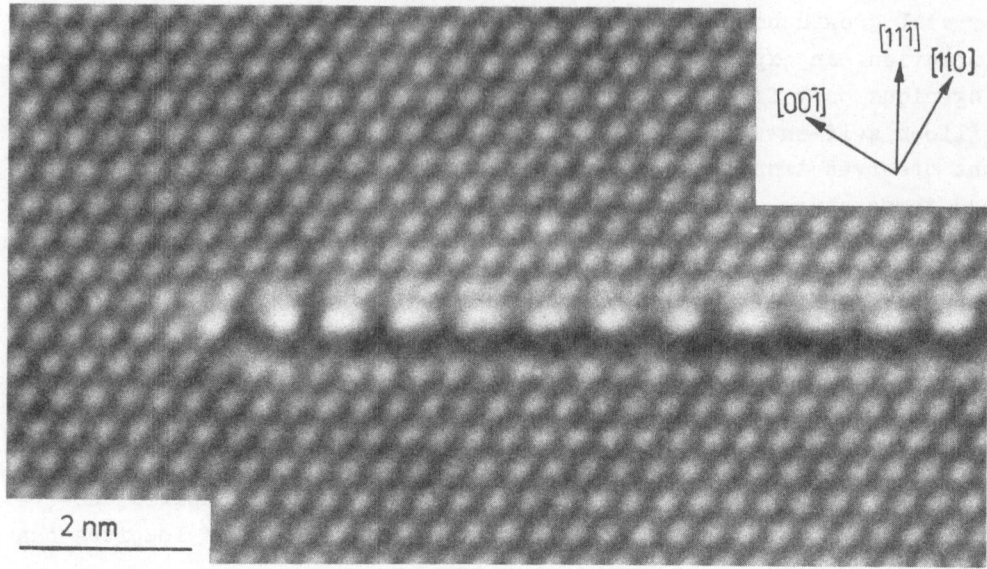

Fig. 5: Pure {111} defect with an atomic-ordered structure, already known from RLD's.

5. Discussion

The significance of {113} defects for the formation of defects during low-temperature annealing was emphasized by Bergholz et al. /11/ and Bourret /2/. According to Bourret /2/ the {113} defects are primary ones. But this interpretation cannot explain the existence of pure {111} defects without the assumption of transformations. In addition, the dimensions of pure {111} defects are in the order of 10nm which corresponds to the dimensions of single {111} segments of loopites. This probably refers to the primary formation of {111} defects. The {111} lattice planes are densely packed planes of the cubic diamond structure and therefore preferred for defect formation. At higher concentrations of point defects (Si_I) particularly, these planes are more favoured than {113} ones, since a higher density of self-interstitials can be inserted. This interpretation is confirmed also by studies of ion-implantation induced defects. Here, the formation of {111} defects (Frank's loops) increases with increasing ion dose (i. e. increasing excess point defect concentration) /12/. At first, {111} defects grow by agglomeration of self-interstitials up to a critical size which is in the order of 10nm /12/, corresponding to the dimensions of pure {111} defects observed here. Following this, a displacement of a/3<111> is supposed to bind the agglomerate into the silicon host lattice (formation of Frank's loop) and to initiate further growth. This appears, however, to be improbable during low-temperature annealing. Instead, further growth can be initiated on {113}. In addition, also a model was proposed for the transformation of {113} defects into {111} defects during nucleation stages by atomic rearrangement /13/. Both models, the primary formation of {111} defects as well as the transformation of {113} defects, probably yield a better interpretation of the existence of {111} defects as well as of complexes containing several {111} and {113} segments.

References

1. A. Bourret, Mat. Res. Soc. Symp. Proc., Vol. 59, Mat. Res. Soc. Pittsburgh, Pa., p. 223 (1986)
2. A. Bourret, Paper presented at the 5th Oxford Conf. on Microsc. of Semicond. 1987, to be published in Inst. Phys. Conf. Ser.
3. T.Y. Tan, R. Kleinhenz, C.P. Schneider, as ref. /1/ p. 195
4. R.C. Newman, J. Phys. C 18, L967 (1985)
5. S. Messoloras, R.C. Newman, R.J. Stewart, J.H. Tucker, Semicond. Sci. Technol. 2, 14 (1987)
6. M. Reiche, O. Breitenstein, phys. stat. sol. (a) 101, K97 (1987)
7. M. Reiche, W. Nitzsche, Proc. 2nd Int. Aut. Meet. Gettering and Defect Engin. in the Semicond. Technol. (GADEST), in press (1987)
8. DIN 50438, part 1 (1978), part 2 (1982), Deutsches Institut für Normung e.V.
9. M. Reiche, W. Nitzsche, Proc. 1st Int. Aut. School Gettering and Defect Engin. in the Semicond. Technol. (GADEST) (ed. H. Richter) Garzau, Akademie der Wissenschaften der DDR, p. 174 (1985)
10. G.S. Oehrlein, J.L. Lindström, S.A. Cohen, Proc. 13th Int. Conf. on Defects in Semicond. (eds. L.C. Kimerling, J.M. Parsey) The Metallurg Soc. of AIME, Warrendale, Pa., p. 701 (1984)
11. W. Bergholz, J.L. Hutchinson, P. Pirouz, J. Appl. Phys. 58, 3419 (1985)
12. D. Baither, H. Bartsch, Doc. Thes., Univ. Jena (1983)
13. H. Bartsch, J. Heydenreich, D. Hoehl, P. Werner, 5th Int. Symp. Structure and Properties of Disloc. in Semicond., Moskau, Abstr. p. 39 (1986)

BAND-EDGE OFFSETS IN SEMICONDUCTOR HETEROJUNCTIONS

H. Heinrich

Institut für Experimentalphysik, Universität Linz

A-4040 Linz, Austria

The band-edge offset (BEO) in semiconductor heterostructures is the most crucial parameter for the interpretation and prediction of the physical properties of such a device. The different experimental procedures for obtaining the BEO are discussed. We have recently proposed a new method, using the electron-beam-induced current (EBIC) as an indication for obtaining flat-band condition when an external bias is applied. Results for GaAs/GaAlAs and PbSe/PbEuSe are given. Alligning deep-impurity levels of transition metals proved to be a valuable tool in predicting BEO's. Recently reported results will be compared with the predictions. On the other hand, the determination of the energetic position of band edges by measuring BEO's can give information where to look for energy levels of as yet unobserved deep impurities. Results for InAs and GaSb are discussed. Based on a combination of transition-metal-energy level alignment and experimental results of BEO's in binary and ternary III-V compounds a level order of valence-band edges in these compounds is proposed. The importance of a fixed reference level formed by transitions metals in the semiconductor is emphasized and recently reported measurements of the deformation potential are reviewed.

1. Introduction

Presently, semiconductor heterojunctions (HJ) are among the most widely studied structures in semiconductor physics and technology. Combinations of HJ's in the form of a quantum well or of superlattices provide a huge amount of interesting pheno-mena in basic science and are fundamental for electronic and optoelectronic devices. For physics as well as for technology the most important parameter of these structures is how the conduction bands (cb) and the valence bands (vb) match at the transition point (or region). For many of the applications it would be desirable to know the band-edge offsets (BEO's) within a few ten meV's. We will not treat the details of the present status of the theoretical approaches for BEO calculations here and will refer the reader to the reviews of Pollmann [1], Kroemer [2], Harrison [3] and Flores and Tejedor [4]. All first principle theories of HJ's as well as the heuristic approaches start from the assumption that the BEO of a given HJ can be obtained when the position of the band edges of the two partners are known on an absolute energy scale. This leads directly to the property of linearity and transitivity [5] and assumes that a canonical order of band edges for different semiconductors may be established according to the bulk properties of the individual materials.

On the other side, Bauer and Sang [6] point out that the complete situation present in reality at the interface has to be taken into account and therefore the BEO cannot be predicted in a unique way for a particular HJ. Experimental data for triples of nearly lattice-matched HJ's (A/B, B/C, C/A) became available recently and demonstrate that transitivity exists within the experimental error. We will treat this in some detail later.

Evidence therefore accumulates that the BEO in certain conditions is indeed a function of the bulk properties of the constituents. We will call such a HJ an "ideal HJ" meaning that it is abrupt, lattice-matched (no strain), with controlled doping up to the transition, and a HJ within one class of semiconductors, e.g. III-V on III-V or II-VI on II-VI. Presently, however, there is only one HJ (GaAs/(Ga,Al)As) for which a consensus on the BEO is obtained from experiments [5] with a few others coming up: (Ga,In)As/InP, GaSb/InAs, (Ga,In)P/GaAs and CdTe/HgTe. We will treat these HJ's in the following in more detail.

Among the heuristic approaches for the prediction of BEO's the electron affinity rule (Anderson model [7]) and the common anion rule [8] had their definite merits in the beginning of the physics of HJ's, but failed when more accurate values of the BEO were demanded [9,10]. The other remaining heuristic approach - the transition-metal (TM) alignment method - passed all the available experimental tests. It has been proposed and applied to many HJ's by Langer and Heinrich [11,12,13] and has been independently reported by Zunger [14]. The predictions of the TM method compare favourable with the results of neutrality-level alignment [15] and the method itself has been put on a theoretically founded base by Tersoff and Harrison recently [16].

In the following paragraph we will give a short summary of experimental methods for the determination of BEO's and will report about the application of EBIC for BEO measurements [17]. We will then compare the results of the TM predictions to recent experimental results of BEO's. An attempt is made to establish a canonical order of the vb's of most of the relevant III-V compounds, by combining experimental results and TM alignment. Finally, we emphasize the importance of a fixed reference level formed by transition metals in the semiconductor and will review recently reported measurements of the deformation potential.

2. Experimental Methods for the Determination of BEO's

The experimental methods for measurements of the BEO have been reviewed in several papers, e.g.: Kroemer [2], Duggan [18], Hickmott [19], Heinrich and Langer [12a], Bauer and Margaritondo [20] and Esaki [21]. In order to demonstrate to the reader the great variety of applied techniques we simply list the different methods here and refer to Refs. [12] and [21] for a critical discussion as well as for the large number of citations of original contributions. Only the most recently reported techniques will be treated in some detail.

a) Experimental determinations from single wells and superlattices.

    i)   Spectroscopic techniques using absorption or emission transitions between quantum levels.

    ii)   Thermoionic emission over barriers formed by "contacting" a layer of wide gap material by a semiconductor of narrower gap.

b) Single heterojunctions.

    i)   External Photoemission Spectroscopy (Ultraviolet or X-ray Photoemission Spectroscopy).

    ii)   Internal Photoemission by exciting charge carriers over the barrier formed by the BEO.

    iii)   Capacitance-Voltage Profiling (CVP) at an depletion layer formed by a Schottky contact on top of a shallow HJ.

    iv)   Charge-transfer techniques.

Recently, Smoliner et al. [22] investigated tunneling from 2D electronic states into continuum states in GaAs/GaAlAs in order to determine the 2D energy level positions and thereby the BEO. Their result for the ratio of cb offset to the difference in the gaps is $\Delta E_c / \Delta E_g = 0.61$.

In principle, the problem of measuring the BEO at a HJ is solved, as soon as the built-in potentials V_{b1} and V_{b2} on both sides of the HJ are determined as indicated in Fig. 1. Traditionally $1/C^2$-V plots are extrapolated for $C \to \infty$ in order to obtain

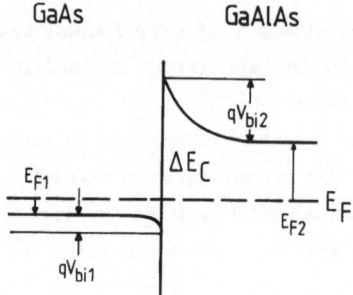

GaAs   GaAlAs

Figure 1. Schematic diagram for the conduction band on an n-n heterojunction. The conduction-band discontinuity ΔE_c and the Fermi levels E_{F1} and E_{F2} on each side of the junction are indicated (after Ref. [17]).

the flatband voltage $V_{b1} + V_{b2}$. However, the information obtained from C-V measurements is too global and does not reflect the microscopic situation at the interface, especially in the presence of interface charges. Therefore, Eisenbeiss et al. [17] applied a new technique, which is based on measuring the EBIC generated at a HJ while simultaneously applying on external bias to the device. The separation of electron-hole pairs generated by an electron beam in the vicinity of an n-n HJ takes place in the depletion region. Applying an external bias (V_{ex}) in the forward direction reduces both types of charges (depletion and accumulation) and therefore reduces the resulting built-in potentials V_{b1} and V_{b2}. When V_{ex} is increased, the flatband condition is obtained for a particular voltage (V_{exo}) and the EBIC disappears. This disappearance of the EBIC is used as an indication that the flatband condition has been reached. In this case $V_{exo} = V_{b1} + V_{b2}$, and if the differences of band-edge energy and the Fermi levels E_{F1} and E_{F2} are known on both sides of the HJ, the BEO (ΔE_c) may be obtained directly from the relation $\Delta E_c = E_{F1} - E_{F2} + qV_{exo}$.

This simple way of evaluation is only valid if no interface charge is present. If an interface charge is present, a compensating space charge region remains, even if V_{ex} is increased further and the EBIC signal cannot be reduced to zero. The degree to which the EBIC signal can be reduced to zero may be taken as a measure of the amount of interface charge and only those samples in which the EBIC signal vanishes may be evaluated.

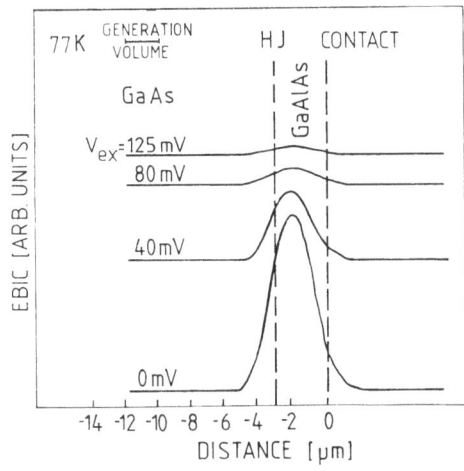

Figure 2. EBIC signal vs. position in the range of the heterojunction and the top contact for 77 K. The epitaxial $Ga_{0.48}Al_{0.52}As$ layer is 2.8 μm thick. With application of an external bias (V_{ex}) the signal is reduced. The diameter of the generation volume (20 kV acceleration voltage of the primary electron beam) is indicated at the top of the diagram (after Ref. [17]).

Fig. 2 gives an example for a scan of the EBIC signal vs. distance across an n-n HJ of GaAs/$Ga_{0.48}Al_{0.52}As$ with V_{ex} as a parameter. With V_{ex} = 125 mV the EBIC signal is practically zero and a ΔE_c of 242 ± 10 meV is evaluated, which gives a ΔE_v = 288 ± 10 meV, equivalent to ΔE_v = 5.5 meV/%Al. In the range of the direct gap

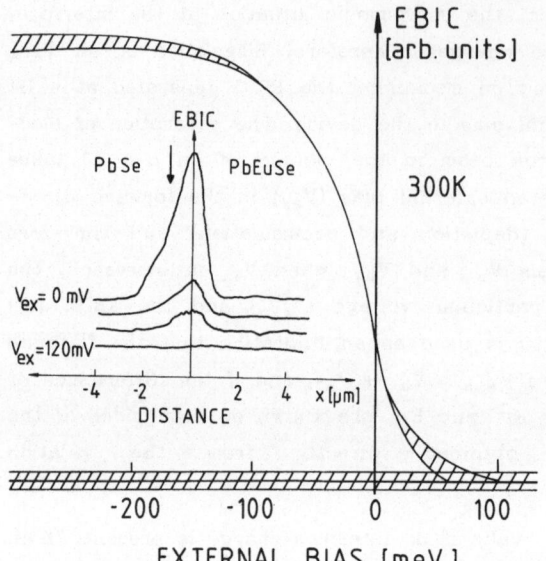

EBIC [arb. units]

300K

EXTERNAL BIAS [meV]

Figure 3. EBIC signal on the PbSe side of a p-p PbSe/Pb$_{0.987}$Eu$_{0.013}$Se HJ vs. bias obtained at a position as indicated by the arrow in the insert. The insert shows the EBIC vs. distance in the range of the HJ with bias as a parameter. The hatched areas indicate the scattering of the data by noise (after Ref. [24]).

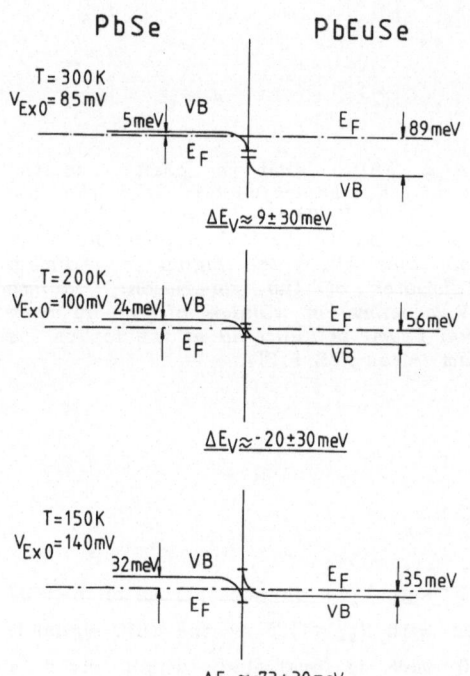

Figure 4. Schematic vb structure of a PbSe/Pb$_{0.0987}$Eu$_{0.013}$Se HJ at three different temperatures (after Ref. [24]).

this gives $\Delta E_c/\Delta E_v = 56/44$.

The flatband voltage at an HJ can be determined also from potential contrast measurements which is in principle very similar to the EBIC method. Potential contrast has been used by Lechner et al [23] with GaAs/Ga$_{0.64}$Al$_{0.36}$As HJ's and $\Delta E_c/\Delta E_v = 66/34$ has been obtained.

Recently, the EBIC method has been applied to investigate BEO's in the interesting structure PbSe/Pb$_{0.987}$Eu$_{0.013}$Se which is important for heterojunction diode lasers [24]. Fig. 3 shows the EBIC signal at a p-p HJ vs. bias. By application of the bias the EBIC signal can be reduced to zero completely, which allows to determine the flatband voltage very precisely. The schematic vb structure at three different temperatures is shown in Fig. 4. The evaluation of ΔE_v depends critically on the Fermi levels on both sides and therefore, on the knowledge of the carrier concentration and the effective masses. For PbEuSe both parameters are known only approximately and therefore, the error margins are estimated to be relatively large (± 30 meV). At room temperature nearly aligned vb's are evaluated ($\Delta E_v = 9 \pm 30$ meV). With decreasing temperatures a pronounced trend to negative values of ΔE_v is observed giving a staggered alignment which means that the wide gap PbEuSe vb is above the PbSe vb. Since the temperature variation of ΔE_v has been measured on the same sample it is more relyable than the absolute value as mentioned above. The IV-VI compounds are known for their strong temperature dependence of the energy gap and since the alignment is strongly asymmetric between PbSe and PbEuSe, it is not too surprising that a variation of ΔE_v with temperature is observed.

All the above mentioned experimental techniques have been applied to investigate GaAs/GaAlAs which is the most widely studied HJ, and also its technology is most advanced. The reported data for $Q_c = \Delta E_c/\Delta E_g$ scatter in a wide range as illustrated in Fig. 5, where they are plotted according to their date of appearance.

Starting from the pioneering work of Dingle [28] in 1974 with $Q_c = 0.85$, a drastic reduction in Q_c was reported in most of the papers of the last years. Even the recent data scatter over the range of Q_c 0.55 to 0.7. A critical evaluation [5] as well as the statistical average over the results reported in the last years puts the most probable value in the range of $Q_c = 0.6$ to 0.65.

Concluding this chapter on BEO's it might be of interest to remark that what happens presently in connection with the measurement of BEO's, has been happening several times in the past of semiconductor physics, whenever a particular property was of interest, e.g. band structure or effective masses; there are experiments from which a certain property can be deduced more or less directly and experiments for whose interpretation an accurate knowledge of that property is essential. One has the impression that in the case of the BEO the device-like techniques, performed on single HJ's belong to the first group, while experiment with quantum wells and superlattices are of the second type.

3. BEO's and Their Relation to TM Impurities

TM's form deep impurity levels in most semiconductors. It has been observed by several authors (e.g. Ledebo and Ridley [29]) that the relative energetic separations

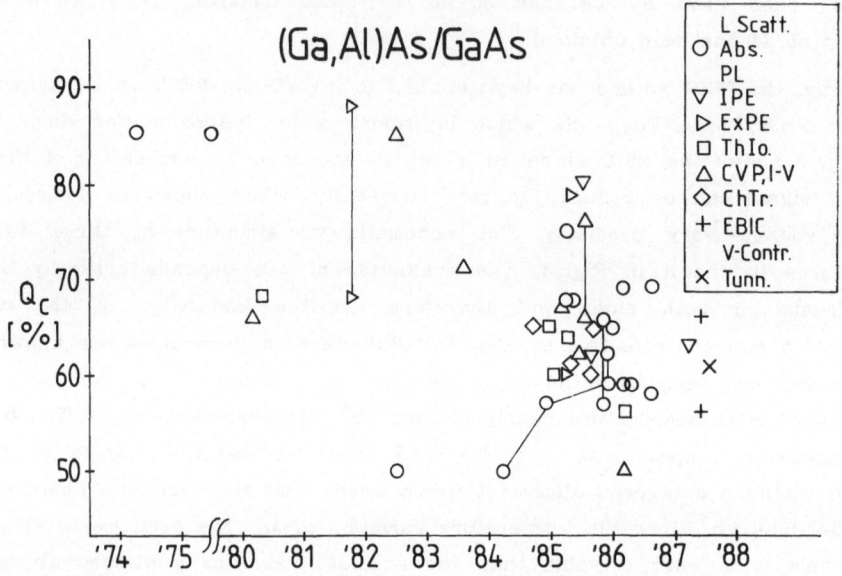

Figure 5. *Experimental data for relative conduction BEO, Q_c, in GaAs/(Ga,Al)As HJ's. When Q_v was reported only, the plot gives $Q_c = 1 - Q_v$. For measured values of ΔE_v obtained in the range of the indirect gap, a linear extrapolation was performed to obtain Q_c. The symbols indicate different experimental techniques: LScatt. - Light Scattering; Abs, PL - Absorption and Photoluminescence; IPE - Internal Photoexcitation; ExPE - External Photoexcitation; ThIo - Thermoionic Transmission over Barriers; CVP, I-V - Capacitance-Voltage Profiling and Current-Voltage Characteristics; ChTr - Charge Transfer; EBIC, V-Contr. - Electron-Beam-Induced Current and Voltage Contrast; Tunn. - Tunneling. The references can be identified from Ref. [12a]. The more recent data are from Refs. [25-27], [17], [23] and [22].*

of different donor or acceptor levels of substitutional TM's are independent of the host material, as long as one stays within one class of semiconductors (III-V or II-VI compounds). When these levels for different semiconductors are plotted on an energy scale with arbitrary origin they are found to coincide when the corresponding vb positions of the individual semiconductors are adjusted (Fig. 6).

Based on this observation and other qualitative arguments, Langer and Heinrich [11] and also Zunger [14] postulated that the aligned TM levels define the relative band-edge positions of the corresponding host semiconductors. Making a statistical averaging procedure in order to minimize the mrs-deviations of all the available data a higher degree of accuracy is obtained. The results are reproduced in Fig. 6. Recently, Tersoff and Harrison [16] showed that the TM method can be put on a firm theoretical basis. Their argumentation is the following: In principle, alignment of the neutrality level determines the BEO in a HJ. This neutrality level is that energy at which the states in the forbidden gap at an interface change from vb-like to cb-like. These states are either called MIGS (metal-induced gap states) when a Schottky contact is under consideration, e.g. [30], or IDIS (induced density of interface states) at a HJ [4]. Mismatching of the neutrality levels results in charge transfer

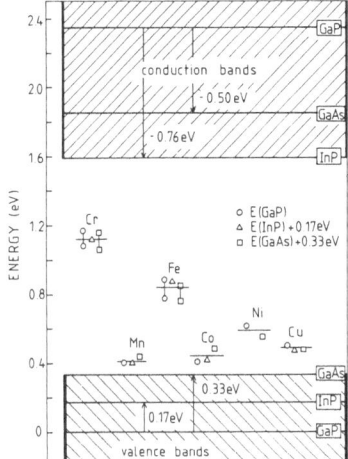 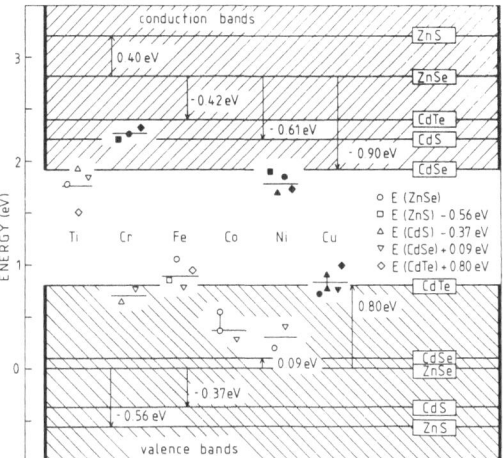

Figure 6. (a) Average energy levels of TM acceptors (-/0) in GaP, InP and GaAs relative to the top of the vb of GaP. They were determined by the appropriate shift of the experimental TM energies in all three compounds so as to minimize the overall mean square deviations. The resulting shifts give the positions of the vb edges relative to GaP. (b) The average energy levels of TM donors (0/+) (open symbols) and acceptors (-/0) (filled symbols) in II-VI compounds. Band-edge shifts are relative to the ZnSe vb edge. The sources of the data can be identified from Ref. [11].

between the two partners which is counteracted by the resulting dipole. Since the density of states in the MIGS is very high the deviation from perfect alignment is usually small. Tersoff and Harrison could then show that the TM levels are in a fixed close relation to that neutrality level. The obtained vb positions by aligning TM levels are therefore identical to the positions obtained by the alignment of the neutrality levels.

The obvious practical advantage of the TM method is that a great number of very precise cation-substitutional impurity levels are known. In the following we will compare the predictions of the TM method with reported experimental results on the few HJ systems which come close to "ideal" HJ's as mentioned above.

3.1 GaAs/GaAlAs

We take the very precise DLTS [31] and photocapacitance data [32] on the Fe^{2+} acceptor level in $Ga_{1-x}Al_xAs$ bulk crystals and obtain the slope of the vb energy as a function of the composition x as

$$\frac{d(E_{vb})}{dx} = (0.453 \pm 0.011) \text{ eV}.$$

According to the TM model, the vb level for a particular x is just the vb offset between GaAs and $Ga_{1-x}Al_xAs$. In the direct gap range this leads to $Q_v = 0.36$; $Q_c = 0.64$, which compares very favourably with $Q_c : Q_v = 62 : 38$ considered by Kroemer [5] to be the most relyable experimental value.

3.2 GaInAs/InP, GaSb/InAs

$Ga_{0.47}In_{0.53}As$ is lattice-matched to InP. However, the few investigations on TM impurities in the binary component InAs of this ternary system do not report about the energetic position of the levels: Cr [33] and Fe [34]. Mn, which was investigated in the quaternary compound $Ga_xIn_{1-x}As_yP_{1-y}$ [35] is too close to the vb as to be considered as a true deep state. The TM method can give only a lower limit for ΔE_v in $Ga_{0.47}In_{0.53}As/InP$: $\Delta E_v > 0.3$ eV (see Ref. [12b]). We are therefore limited to rely on the reported experimental data for the BEO. Let us restrict for simplicity on the lattice-matched ternary compound. For this composition Forrest et al. [36] report from capacitance-voltage profiling about 61% of ΔE_g in the vb, a value recently confirmed by Lang et al. [37] ($Q_v = 0.58$) by admittance spectroscopy. Temkin et al. [38] suggest $Q_v = 0.50$ based upon the analysis of absorption spectra of rectangular quantum wells (a configuration not very sensitive to the BEO determination [39]). However, a recent publication by Skolnick et al. [40] based on the same experimental technique, completely confirms the data by Forrest et al. and Lang et al., giving $\Delta E_c = 235 \pm 20$ meV and $\Delta E_v = 380$ meV. Finally, Brunnemeier et al. [41] opt for $Q_v = 0.4$ from threshold of emission. This situation is illustrated in Fig. 7, taking the value of $Q_v = 0.61$ ($\Delta E_v = 0.35$ eV) as the most trustworthy one.

Starting from the level order of cb and vb as taken from Fig. 6, a linear dependence of the vb of (Ga,In)As is plotted, giving $Q_v = 0.61$ for the $Ga_{1-x}In_xAs/InAs$ HJ at $x = 0.53$. The total nonlinearity of the gap of (Ga,In)As as a function of the composition is accommodated in the cb. In this way the vb of InAs is located 0.37 eV above the vb of GaAs. One should remember, however, that this value depends entirely on the choice of Q_v. Adopting the results of one of the other authors results in a shift of the InAs vb as indicated by the hatched region.

Recently, BEO values for GaSb/InAs have been reported, showing a complete energetic separation of the energy gaps (type II). Claessen et al. [42] and references therein find from optical absorption that this separation is 150 meV. Gualtieri et al. [43] report from X-ray photoemission-core-level spectroscopy $\Delta E_v = 0.51 \pm 0.1$ eV, a result which gives very similar alignment (Fig. 7).

For comparison the TM levels of Fig. 6 are plotted also in Fig. 7. Obviously, Fe and Cr are the most likely candidates to form levels in the gap of InAs. Indeed, Fe has been found as an EPR–active center [34] and Cr has been reported to be a compensating donor in p-type InAs [33]. For GaSb, V (and also Ti) are promising candidates to form deep levels in the gap. Experimentally determined BEO's can therefore provide information where to look for TM levels in semiconductors where non have been observed so far.

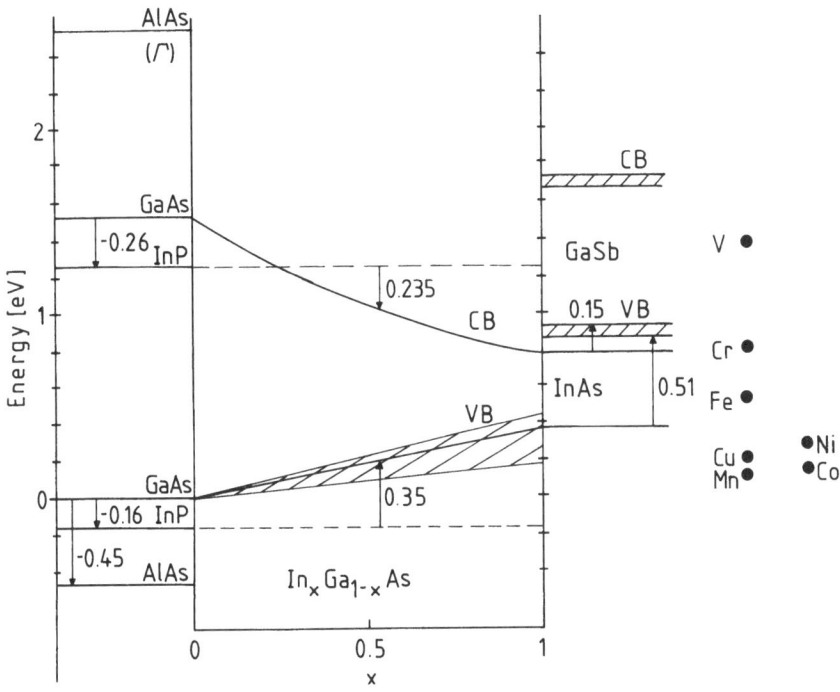

Figure 7. Band-edge diagram for the $In_x Ga_{1-x} As$ compound, assuming a linear dependence of the vb energy on x. The arrow indicates the experimental value of ΔE_v, in this particular case $Q_v = 0.61$ with $\Delta E_v \approx 0.35$ eV [36,37], recently confirmed by Ref. [40] ($\Delta E_c = 0.235$ eV, $\Delta E_v = 0.38$ eV). Band edges for GaSb are shown as obtained from GaSb/InAs HJ's [42,43]. For comparison the position of TM levels are plotted as given in Fig. 6. The value for V is from Ref. [44]. The hatched areas illustrate the variation of the reported data by different authors.

3.3 GaAs/(In,Ga)P

Among the lattice-matched HJ's of III–V compounds GaAs/In$_{0.5}$(Ga$_{1-x}$Al$_x$)$_{0.5}$P is a very interesting one, because ΔE_g is large and the discontinuity in the vb is larger than in the cb. HJ's of the ternary type GaAs//In$_{0.5}$Ga$_{0.5}$P can be compared with the predictions of the TM method and very accurate experimental results became available recently as obtained by the capacitance-profiling method [45,46]. Rao et al. [45] report $\Delta E_v = 0.24$ eV $\pm$ 10 meV and Watanabe et al. [46] give $\Delta_v = 0.3$ eV. The value predicted by the TM method is 0.245 eV assuming a linear varation of the vb energy between InP and GaP (Fig. 6). The agreement is very good and, furthermore, it is obtained for a HJ of different anions, a point mentioned recently by Tersoff and Harrison [16].

3.4 CdTe/HgTe

At present no consensus exists on the BEO in CdTe/HgTe. The earliest data suggest that the BEO is about 50% of the CdTe energy gap. In contrast, spectroscopic studies of multiquantum wells indicate an almost perfect vb alignment [47]. External photoemission experiments [48] gave much larger BEO's equal to $\Delta E_v = 0.35 \pm 0.06$ eV. This situation has been "confirmed" very recently: $\Delta E_v = 0.04$ eV obtained from magnetooptical studies in CdTe/Hg$_{0.96}$Mn$_{0.04}$Te [49] and $\Delta E_v = 0.36$ eV as obtained from XPS [50].

Since all impurity states in HgTe are resonant, one must be very cautious with regard to TM energy-level positions. There is, however, quite strong evidence [51] that the Fe^{2+} localized level lies about 0.2 eV below the top of the vb of HgTe. Since the same level is [52] about 0.15 eV above the vb of CdTe, a vb BEO of about 0.3 − 0.4 eV is expected in good agreement with the photoemission results. The situation is examplified in Fig. 8. A vb BEO of approximately this order of magnitude has been obtained also by most of the recent theoretical calculations as reported by Tersoff [10], Zoryk et al. [53], Zunger et al. [54] and Van de Walle et al. [55]: 0.50 eV, 0.37 eV, 0.35 eV, and 0.28 eV, respectively.

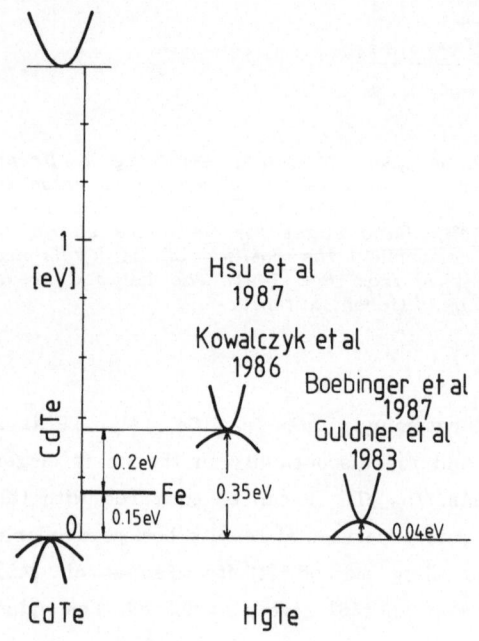

Figure 8. Position of the vb's of CdTe and HgTe according to different experimental investigations. The position of the Fe^{2+} level relative to the CdTe vb and HgTe vb is indicated.

4. Transitivity and Valence-Band Positions of III-V Compounds

In this chapter we make the attempt to establish a canonical order of the vb positions of most of the important binary III-V compounds. Such an order is definitely of interest for the threefold connection of heterojunctions - TM levels - and Schottky barriers, as well as for photoemission and whenever the relative position of bands in semicoductors are of interest. A fixed level order also implies transitivity of the BEO in a triple of HJ's. Within the binary III-V compounds no triple can be found for which lattice matching is fulfilled. To overcome this difficulty one has to include either ternary compounds or other semiconductors for the formation of HJ's. Results for both cases became available recently. Katnani and Bauer [56] find from core-level spectroscopy:

$$\Delta E_v(GaAs/AlAs) + \Delta E_v(AlAs/Ge) + \Delta E_v(Ge/GaAs) = 0.005 \pm 0.15 \ eV$$

and Watanabe and Ohba [46] obtain from capacitance profiling:

$$\Delta E_c(GaAs/In_{0.5}Ga_{0.5}P) + \Delta E_c(In_{0.5}Ga_{0.5}P/In_{0.5}Al_{0.5}P) + \Delta E_c(In_{0.5}Al_{0.5}P/GaAs) = 0.01 \ eV.$$

Both cases confirm the validity of transitivity. Another obvious candidate for testing transitivity is GaAs/InGaP/AlAs, however, to our knowledge only two of the three HJ's have been investigated sofar.

When trying to establish an order of vb edges for a large number of III-V compounds one is running into the problem that only a few pairs of lattice-matched HJ's exist between binary compounds and even including the ternary compounds does not help to establish this order unambigously. We are therefore combining the available experimental results with the results of the TM method according to the following procedure.

i)   We start from vb positions as given by TM alignment.

ii) Experimental results for lattice-matched binary and ternary compounds are considered only.

iii) A linear dependence of the vb edge on composition is assumed in ternary compounds.

In this way results are obtained as shown in Fig. 9. Level positions from TM alignment are drawn as full lines, experimentally obtained values are shown dashed, and results obtained by a combination of both are drawn dash-dotted. Hatched areas indicate the spread of experimental results. Starting from the vb of GaAs (which was set equal to zero) the positions for InAs and GaSb are drawn as obtained in Fig. 7, already. The values for InAs and therefore for GaSb depend on the joice of the most relyable experimental results for GaInAs. As discussed in Section 3.2, the results obtained consistently by capacitance-voltage profiling and admittance spectroscopy [36,37] are used here. For $In_{0.5}Ga_{0.5}P$ [45,46] and $In_{0.5}Al_{0.5}P$ [46] results discussed in Section 3.3 are plotted. InP and AlP are expected to be symmetrically positioned with respect to InAlP. This puts the AlP vb at 1.08 eV below GaAs. The estimated error margin is about ±0.1 eV as obtained from a combination of the error margins of InP and InAlP.

5. On the Importance of a Stable Reference Level

Obviously, energy levels of substitutional TM's are independent from the host material. This has been proved both by theory and experiment as reviewed above. The "true" reference level for aligning HJ's and forming Schottky barriers is of

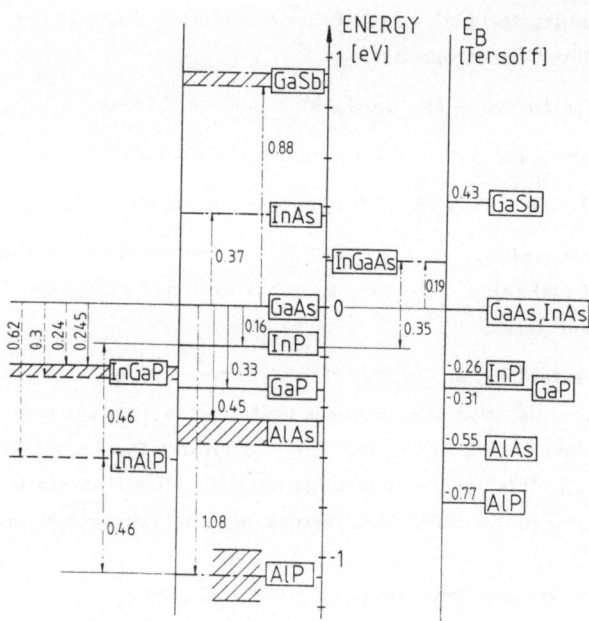

Figure 9. Proposed vb-edge positions of III-V compounds as obtained by combining TM alignment and experimental results of lattice-matched binaries and ternaries. TM alignment - full lines; experimental - dashed lines; combination of both - dash-dotted. Hatched areas indicate the spread of experimental results. For comparison the vb positions as obtained by neutrality level alignment are shown on the right side (Tersoff [10]).

course the charge-neutrality level, however, TM levels are linked within a constant to that neutrality level. At present TM levels are the most accurate way to localize the neutrality level by experiment. On the other side, this close relation allows to use TM levels as an energy-reference point within the bulk of the semiconductor, completely independent of surface properties as it is the case e.g. for the vacuum level. With a TM level as a fixpoint, individual variations of the band edges become measurable by altering the external conditions. Langer and Heinrich [13] proposed that pressure measurements of the TM energy levels should yield together with the energy-gap-pressure dependence the individual pressure coefficents for both the valence and the conduction bands, and therefore, the deformation potentials. Very recently this procedure has been applied to determine the deformation potentials in GaAs and InP by Nolte et al. [57]. Their results for GaAs with (111)-uniaxial stress are reproduced in Fig. 10. These results provide evidence that the pressure derivative is the same for different TM's (Ti and V) and also for different charge states of the impurities, as it should be the case when the TM levels form an universal reference level. The obtained deformation potentials for the bottom of the cb's of GaAs and InP are a_c = -9.3 ± 1 eV and a_c = -7 ± 1 eV, respectively.

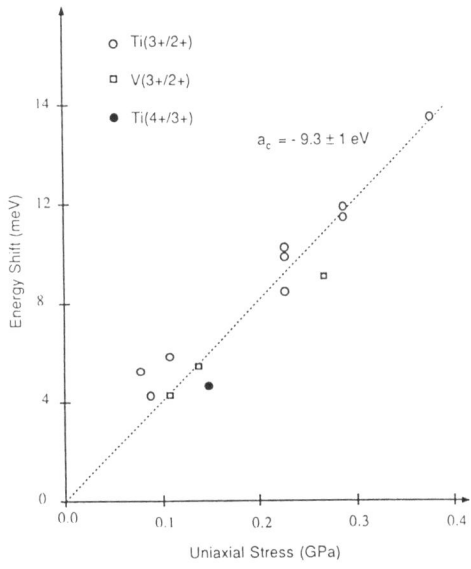

Figure 10. Energy shifts of DLTS peaks (after Ref. [57]) as a function of ⟨111⟩ oriented uniaxial stress for Ti and V of different charge state. Within 30% an universal pressure derivative was observed.

Acknowledgement

The continous and helpful discussions with my colleague J.M. Langer is highly appreciated. Work supported by "Fonds zur Förderung der wissenschaftlichen Forschung", Austria.

References

[1]   J.C. Pollman, in Festkörperprobleme: Advances in Solid State Physics, ed. by J. Treusch (Vieweg, Braunschweig, 1980), Vol. XX, p. 117

[2]   H. Kroemer, Surf. Sci. **132**, 543 (1983), and in: Proc. of the NATO Advanced Study Institute on Molecular-Beam Epitaxy and Heterostructures, Erice, Sicily, 1983, ed. by L.L. Chang and K. Ploog (Martinus Nijhoff, The Netherlands, 1984), p. 331, and J. Vac. Sci. Technol. **B2**, 433 (1984)

[3] W.A. Harrison, Proc. 4th Int. Winterschool, Mauterndorf, ed. by F. Kuchar, G. Bauer and H. Heinrich, Solid State Sciences, Vol. 67, p. 62 (Springer, Berlin, 1986)

[4] F. Flores and C. Tejedor, J. Phys. C, Solid State Phys. 20, 145 (1987)

[5] H. Kroemer, Surf. Sci. 174, 299 (1986)

[6] R.S. Bauer, H.W. Sang, Jr., Surf. Sci. 132, 479 (1983)

[7] R.L. Anderson, Solid State Electronics 5, 341 L(1962)

[8] J.O. McCaldin, T.C. McGill and C.A. Mead, Phys. Rev. Lett. 36, 56 (1976)

[9] D.W. Niles and G. Margaritondo, Proc. 18th Int. Conf. on the Phys. of Semicond. Stockholm, O. Engström ed., World Scientific, p. 187 (1987), Phys. Rev. B 34, 2923 (1986)

[10] J. Tersoff, Phys. Rev. Lett. 56, 2755 (1986)

[11] J.M. Langer and H. Heinrich, Phys. Rev. Lett. 55, 1414 (1985), Physica 139B, 444 (1985)

[12] H. Heinrich and J.M. Langer, in Festkörperprobleme: Advances in Solid State Physics (Vieweg, Braunschweig, 1986), Vol. 26, p. 251; and Proc. 4th Int. Winterschool, Mauterndorf, in: Two Dimensional Systems: Physics and Devices (Springer Series in Solid State Sciences, Vol. 67, p. 83, 1986), ed. by G. Bauer, H. Heinrich and F. Kuchar

[13] J.M. Langer and H. Heinrich, Proc. 18th Int. Conf. on the Phys. of Semicond. Stockholm, O. Engström ed., World Scientific, p. 175 (1987)

[14] A. Zunger, Ann. Rev. Mater. Sci. 15, 411 (1985)

[15] J. Tersoff, Phys. Rev. Lett. 56, 675 (1986)

[16] J. Tersoff and W. Harrison, Phys. Rev. Lett. 58, 2367 (1987)

[17] A. Eisenbeiss, H. Heinrich, J. Opschoor, R.P. Tijbury and H. Preier, Appl. Phys. Lett. 50, 1583 (1987)

[18] G. Duggan, J. Vac. Sci. Technol. B3 (4), 12241 (1985)

[19] T.W. Hickmott, Proc. 4th Int. Winterschool, Mauterndorf, in: Two Dimensional Systems: Physics and Devices (Springer Series in Solid State Sciences, Vol. 67, 1986), ed. by G. Bauer, H. Heinrich and F. Kuchar

[20] R.S. Bauer and G. Margaritondo, Physics Today, January 1987, p. 27

[21] L. Esaki, IEEE J. Quantum Electronics QE-22, 1611 (1986)

[22] J. Smoliner, R. Lassnig, E. Gornik, G. Weimann and K. Ploog, Proc. 3rd Int. Conf. on Modulated Semicond. Structures, Montpellier 1987, in print

[23] A. Lechner, M. Kneidinger, K. Lübke and H. Thim, Inst. Phys. Conf. Ser. No. 83: Chapt. 4, p. 267 (1987)

[24] H. Heinrich, A. Eisenbeiss, C. Panhuber, H. Preier and Norton, to be published

[25] J. Menendez, A. Pinczuk, A.C. Gossard, J.M. English, D.J. Werder and M.G. Lamont, J. Vac. Sci. Technol. B(4), 1041 (1986)

[26] B.A. Wilson, P. Dawson, C.W. Tu and R.C. Miller, J. Vac. Sci. Technol. B(4), 1037 (1986)

[27] M.A. Hause, M.A. Emanuel, S.C. Smith, J.J. Coleman and G.E. Stillman, Appl. Phys. Lett. **50**, 404 (1987)

[28] R. Dingle, A.C. Gossard and C.H. Henry, Phys. Rev. Lett. **33**, 827 (1974)

[29] L.A. Ledebo and B. Ridley, J. Phys. **C15**, L961 (1982)

[30] J. Tersoff, Phys. Rev. Lett. **52**, 465 (1984)

[31] D.V. Lang, R.A. Logan, L.C. Kimerling, in Physics of Semiconductors, Proc. of the 13th Int. Conf., Rome 1976, ed. by F.G. Fumi, Tipografia Marres, Rome 1974, p. 615

[32] Z.G. Wang, I.A., Ledebo, H.G. Grimmeiss, J. Appl. Phys. **56**, 2762 (1984)

[33] E.M. Omel'yanovskii, V.I. Fistul', L.A. Balagurov, V.S. Ivleva, V.V. Karataer, M.G. Mil'vidskii and A.N. Popkov, Sov. Phys. Semicond. **9**, 381 (1975)

[34] T.L. Estle, Phys. Rev. **136**, A1702 (1984)

[35] P.S. Whitney and C. Fonstad, J. Appl. Phys. **57**, 4663 (1985)

[36] S.R. Forrest, P.H. Schmidt, R.B. Wilson, M.L. Kaplan, Appl. Phys. Lett, **45**, 1199 (1984)

[37] D.V. Lang, M.B. Panish, F. Capasso, J. Allam, R.A. Hamm and A.M. Sergent, Appl. Phys. Lett. **50**, 736 (1987)

[38] H. Temkin, M.B. Panish, P.M. Petroff, R.A. Hamm, J.M. Vandenberg, S. Sumski, Appl. Phys. Lett. **47**, 394 (1985)

[39] G. Duggan, J. Vac. Sci. Technol. **B3**, 1224 (1985)

[40] M.S. Skolnik, L.L. Taylor, S.J. Bass, A.D. Pitt, D.J. Mowbray, A.G. Cullis and N.G. Chew, Appl. Phys. Lett. **51**, 24 (1987)

[41] P.E. Brunnemeier, D.G. Deppe and N. Holonyak, Jr., Appl. Phys. Lett. **46**, 755 (1985)

[42] L.M. Claessen, J.C. Maan, M. Altarelli, P. Wyder, L.L. Chang and L. Esaki, Phys. Rev. Lett. **57**, 2556 (1986)

[43] G.J. Gualtieri, G.P. Schwartz, R.G. Nuzzo, R.J. Malik and J.F. Walker, J. Appl. Phys. **61**, 5337 (1987)

[44] W. Ulrici, L. Eaves, K. Friedland, D.P. Halliday and J. Kreißl, Defects in Semicond. (von Bardeleben, ed.) Materials Science Forum Vol. 10-12, 639 (1986) and references therein

[45] M.A. Rao, E.J. Caine, H. Kroemer, S.I. Long and D.I. Baci, J. Appl. Phys. **61**, 643 (1987)

[46] Miyoko O. Watanabe and Yasuo Ohba, Appl. Phys. Lett. **50**, 906 (1987)

[47] Y. Guldner, Proc. 3rd Int. Winterschool, Mauterndorf (Kuchar, Bauer, Heinrich, ed.) Springer Series in Solid State Sciences, Vol. 53, p. 200 (1984)

[48] S.P. Kowalczyk, J.T. Cheung, E.A. Kraut and R.W. Grant, Phys. Rev. Lett. **56**, 1605 (1986)

[49] G.S. Boebinger, Y. Guldner, J.M. Berroir, M. Voos, J.P. Vieren and J.P. Faurier, Proc. 3rd Int. Conf. on Modulated Semicond. Structures, Montpellier 1987, in print

[50] C. Hsu, Trou Minh Duc and J.P. Faurier, Proc. 3rd Int. Conf. on Modulated Semicond. Structures, Montpellier 1987, in print

[51] W. Dobrowolski, K. Dybko, A. Mycielski, J. Wrobel, S. Piechota, M. Palczewska, H. Szymczak and Z. Wilamowski, Proc. 18th Int. Conf. on the Phys. of Semicond. Stockholm, O. Engström ed., World Scientific, p. 1743 (1987)

[52] K. Lischka, G. Brunthaler and W. Jantsch, J. Cryst. Growth **72**, 355 (1985)
A recent reevaluation of CdTe:Fe including lattice relaxation puts the Fe^{2+}-level at 0.35 eV above the vb edge in CdTe:
W. Jantsch, G. Brunthaler and G. Hendorfer, Defects in Semiconductors (v. Bardeleben ed.), Materials Science Forum, Vol. 10-12 (1986), p. 515

[53] A. Zoryk and M. Jaros, Appl. Phys. Lett. **50**, 1191 (1987)

[54] A. Zunger and Su-Hani Wei, J. Vac. Sci. Technol., to be published

[55] C.G. Van De Walle and R.M. Martin, J. Vac. Sci. Technol., to be published

[56] A.D. Katnani and P.S. Bauer, Phys. Rev. **B33**, 1106 (1986)

[57] D.D. Nolte, W. Walukiewicz and E.E. Haller, Phys. Rev. Lett. **59**, 501 (1987)

DEFECT DYNAMICS IN CRYSTALLINE AND AMORPHOUS SILICON

Sokrates T. Pantelides

IBM Thomas J. Watson Research Center, Yorktown Heights, NY 10598

Extended Abstract

In crystalline Si, it has long been recognized that the two primitive conjugate intrinsic defects are the vacancy and the self-interstitial and that both can potentially mediate self-diffusion and impurity diffusion. Their detailed properties, their relative importance and their precise role in observed phenomena, however, have been the subject of contrasting speculations and much controversy. At cryogenic temperatures ($\sim$ 4K), the vacancy was unambiguously identified in the 1960's by Watkins after electron irradiation. Its migration energy was measured to be of order 0.3 eV depending slightly on the charge state. The vacancy's conjugate partner, however, the self-interstitial remained undetected. The detection of interstitial Al impurities, which had been present as substitutional dopants, was indirect evidence that self-interstitials can migrate rapidly, most likely athermally. At high temperatures (over 800° C), The self-diffusion coefficient was measured by many authors to have an activation energy of order 4-5 eV. If self-diffusion is mediated by a defect, the activation energy is equal to the sum of formation and migration energies. Independent measurements of formation energies were lacking, however, for both vacancies and interstitials. In view of the very small migration energies measured at cryogenic temperatures, one could conclude that, if either defect mediates self-diffusion, its formation energy ought to be of order 4 eV. A variety of indirect evidence, however, and crude theoretical estimates led to widespread belief that the formation energy of the dominant defect is only 2.5 eV, requiring a migration energy of order 2 eV, significantly larger than measured at cryogenic temperatures. At the same time, there was widespread disagreement over whether the vacancy or self-interstitial was the dominant defect, each side invoking indirect experiments, often the same experiment, to support their point of new. The drastic change in migration energy from cryogenic to high temperatures was usually attributed to a change in the atomic structure of the defect, a "spreading out" or delocalization. This notion was actually invoked back in 1968 by Seeger and Chik as an explanation of the observed large pre-exponential of the self-diffusion coefficient, which implies large entropy. All these ideas remained speculative, however, as no atomistic description of the "spreading out" was determined.

Parameter-free calculations by Car, Kelly, Oshiyama and Pantelides, first reported in 1984 and updated in 1985, determined that vacancies and self-interstitials have comparable formation energies, of order 4-6 eV, depending on the change state. These large formation energies, as compared to 2.5 eV, yielded a simple reconciliation of the cryogenic and high-temperature data without the need to invoke a drastic change in migration energies. The calculations also established that both defects play a role in self-diffusion. Since then, similar theoretical values for formation energies were obtained independently by others. More significantly, in 1986 Dannefaer, Mascher and Kerr reported a direct measurement of the formation energy of the neutral vacancy. Their value, 3.6 ±0.2 eV, is in excellent agreement with the theoretical value of 3.8 ± 0.5 eV. A review of the literature of the 2-3 years reveals that the views that were heavily debated before 1984 have disappeared, indicating quiet acceptance of the understanding that was provided by the theoretical calculations and the experimental confirmation of a large formation energy for the vacancy. More recently, Pandey has reported calculations of the activation energy for a direct exchange of two neighboring Si atoms, i.e., self-diffusion with defects. The activation energy, ∼ 4.5 eV is similar to the activation energies for vacancy and interstitial-mediated self-diffusion. In the absence of entropy calculations for these three processes, their relative importance is not known.

In addition to reconciling cryogenic and high-temperature data, the theoretical calculations also provide a detailed description of the properties of the elusive self-interstitial. In particular, they established that there are several paths which support athermal migration for the self-interstitial. All of these paths involve the high-symmetry tetrahedral interstitial site. Athermal migration is achieved via the successive capture and emission of electrons (Bourgoin-Corbett mechanism).

Further theoretical calculations by Car, Kelly, Oshiyama and Pantelides addressed the question of impurity diffusion, again establishing that both vacancies and self-interstitials participate. This work also established that the preferred mode for self-interstitial mediation is the kick-out mechanism. The relative importance of the various mechanisms for each dopant still remain controversial, however.

In amorphous silicon (a-Si), for the last twenty years, it was universally believed that the dangling bond (threefold-coordinated Si) is the dominant defect giving rise to an EPR signal with g = 2.0055. Overcoordination was either overlooked completely or dismissed as impossible. Recently, the present author noted that though the vacancy in crystalline Si (c-Si) consists of four threefold-coordinated Si atoms, the tetrahedral self-interstitial consists of four fivefold-coordinated Si atoms. It was noted further that threefold- and fivefold- coordinated Si atoms are in principle the intrinsic primitive conjugate defects in a-Si and should be examined on an equal footing as possible candidates for the paramagnetic center (known as the D center). The similarity of the formation energies of vacancies and self-interstitials in c-Si lent support to the notion that both primitive defects a-Si might be present and play a role in determining the observed properties of the material.

An elementary calculation of the electronic structure of the fivefold-coordinated Si revealed that, just like the threefold-coordinated Si, it has a state in the gap containing an unpaired electron when the defect is neutral. Its wave function was found to be delocalized over the five neighbors of the central atom, with one of the atoms typically being dominant. This state was labeled "floating bond", to parallel "dangling bond". Careful examination of the EPR hyperfine data led to the conclusion that they favor floating bonds being the paramagnetic defect.

An important property of the floating bond is that it can migrate quite efficiently in the amorphous network: The extra bond is simply passed from one atom to another while individual atoms change their positions only slightly. Dangling bonds can in principle migrate by a similar bond-switching mechanism, but higher energies are probably involved because of the need to maintain a microvoid on the front side. Interconversion of a dangling bond into a floating bond can enhance its migration.

By assuming a self-mobile intrinsic defect, it is possible to deduce a set of point defect reactions in terms of which many observations can be explained in an elegant, microscopic and systematic way. Until now, atomistic descriptions of the observed phenomena have been lacking or provided unsatisfactory pictures. For example, an atomistic description of the annealing of D centers has not been available in terms of the conventional concepts of immobile dangling bonds. A mobile D center however, can participate in the reaction

$$D + SiH \rightleftarrows \square + H_i \qquad (1)$$

where SiH stand for and Si-H bond, $\square$ stands for a fourfold-coordinated network, and Hi for interstitial hydrogen. The latter can then kill another D center :

$$D + H_i \rightleftarrows SiH \qquad (2)$$

Note that, through these two reactions, D centers destroy themselves while the density of SiH bonds is conserved in agreement with observations. In pure a-Si (without H), D center annealing is significantly slower, reflecting the absence of reactions (1) and (2). The only mechanism for annealing in that case is the mutual annihilation between dangling and floating bonds, i.e.

$$DB + FB \rightleftarrows \square \qquad (3)$$

which is quite slower because the concentration of both DBs and FB's are typically smaller than those of SiH in a-Si:H.

Reaction (1) and (2) also mediate H diffusion. Reaction (1) which is equivalent to the kick-out mechanism for impunity diffusion in c-Si, occurs with a smaller activation energy if the D centers already exist. In that case, H is kicked out of an SiH bond. In reaction (2) on the other hand, H_i is created by the direct break up of SiH bonds, which costs more energy. Thus, reaction (1) provides a natural explanation of the observed low-temperature H diffusion.

Reaction (3) in the reverse direction represents the creation of "Frenkel" pairs. This reaction is likely to be enhanced by band-gap light which can provide the energy needed to created the pair. The present author has proposed that this process underlies the Staebler-Wronski effect, i.e., the creation of excess D centers by prolonged illumination. Examination of relevant data has led this author to assign the luminescence peak at ~ 0.9 eV to dangling bonds.

Finally, migrating floating bonds provide a natural explanation for doping. Impurities such as phosphorus or boron normally optimize their energy in threefold coordination. A fraction of them, however, exist in fourfold coordination which results in doping. In other words, overcoordination is essential for doping. In terms of floating bonds, the existence of fourfold-coordinated P is a natural consequence of the reaction

$$P_3 + FB \rightleftarrows P_4 \tag{4}$$

where the subscript denotes coordination. A similar reaction holds for boron. Doping, of course, moves the Fermi level closer to one or the other band edge which leads to a lowering of the formation energy of charged defects. An increase in the density of such defects follows, as observed.

This work was supported in part by ONR contract No. N00014-84-C-0396.

References to all the original work mentioned above can be found in two recent reviews by the present author:
S. T. Pantelides, "Defect Reactions and Atomic Diffusion in Silicon" in Materials Issues in Silicon Integrated Circuit Processing", edited by M. Wittmer, J. Stimmell and M. Strathmann, (Materials Research Society, Pittsburgh, 1986), p. 147; and S. T. Pantelides, "Defect Dynamics and the Properties of Amorphous Silicon--A New Perspective", in Amorphous Silicon Semiconductors-- Pure and Hydrogenated, edited by D. Adler, Y. Hamakawa, A. Madan and M. Thompson, (Materials Research Society, Pittsburgh, 1987).

On the Diffusion of Oxygen in a Silicon Crystal

Lawrence C. Snyder, James W. Corbett,

Peter Deák[+] and Rongzhi Wu

State University of New York, 1400 Washington Ave., Albany, N.Y. 12222

1. INTRODUCTION.

Gosele and Tan[1] discussed the relation of oxygen diffusion to thermal donor formation at 450 °C in crystalline silicon. They found the accepted diffusion constant for interstitial oxygen (O_1) in silicon to be too slow by several orders of magnitude to explain the rate of oxygen thermal donor formation. They were forced to postulate a second diffusion mechanism for oxygen in silicon. They postulated this to be a rapidly diffusing oxygen dimer (O_2), a gas-like molecular oxygen without bonds to silicons. They estimated the diffusion constant of O_2 to be 10^8 times that of O_1.

Stavola has found that the diffusion rate of interstitial oxygen estimated from the relaxation kinetics of stress induced dichroism in Czochralski grown silicon crystals can be several orders of magnitude greater, depending on the thermal history, than that observed for isolated interstitial oxygen.[2] For a silicon crystal that received an oxygen dispersing heat treatment of 20 hours at 1350 °C, the measured relaxation times yield a diffusivity that extrapolates well to the mass transport measurements made at higher temperatures. The oxygen diffusivity is two orders of magnitude greater for crystals which have received an "oxygen stabilization" heat treatment at 900 °C for 2 hours. Stavola and Snyder[3] postulated that the nominally isolated oxygen in the crystal is loosely clustered, and that interactions between oxygen atoms, primarily through strain, might effect the saddle point configuration for diffusion and modify the jump rate.

In this work we have attempted to follow up the proposal of Gosele and Tan[1] that an oxygen dimer provides a rapid diffusion path. We first have searched for structures by which two interstitial oxygen atoms bind in a silicon crystal to form a dimer. Then we have computed the activation energy for diffusion of an isolated interstitial oxygen atom and have investigated whether the dimer provides a path for a more rapid diffusion of oxygen in a silicon crystal.

2. THEORETICAL BACKGROUND

Ab initio SCF calculations (some of which are published in ref. 4) on $Si_5H_{12}O_n$ clusters, with n = 0 to 4 , suggested that interstitial oxygen atoms tend to bond to a common silicon atom. However, we now know that these clusters were too small to quantitatively investigate questions of oxygen clustering in a silicon crystal.

In exploratory molecular cluster[5] and cyclic cluster[6] computations using semiempirical quantum chemical methods to study the chemistry of oxygen in silicon, we have concluded that the modified intermediate neglect of differential overlap (MINDO/3) method developed by Dewar and coworkers,[7] and using the Si-O bonding parameters developed by Edwards and Fowler[8] gives the best description of the molecular and solid state properties of silicon. We employ MINDO/3 and those oxygen parameters in the work reported here. We have also concluded that to simulate defect chemistry on a silicon lattice, one must allow relaxation of the positions of both first and second neighbor atoms of the defect in the lattice.[9]

The work we report here is part of a larger body of research in which we are studying the chemistry of up to five interstitial oxygen atoms in order to represent the core of the 450 °C oxygen thermal donor.[3,4] In that work and here we have adopted the reference $Si_{47}H_{60}$ cluster depicted in Figure 1a . In this cluster all Si-Si bond lengths are initially taken to be 2.37222 Angstroms and all bond angles to be tetrahedral. We use Si-H bonds to tie up silicon valencies on the perimeter of the cluster. All Si-H bonds are given a length of 1.47422 Angstroms. These bondlengths are those which minimize the total energy of the Si_5H_{12} molecule when all bond angles are assumed to be tetrahedral. In the work reported here the positions of all silicon atoms not bonded to hydrogen are optimized in all clusters containing defects. Therefore we have also optimized their positions as depicted in Figure 1b and subject to C_{2v} symmetry in the reference cluster. For the optimized reference cluster we compute a heat of formation of -5333.463 eV. The Koopmans theorem ionization potential of the reference cluster is 7.2 eV.

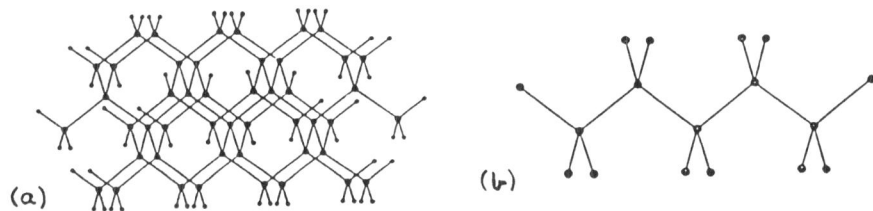

Figure 1: (a) The reference cluster $Si_{47}H_{60}$.
(b) The optimized positions of the 17 central silicon atoms.

Our research plan is to introduce one or two oxygen atoms into the central plane of the reference cluster, and then compute the structures of minimum energy with respect to the positions of oxygen atoms and of all the silicon atoms not having perimeter Si-H bonds.

3. THE POTENTIAL FOR DIFFUSION OF INTERSTITIAL OXYGEN

Our computed structure for interstitial oxygen in the molecular cluster $Si_{47}H_{60}O_i$ is shown in Figure 2a. The silicon atoms whose positions were varied are shown there as large dark circles. Their respective lattice postions are shown as small light circles. The interstitial oxygen atom is shown as a circle of intermediate size. Its bonds with silicons are computed to be linear with the MINDO/3 method used here. This is a usual result of small basis calculations on disiloxan as well. The computed heat of formation of the interstitial oxygen cluster is -5647.737 eV. The computed ionization potential is 7.1 eV.

We have assumed that the transition state for the diffusion of interstitial oxygen between two sites having a common silicon center is the C_{2v} ylid structure.[3] Our computed structure of minimum energy is depicted in Figure 2b. The computed heat of formation is -5645.247 eV. The computed ionization potential is 5.6 eV.

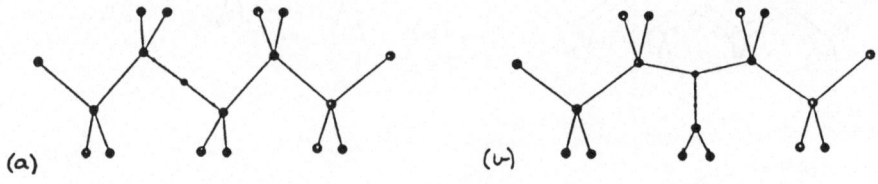

(a)   (b)

Figure 2: (a) The interstitial oxygen cluster. (b) The C_{2v} ylid structure.

Our estimate of the activation energy for diffusion of an isolated interstitial oxygen (the interstitial oxygen monomer) in silicon is the computed difference in energy of the interstitial oxygen and the C_{2v} ylid clusters: that is +2.49 eV. This is surprisingly close to the observed 2.53 eV activation energy for oxygen diffusion.[10] We take take this result to give some credibility to our subsequent prediction of the potential for diffusion of an interstitial oxygen dimer.

4. THE BINDING OF TWO INTERSTITIAL OXYGEN ATOMS TO FORM A DIMER (O_2)

The lowest energy structure we have computed with two oxygen atoms added to the reference cluster, $Si_{47}H_{60}O_2$, is depicted in Figure 3a. We call this the C_{2v} di-interstitial oxygen structure, because two interstitial oxygen atoms are bonded to a common silicon atom. This cluster has a computed heat of formation of -5962.114 eV. The computed ionization potential is 7.0 eV. This C_{2v} di-interstitial oxygen structure is binding by 0.1 eV with respect to two isolated interstitial oxygen atoms. Specifically, the transformation of two interstitial oxygen clusters to one C_{2v} di-interstitial oxygen cluster plus one reference cluster is computed to be exothermic by 0.1 eV.

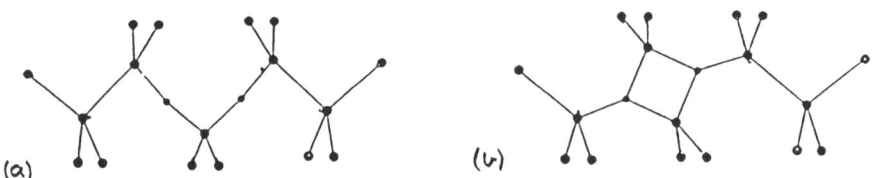

Figure 3: (a) The C_{2v} di-interstitial oxygen structure.
(b) The four-member ring structure.

We have found a second $Si_{47}H_{60}O_2$ structure computed to be a local minimum in energy only 0.05 eV higher than the C_{2v} di-interstitial oxygen cluster. We call this the four-member ring structure because it contains a four-member ring with alternating oxygen and silicon atoms as depicted in Figure 3b. Both oxygen atoms are trivalent, as in the hydronium ion, but bonded to three silicon atoms. This is in contrast to the proposal by Gosele and Tan[1] of a rapidly diffusing O_2 dimer without bonds to the silicons.

Because of the high symmetry of the four-member ring structure, its bonds can be broken in two ways leading to two C_{2v} di-interstitial oxygen structures displaced on the lattice relative to each other. Thus the four-member ring is on a diffusion path of the di-interstitial oxygen dimer.

We note that there are four sigma antibonds from silicon to oxygen atoms of the four-member ring. These antibonds form a manifold of four orbitals which are nearly degenerate. One of these is a major component of the highest filled molecular orbital of the four-member ring cluster. We expect configuration interaction and the corresponding electron correlation to be more prominent in the four-member ring structure than for the C_{2v} di-interstitial oxygen structure. We would not be surprised if *ab initio* electronic structure calculations which give a more explicit account of electron correlation show the four-member ring to be more stable and the binding energy greater.

We also note that the computed ionization potential for the four-member ring is 5.1 eV. This is the lowest for the structures considered here. The computed ionization potential for a 54 silicon atom cyclic cluster described with MINDO/3 is 5.7 eV. This corresponds to the approximate location of the top of the valence band of silicon in the MINDO/3 approximation. The highest filled orbital of our four-member ring appears to be in the gap of the silicon crystal. The low ionization potential of the four-member ring may provide a mechanism[11] for ionization to assist its diffusion.

5. A POTENTIAL FOR DIFFUSION OF THE INTERSTITIAL OXYGEN DIMER

The four-member ring structure is intermediate between two C_{2v} di-interstitial oxygen structures displaced in the lattice. We have constructed a potential surface for the diffusion of a pair of oxygen atoms by drawing a path for the two silicon and the two oxygen atoms of the four-member ring which carries them to their locations in the C_{2v} di-interstitial oxygen structure. In constructing this path we have attempted to maintain nearly constant Si-O bond lengths in those bonds which are preserved in this process. Seven intermediate configurations of these four atoms on the assumed trajectory are shown in Figure 4a. along with stick figures of the initial and final configurations. For each position of the atoms of the four-member ring, the location of all other silicon atoms were optimized (except of the ones with Si-H bonds). The potential surface connecting the total energy for points of this path is drawn in Figure 4b.

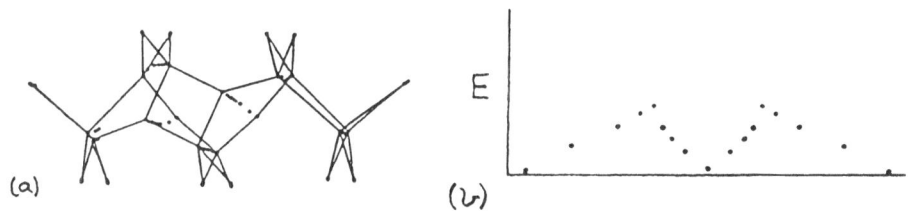

Figure 4: (a) Location of computed path from the four-member ring structure to the C_{2v} di-interstitial oxygen structure. (b) Plot of computed energy over the path in (a).

The activation energy for diffusion along this trajectory is computed to be 1.36 eV. The transition state is closer to the four-member ring structure. We may not have found the trajectory having the lowest activation energy, but we believe we are close to it. The value 1.36 eV must be regarded as an upper limit for the MINDO/3 activation energy for diffusion of the oxygen dimer.

It is not obvious why the four-member ring is more stable relative to the di-interstitial oxygen than the C_{2v} ylid is relative to the single interstitial oxygen for the monomer. However, a major contributing factor may be the stability of the highest filled molecular orbital of the four-member ring. This molecular orbital is the most stable linear combination of the four Si-O antibonding σ orbitals of the four-member ring.

6. THE RELATIVE RATES OF DIFFUSION OF INTERSTITIAL OXYGEN MONOMER AND DIMER

If one assumes that the prefactors in the rate equation for diffusion of the interstitial oxygen monomer and dimer are the same, then one can calculate the ratio of their diffusion constants from the difference of the two activation energies of diffusion, ΔE_{O_1} and ΔE_{O_2}:

$$\frac{D_{O_2}}{D_{O_1}} = \exp[-\frac{1}{RT}(\Delta E_{O_2} - \Delta E_{O_1})]$$

Using the computed values, $\Delta E_{O_1} = +2.49$ eV and $\Delta E_{O_2} = +1.36$ eV, at a temperature of 450 °C with R = 8.6 10^{-5} eV we obtain

$$\frac{D_{O_2}}{D_{O_1}} = 8.2 \ 10^7$$

This is close to the original estimate by Gosele and Tan[1] that the diffusion constant of O_2 is 10^8 greater than that of O_1.

7. RELATION TO OTHER PHENOMENA

It is now possible to give a tentative explanation of several previously puzzling phenomena. In silicon crystals which were given an "oxygen stabilization" heat treatment, all of the interstitial oxygen of the sample showed the enhanced relaxation rate of stress induced dichroism. We propose that the oxygen stabilization generates a small concentration of the rapidly diffusing dimer. Collisions of this dimer with the interstitial oxygen monomer may become an important mechanism for the relaxation of dichroism of the monomer. Thus the activation energy for relaxation of the stress-induced dichroism of the monomer, which is observed to be 1.9 eV after "oxygen stabilization" treatment[3], approaches the activation energy for diffusion of the dimer.

Spaeth[12] has observed the electron nuclear double resonance (ENDOR) spectra of 450 °C oxygen thermal donors in silicon. He observed the ^{29}Si superhyperfine interactions for up to seven neighbor shells of four thermal donors. These distinct thermal donors are believed to differ in the number of oxygen atoms contained. From the analysis of the ENDOR each was found to have C_{2v} symmetry. Spaeth concluded that his results require that the addition of (oxygen) atoms must occur pairwise. This observation would be consistent with the addition of the rapidly diffusing oxygen dimer.

References

+ *Permanent address*:Physical Institute of the Technical University, Budapest H-

1111 Budafoki ut 8. Hungary.

(1) U. Gosele and T. Y. Tan, Appl. Phys. **A28**, 79 (1982).
(2) M. Stavola, J. R. Patel, L. C. Kimerling, and P. E. Freeland, Appl. Phys. Lett. **42**, 73 (1983).
(3) M. Stavola and L. C. Snyder, in *"Defects in Silicon"*, edited by W. M. Bullis and L. C. Kimerling (The Electrochem. Soc., Pennington, 1983) p.61.
(4) L. C. Snyder and J. W. Corbett, in *"Oxygen, Carbon, Hydrogen and Nitrogen in Silicon"*, eds. S. V. Pearton, J. W. Corbett, J. C. Mikkelson Jr., and S. J. Pennycock (Material Res. Soc., Pittsburgh, 1986) p.207.
(5) P. Deak, L. C. Snyder, R. K. Singh, and J. W. Corbett, Phys. Rev. B *in print* (1987).
(6) P. Deak and L. C. Snyder, Phys. Rev. B *in print* (1987).
(7) R. C. Bingham, M. J. S. Dewar, and H. C. Lo, J. Am. Chem. Soc. **97**, 1285 (1975).
(8) A. H. Edwards and W. B. Fowler, J. Phys. Chem. Solids **46**, 841 (1985).
(9) P. Deak, L. C. Snyder, and J. W. Corbett, Phys. Rev. B *to be published*
(10) J. C. Mikkelsen, Jr., in *"Oxygen, Carbon, Hydrogen, and Nitrogen in Silicon"*, eds. S. V. Pearton, J. W. Corbett, J. C. Mikkelson Jr., and S. J. Pennycock (Material Res. Soc., Pittsburgh, 1986) p.19.
(11) J. C. Bourgoin and J. W. Corbett, Phys. Lett. **38A**, 135 (1972).
(12) J. Michel, J. R. Niklas, J.-M. Spaeth, and C. Weinert, Phys. Rev. Lett. **57**, 611 (1986).

HEXAGONAL SITE INTERSTITIAL RELATED STATES IN SILICON

G. Papp

Department of Theoretical Physics, Attila József University,
Szeged, Aradi Vértanúk tere 1., HUNGARY

P. Boguslawski

Institute of Physics, Polish Academy of Sciences,
Warsaw, POLAND

A. Baldereschi

Institut de Physique Appliquée, Ecole Polytechnique Fédérale,
Lausanne, SWITZERLAND
Istituto di Fisica Teoretica, Universitá di Trieste, ITALY

The Zn, Si and S interstitials in silicon are studied at the
hexagonal site with the large unit cell pseudopotential
method. The electronic energy scheme and charge density are
investigated. The origin of the impurity related states
stems from the interactions between the atomic orbitals of
the interstitials and the nearest-neighbour bonding orbitals
of the host crystal. The energy levels are strongly affected
by the impurity potential.

The different kinds of impurities in Si at the tetrahedral
site have been studied extensively [1-5]. The results have shown
that its energy level scheme shows chemical trends from one
impurity to another. While the existence of these trends at the
tetrahedral site is well established by now, hexagonal site
results [6-8] are not yet examined systemathically. In the
present work we select the hexagonal interstitial site in Si.
The electronic structures for Zn and S are given and compared
with our earlier self-interstitial results [9].

We use the large unit cell approach [10], and assume that
the impurities are distributed periodically. The cell – centered
on the impurity – contains sixteen host atoms. The energy bands
are calculated with the self-consistent pseudopotential method
using the Appelbaum-Hamann pseudopotential for Si [11], and the
ion potential for impurities was chosen as in [12]:

$$v(r)=-2Z/r \ \mathrm{erf}(r/r_0)+a/r_0\exp(-(r/r_0)^2)+r^2b/r_0^3\exp(-(r/r_0)^2), \quad (1)$$

where the parameters are in Table I.

Table I.
Parameters for ion potentials

	r_0	a	b
Zn	1.5	3.37	-2.96
S	1.0	15.0	-7.50

The exchange part of the potential was treated by the Xα approximation with $\alpha = 0.76$.

We limit ourselves the neutral interstitials and fill the lowest 33 and 35 bands with two electrons for Zn and S, respectively.

The effect produced by the interstitials are studied with the Hamiltonian

$$H(x,r)=p^2/(2m)+U_0(r)+x(U(r)-U_0(r)) , \qquad (2)$$

where $U_0(r)$ and $U(r)$ are the self-consistent crystal potentials of perfect Si and of Si with superlattice of interstitials, respectively. The parameter $x(0\leq x\leq 1)$ switches on the impurity potential. Impurity related states are identified by direct inspection of the Bloch functions as those with appreciable amplitude near the impurity.

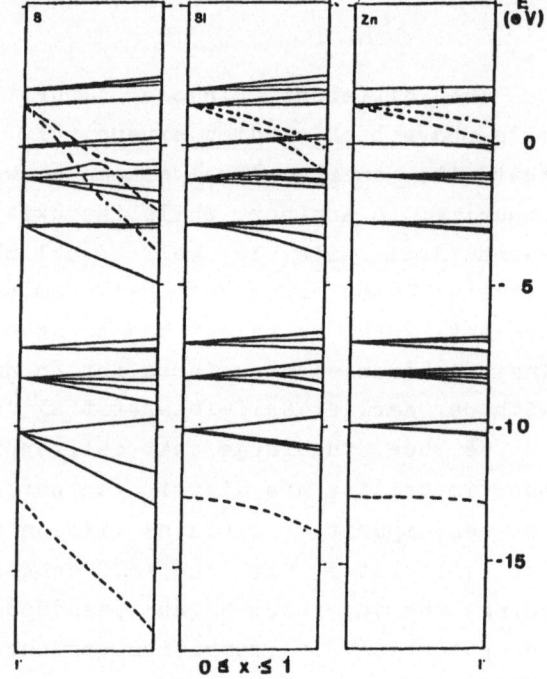

Let us proceed to the results. In Figure 1. we presented the energies of the Hamiltonian (2) at the Γ point as function of x. The results clearly show that at the hexagonal site two levels are pulled down from the conduction band and drop below the valence band edge for S (left) and remain in the gap for Zn (right). The dotted lines are s-like, and the dashed--dotted lines are p_z-like states. Furthermore a bound hyperdeep-like state is formed below the valence band.

As we have shown for the self-interstitial [9],

Figure 1. Band energies at the Γ point when the impurity potential is switched on for Zn (left), for Si (middle) and for S (right) impurity. The zero of energy is set at the top of the valence band of pure Si.

the main difference between the tetrahedral and the hexagonal site is the existence of a p_z-like state. The appereance of this state below of the conduction band is due to symmetry, as there is no such combination of the nearest-neighbour bonds that belongs to the same irreducible representation of D_{3d} point group as the p_z one. Therefore it is not repelled towards higher energy by the nearest-neighbour bonds.

The other feature of the impurity related states is, that the isotropic part of impurity potential is sufficient to produce them. The Figure 2. contains the impurity potential (lower curves) and the isotropic part of the impurity potentials (upper curves). We can see from Figure 1. that the deepness of the p_z- -like states for the diverse impurities are very different. For example for S impurity the p_z-like state is lower than s-like resonance, while for Zn the results remain the same as for the self-interstitial, i.e. the p_z-like state is of higher energy than the s-like one. From the two Figures we can conclude, that the deepness of the p_z-like state is strongly correlated – much more than the s-like one – to the strength of the potential.

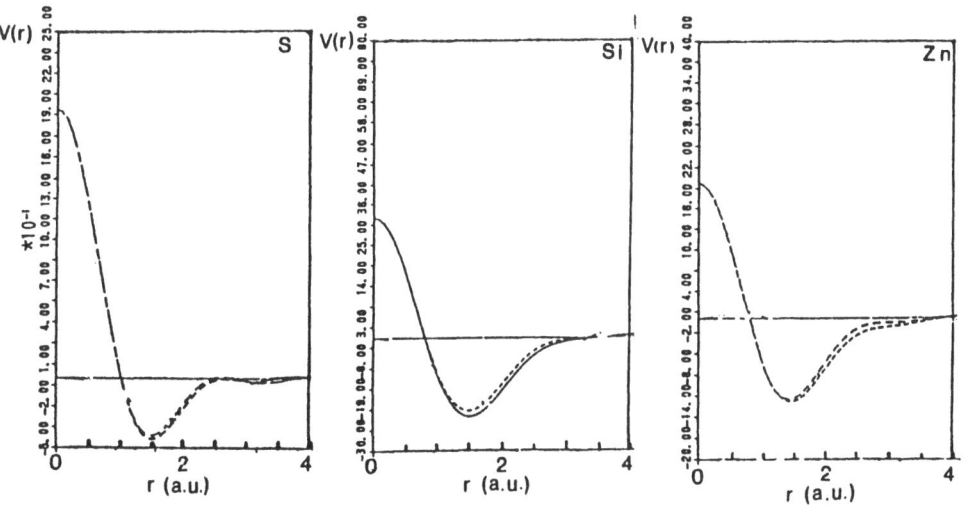

Figure 2. The self-consistent impurity potentials (lower curves) and their isotropic part (upper curves) for S (left), for Si (middle) and for Zn (right).

Sankey and Dow [1] studied the different kinds of impurities in Si at the tetrahedral site and were led to a simple tight--binding model for the impurity states. They emphasized the importance of the hyperdeep state. Figure 3. contains the charge density contour plots of these states for the impurities at the hexagonal site.

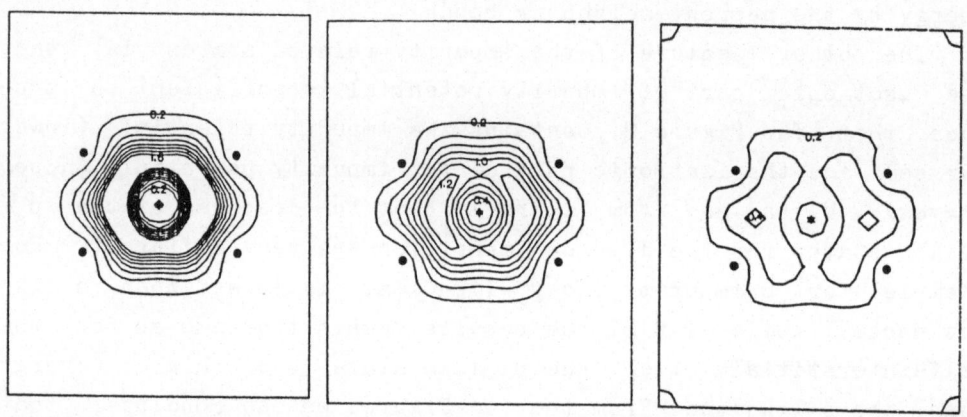

Figure 3. Electron density contour plot in a (211) plane of the hyperdeep states for S (left), for Si (middle) and for Zn (right). The interstitial and host atoms are indicated by stars and dots, respectively. Contour values are given in units of one electron/Si unit cell.

To clear up the origin of the impurity related states we have decomposed their wawefunctions with projections onto the valence and the conduction subspaces of perfect Si:

$$\Psi = P_v \Psi + P_c \Psi = \Psi_v + \Psi_c, \tag{3}$$

where P_v and P_c are projectors onto the valence and the conduction band, respectively. In all cases we have found that Ψ_v consists of the host crystal bonding orbitals and Ψ_c corresponds to the atomic orbitals of the interstitials orthogonalized to the host crystal valence state. Figure 4. illustrates this feature for the hyperdeep state of S impurity. The other impurity related states show the same origin but for the s-like resonance the contribution of the conduction band is much more pronounced than that of the valence band (see Figure 5., where there is no contribution from the valence band with this resolution).

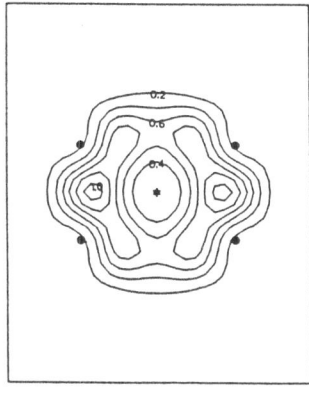 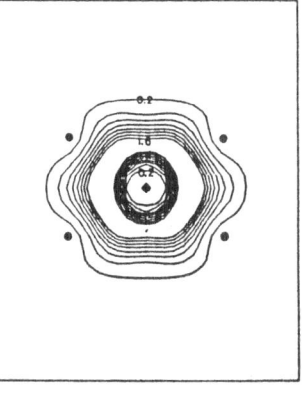 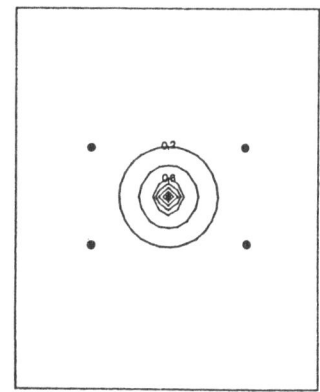

Figure 4. Electron density contours in a (211) plane of the hyperdeep state of S (middle) and its projection onto the valence (left) and onto the conduction (right) band. See Figure 3. for contour values.

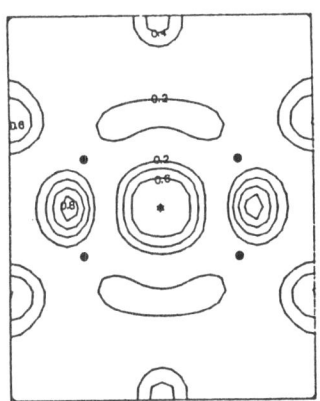 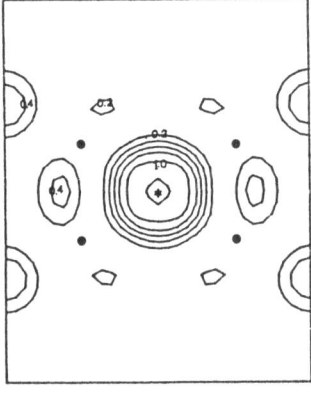

Figure 5. Electron density contours in a (211) plane of the s-like resonance of Zn (left), and its projection onto the conduction (right) band. See Figure 3. for contour values.

As a conclusion we can say, that the features mentioned above can be understood if we apply the model proposed by us previously [9]. The hyperdeep and s-like states are bonding and antibonding combinations of the s function of impurity and the total symmetric combination of the six nearest-neighbour bonds, respectively, and the p_z-like state stems from the relevant atomic state only.

REFERENCES

1. H.P. Hjalmarson, P. Vogl, D.J. Wolford, & J.D. Dow, Phys. Rev.Lett. <u>44</u>, 810 (1980); O.F. Sankey, & J.D. Dow, Phys.Rev. <u>B27</u>, 7641 (1983).

2. M. Lanoo, & J. Bourgoin, Point Defect in Semiconductors I. Springer Verlag, Berlin (1981).

3. C. Weigel, Inst.Phys.Conf. Ser.No. <u>46</u>, 186 (1979).

4. E. Kauffer, P. Pecheur, & M. Gerl, Revue Phys.Appl. <u>15</u>, 849 (1980).

5. S.T. Pantelides, I. Ivanov, M. Scheffler, & J.P. Vigneron, Physica <u>116B</u>, 18 (1983); J.P. Vigneron, M. Scheffler, & S.T. Pantelides, Physica <u>117B&118B</u>, 137 (1983).

6. G.A. Baraff, M. Schlüter, & G. Allan, Phisica <u>116B</u>, 76 (1983); Phys.Rev. <u>B27</u>, 1010 (1983); Phys.Rev.Lett. <u>50</u>, 739 (1983).

7. Y. Bar-Yam, & J.D. Joannopoulos, Phys.Rev.Lett. <u>52</u>, 1129 (1984)

8. R. Car, P.J. Kelly, A. Oshiyama, & S.T. Pantelides, Phys. Rev.Lett. <u>52</u>, 1814 (1984).

9. P. Boguslawski, G. Papp, & A. Baldereschi, Solid State Comm. <u>52</u>, 155 (1984)

10. Several authors have used the large unit cell approach to study impurity problems. See, for example: S.G. Louie, M. Schlüter, J.R. Chelikowsky & M.L. Cohen, Phys.Rev. <u>B13</u>, 1654 (1976); U. Lindefelt, J.Phys. <u>C11</u>, 85 and 3651 (1978); W.E. Pickett, M.L. Cohen & C. Kittel, Phys.Rev. <u>B20</u>, 5050 (1979); D. Vanderbilt & J.D. Joannopoulos, Phys.Rev.Lett. <u>49</u>, 823 (1982); P. Dzwig, M.G. Burt, J.C. Inkson & V. Crum, J.Phys. <u>C15</u>, 1187 (1982); R.A. Evarestov, Kvantovohimicseszkie metodi v teorii tverdovo tela, Izdatelsztvo Leningradszkovo Universiteta (1982)

11. J.A. Appelbaum, & D.R. Hamann, Phys.Rev. <u>B8</u>, 1777 (1973)

12. J. Bernholc, S.T. Pantelides, N.O. Lipari, & A. Baldereschi, Solid State Commun. <u>37</u>, 705 (1981) ; J. Bernholc, N.O. Lipari, S.T. Pantelides, & M. Scheffler, Phys.Rev. <u>B26</u>, 5706 (1982).

The Diffusion and Electronic Structure of Hydrogen in Silicon

Peter Deák[+] , Lawrence C. Snyder and James W. Corbett

State University of New York 1400 Washington Ave. Albany, N.Y. 12222

1. INTRODUCTION.

The current interest in the properties of hydrogen in silicon derives mainly from its ability to electrically passivate dangling bonds and substitutional acceptor and donor impuritities.[1-4] However, the behavior of hydrogen away from defect centers is also far from being understood.[5] Based on earlier calculations,[6,7] it is generally assumed that in the absence of other defects most of the hydrogen can be found in molecular form embedded in the lattice at the tetrahedral intersitial (T_d) site. The activation energy for the diffusion of the molecule was calculated to be 0.95 eV in ref. 6 and 2.7 eV in ref. 7. The dissociation energy was found to be about 1.6 eV in ref. 7. Based on the latter results most of the hydrogen is tought to be "trapped" by molecule formation at lower temperatures.

Different calculations predict different minimum energy positions for atomic hydrogen, generally 1 -2 eV higher in energy than in the molecular form. An early EHT (Extended Huckel Theory[8]) calculation[9] and the CNDO (Complete Neglect of Differential Overlap[10]) calculation in ref. 6 predicted the T_d site as minimum. The MNDO (Modified Neglect of Diatomic Overlap[11]) calculation in ref. 7 results in a minimum energy postion for atomic hydrogen at the so called M site (see Figure 1). The empirical tight binding calculation of Johnson et al.[4] found the antibonding (AB) site (along the <111> direction on the opposite side of a Si-Si bond) as minimum. Only the calculation in ref. 6 included lattice relaxations. The first neighbor shell around the hydrogen was allowed to relax in a large cluster but no significant displacement has been found. In a more recent calculation DeLeo and Fowler used a small cluster with attached springs to simulate the force-field of the crystalline background more effectively. Their MNDO calculation[12] predicted the bond centered interstitial (BC) site to be the most stable configuration for atomic hydrogen; the neighboring two silicon atoms being pushed back considerably, by about 0.4 A.

The activation energy for the diffusion of atomic hydrogen is predicted to be very low in these calculations: 0.05 eV via the hexagonal (Hex) site in ref. 6 and 0.32 eV via the C site (see Fig. 1) in ref. 7 . In comparison, the high temperature experimental activation energy for diffusion is 0.48 eV.[13] The presence and diffusion of molecular hydrogen has not been verified by direct experiment. The low temperature diffusion experiments support neither the assumption of molecular diffusion nor of an activation energy around 2.5 eV.[14] The strongest indirect evidence for the presence of H_2 molecules is the absence of hydrogen related EPR activity.[5] There is experimental indication however that there may be two diffusing species: atomic hydrogen and the proton[15-17]

2. COMPUTATIONAL METHOD.

We have performed calculations using an open shell version[18] of the MINDO/3 (Modified Intermediate Neglect of Differential Overlap[19]) method which we have found to be superior to CNDO or MNDO in a recent comparison on crystalline silicon[20] and for molecular clusters containing various defects.[21] The environment of hydrogen has been simulated by an Si_{32} cyclic cluster. The details of the cyclic cluster model we use can be found in ref. 20. The Si_{32} cyclic cluster provides the one-electron eigenvalues at the Γ, X, and $\Sigma[1/2,1/2,0]$ points of the

Brillouin-zone of the perfect crystal. First to fourth neighbor interactions are considered only. The periodicity of the defects is > $2a_0$, where a_0 is the lattice parameter. Semiempirical methods based on the neglect of overlap integrals between different atomic orbitals grossly overestimate the stability of states with high atomic s-orbital contribution. As a result, the Γ'_2 state of the conduction band, which is antibonding between any two pair of neighbor silicon atoms, lies much lower in energy than the valleys around X_1. Upon introducing a defect into the system there is a high probability of a Γ'_2-like state being occupied which leads to unreal results[20]. We have applied a correction matrix to the MINDO/3 core-hamiltonian to simulate the effect of neglected overlap on the symmetric antibonding combination of atomic s-orbitals. Details of the procedure will be given in a forthcoming publication.[22] The sole effect of the correction is the Γ'_2 level being pushed up in energy to its expected position above Γ_{15}. The results of the perfect cluster calculation are: lattice parameter 5.64 A, binding energy 3.73 eV/atom, maximum phonon frequency 491.5 cm^{-1}, ionization treshold 5.77 eV, and an indirect gap of 1.59 eV (calculated in the frozen core approximation). In comparison the experimental values are 5.43 A , 4.68 eV , 518.2 cm^{-1} , 5.45 eV , and 1.12 eV respectively (refs. 23-27).

In the geometry optimization procedure the Davidon-Fletcher-Powell algorithm[28] has been used to find the minimum of the total energy with respect to the positions of the hydrogen atom and of the silicon atoms. The convergence problem in the individual SCF calculations could be overcome by the dynamic damping procedure of Zerner and Hehenberg.[29]

3. RESULTS AND DISCUSSION.

We have performed calculations to explore the potential surface of neutral atomic hydrogen in the silicon lattice. In a first step the potential surface of hydrogen has been maped in a rigid lattice by allowing it to move along various symmetry lines. The local minima we have found indicated a less than 0.03 eV activation energy for the hydrogen diffusing along a path which follows the chain of silicons in the (110) plane at a distance of about 1.5 A from the silicon bonds.

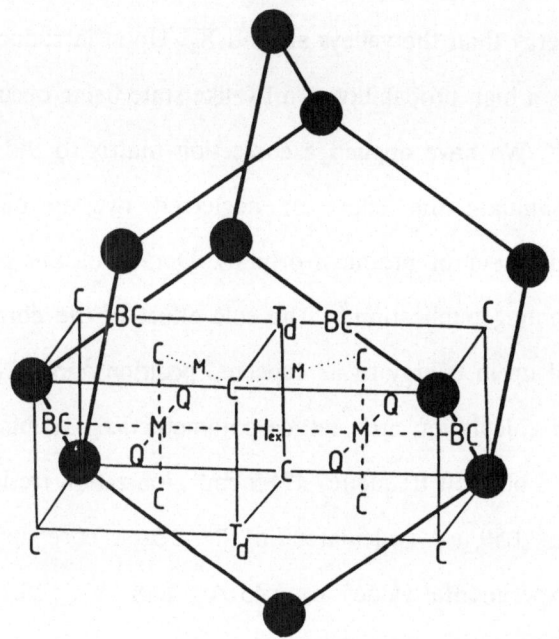

Figure 1.: Symmetric interstitial sites in the silicon lattice.

To explore the effect of lattice relaxation, in consecutive calculations the hydrogen atom has been placed at the vicinity of all of the symmetric interstitial sites displayed in Figure 1. Then the hydrogen and two shells of silicon neighbors have been allowed to relax freely. Two wells have been found in the total energy surface, both accompanied with considerable lattice relaxation. We note that fixing the second neighbor silicon atoms largely reduces the relaxation of the first neighbors and actually prevents the hydrogen from moving into these wells. Both wells are the consequence of an essentially covalent chemical bond established between the hydrogen and one or two neighboring silicons at the cost of Si-Si bonds. We find that the diffusion barrier arises almost solely from the rearrangment of the lattice.

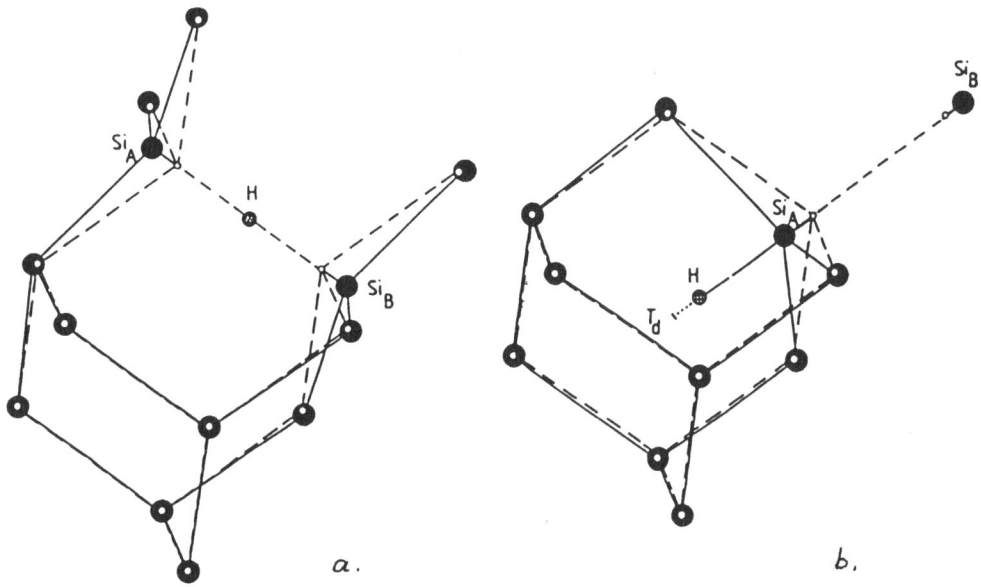

Figure 2.: Relaxation around the BC and AB sites. Dashed lines and empty circles mark the unrelaxed lattice.

The most stable configuration is the BC site (Figure 2.a). The calculated heat of solution is 1.98 eV, in fair agreement with the experimental 1.88 eV.[13] The two silicon neighbors of the hydrogen are pushed back symmetrically along the <111> line by 0.42 A, and a three-center bond is formed with the hydrogen located at 1.64 A from each of the two silicons. From silicon chemistry it is known[30] that in a silane molecule interacting with a silyl radical the bridging hydrogen is located asymmetrically between the neighboring silicon atoms in the neutral charge state , while it is symmetric in the single positive charge state. We have investigated the question of symmetry in the BC site very carefully for the neutral H atom in the lattice. The minimum energy occurs at a symmetric position, apparently as a consequence of the restraining effect of the crystalline environment. The bonding combination of the H *1s* orbital with the orbitals of the neighboring silicons has an energy deep in the valence band while the antibonding combination is above the conduction band. The extra electron goes into an antibonding combination of the orbitals of the two silicon atoms. This level is pulled down from the conduction band and becomes practically degenerate with the highest lying bonding orbitals at -5.79 eV. Since a MINDO/3 band structure calculation[31] predicts the valence band edge at -5.4 eV, this level must fall into the valence band. Therefore, although 90 % of the unpaired spin is localized on the *p*-orbitals of these two silicons, it may not be seen by the EPR, and it is also electrically inactive. We have calculated the force constant for the hydrogen atom oscillating along the <111> line between the fixed silicon atoms (bond-stretch) and also for the perpendicular motion in the (110) plane. The potential well for the stretching mode is rather anharmonic. The vibrational wave number in the harmonic approximation is 784 cm^{-1}. This value is fairly close to a prominent band at 810 cm^{-1} associated with hydrogen in silicon at low temperature.[32,33] The wave number of the bending mode is 697 cm^{-1}.

The other potential well is centered at 0.42 A from the T_d site in the <111> direction, opposite to an Si-Si bond (between T_d and Q on Fig.1). We will denote this point as the AB (antibonding) site. It is a local minimum, 0.93 eV above the energy of the BC site. The nearest silicon relaxes toward the hydrogen (to an Si-H distance of 1.51 A) and becomes almost coplanar with its three silicon neighbor (Figure 2.b). There is a strong bonding between the hydrogen and the nearest silicon, while the bond between the two silicons along <111> is greatly weakened. About 76 % of the unpaired spin is localized on the silicon opposite to the Si-H bond, 69 % on its *p*-orbitals. The extra electron occupies a dangling bond-like α-spin orbital in weak interaction with the neighboring H-Si complex. The one-electron energy is -6.82 eV, i.e., well in the valence band. The corresponding empty β-spin orbital has an energy level below the conduction band edge (see Figure 4). This state should behave as a deep single acceptor. The hydrogen related stretching vibration has a wave number of 2223 cm^{-1}, close to the high wave number end, 2210 cm^{-1}, of the hydrogen related IR spectrum at higher temperatures.[32-34]

The two sites we have found are consistent with the channeling results of Nielsen[35] who found 80 % of the hydrogen at about 0.3 A off the BC site toward the C site and 20 % near the T_d site in the <111> direction. The soft bending mode we calculate for the BC site gives an average displacement of 0.1 A from the BC site toward the C point.

In exploring the diffusion path of atomic hydrogen we have considered five possible routes (Figure 3). Around a BC site there are three other inequivalent BC sites along the chair-form hexagon of the silicon atoms. Let us denote them as BC_1, BC_2 and BC_3 in the order of increasing distance from the starting BC_0 site.

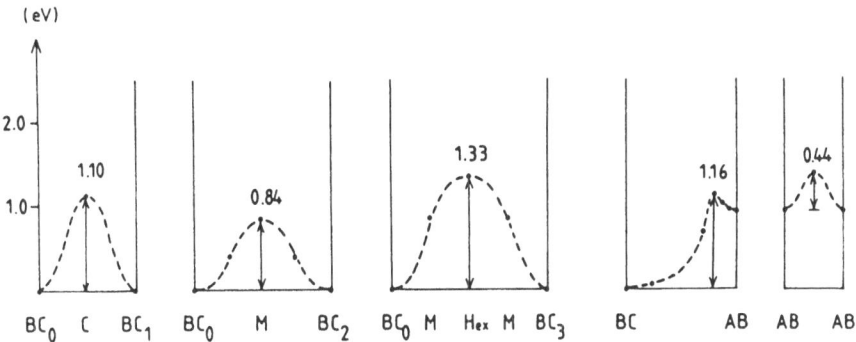

Figure 3.: Energy barriers along different diffusion paths of hydrogen.

For a motion of H from BC_0 to BC_1 in the (110) plane spanned by three silicon atoms the maximum energy should occur on the <001> line due to symmetry reasons. Prescribing C_{2v} symmetry in the optimization we have found a local minimum along that line very close to the C point. At this site the hydrogen is bonded to the silicon atom on the <001> axis only. The total energy is 1.10 eV higher than at the BC site, where it is bonded to two silicons. Another possibility is via the AB site through two planes rotated by 120°. We studied this motion by minimizing the energy with respect to the position of the H atom along lines enclosing increasing angles from 0 to 70.56° with the Si-Si bond. The turning point occurs at about 55° with an energy of 1.16 eV above that of the BC site and 0.22 eV above that of the AB site. It is characterized by the breakup of the bond between the H and the further lying Si neighbor.

The motion between two nearest BC sites is accompanied by a considerable rearrangement of the neighbor silicon atoms. This leads to the relatively high diffusion barriers. Moving the hydrogen to second and third neighbor BC sites leads through more "open" regions. An obvious route would be via the BC_0 - M - Hex - M - BC_3 line parallel with the <110> direction. Calculations at these points give a maximum energy at the hexagonal intersitial site with 1.33 eV above the energy of the BC point. The increase in energy is due to the almost complete break-up of H-Si bonds. The unpaired spin in this case is localized almost entirely on the hydrogen. On the other hand, the M point is also the midpoint between second neighbor BC sites. Moving the hydrogen along that line keeps it bonded to one or two of the same four silicon atoms, A,B,C and D. These do not have to rearrange very much while the bonding transforms from A-H-B to H-B to B-H-C to H-C to C-H-D. The maximum of the total energy occurs at M, 0.84 eV above the energy of the BC site. This is the lowest energy barrier we have found between any BC sites. We note, that the relative energies in this diffusion study are only approximate, for the choice of the degreee of freedom for lattice relaxation is somewhat ambigous at different hydrogen positions.

We have also calculated the energy barrier between two nearest AB sites in the (110) plane with the hydrogen crossing the <001> line. We have found another local minimum along that line 0.4 A from T_d. The total energy relative to the AB sites is 0.44 eV.

The summary of these diffusion studies would suggest that at low termperatures when practically all of the hydrogen atoms occupy a BC site the activation energy for diffusion is 0.84 eV (comparable to the barrier for H_2 diffusion as given by Mainwood and Stoneham[6]) and the hydrogens are essentially trapped. At high temperatures some part of the total hydrogen concentration may occupy AB sites and diffusion flow can be started with an activation energy of 0.44 eV, in fair agreement with the high temperature experimental value of 0.48 eV[13].

To check the validity of a trapping mechanism of hydrogen in an atomic form at the BC site, we have to compare it with the likelihood of molecule formation. We have performed calculations on a H_2 molecule in the Si_{32} cyclic cluster. We started the calculation with the bond-center of the molecule on the T_d site and allowing the hydrogens and two shells of silicon neighbors to move freely. The molecule did not dissociate and did not move off the T_d site either. The <111> and <001> orientations are almost degenerate in energy. The energy per one hydrogen atom relative to the lattice is 0.12 eV higher than for a single atom at the BC site. This implies that the BC site is at least competing with the molecule formation in trapping the hydrogens.

There are indications that the hydrogen atom may become ionized and diffuses as a proton. This would require a hydrogen related donor level in the gap.

In Figure 4. the electronic structures related to different H positions are shown.

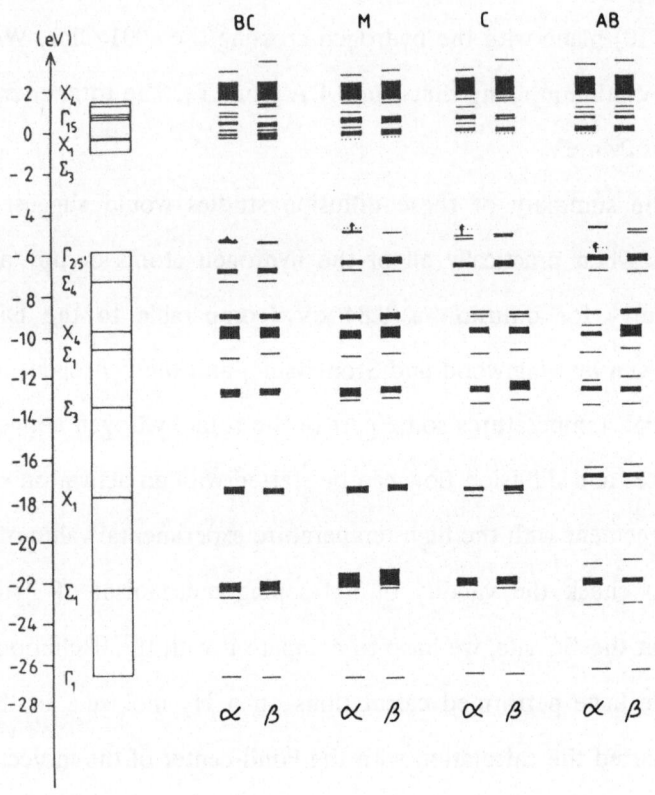

Figure 4.: Electronic structure of the Si_{32}:H system at different hydrogen positions.

No gap levels have been found with the hydrogen at the hexagonal site, while a single acceptor state was found with connection to the AB site, in agreement with ref. 4. Comparing the energy position of the singly occupied Si-Si antibonding orbital at the BC site to the position of the valence band edge of a MINDO/3 calculation (or for that matter to the valence band edge provided by the perfect cluster), this level must fall into the valence band. However, as the hydrogen moves between BC sites through the M or the C point this level moves up and may emerge from the valence band as a deep donor-like state. Our calculation indicates that these states may be ionized only in p-type material.

References

+ *Permanent address*:Physical Institute of the Technical University, Budapest H-1111 Budafoki ut 8. Hungary.

(1) J. I. Pankove, D. E. Carlson, J. E. Berkeyheiser & R. O. Wance, Phys. Rev. Lett. **51**, 2224 (1983).
(2) G. G. DeLeo & W. B. Fowler, J. Electron. Mater. **14a**, 745 (1985).
(3) G. G. DeLeo & W. B. Fowler, Phys. Rev. B **31**, 6861 (1985).
(4) N. M. Johnson, C. Herring & D. J. Chadi, Phys. Rev. Lett. **56**, 769 (1986).
(5) S. J. Pearton, J. W. Corbett & T. S. Shi, Appl. Phys. **43**, 153 (1987).
(6) A. Mainwood & A. M. Stoneham, J. Phys. C **17**, 2513 (1984).
(7) T. S. Shi, S. N. Sahu, J. W. Corbett & L. C. Snyder, Scientia Sinica **27**, 98 (1984).
(8) R. Hoffman, J. Chem. Phys. **39**, 1397 (1963).
(9) V. A. Singh, C. Weigel, J. W. Corbett & L. M. Roth, Phys. Stat. Sol. **B81**, 637 (1977).
(10) J. A. Pople & D. L. Beveridge, *Approximate Molecular Orbital Theory*, McGrew-Hill, New York (1970).
(11) M. J. S. Dewar & W. Thiel, J. Am. Chem. Soc. **99**, 4899 (1977).
(12) G. G. DeLeo & W. B. Fowler, Bulletin of the Am. Phys. Soc. **32**, 841 (1987).
(13) A. Van Wieringen & N. Warmoltz, Physica **22**, 55 (1978).
(14) N. M. Johnson, *private communication*
(15). M. Capizzi & A. Mittiga, *presented at the 3rd Intl. Conf. on Shallow Impurities*, Trieste July (1986).
(16) N. M. Johnson, C. Herring & D. J. Chadi, *Proceedings of the 18th Intl. Conf. Physics in Semiconductors*, Stockholm, Aug. (1986).
(17) S. T. Pantelides, *ibid.*
(18) P. K. Bischof, J. Am. Chem. Soc. **98**, 6844 (1976).
(19) R. C. Bingham, M. J. S. Dewar & H. C. Lo, J. Am. Chem. Soc. **97**, 1285 (1975).
(20) P. Deak, L. C. Snyder, R. K. Singh & J. W. Corbett, Phys. Rev. B *in print* (1987).
(21) P. Deak & L. C. Snyder, Phys. Rev. B *in print* (1987).
(22) P. Deak & L. C. Snyder, *to be published*.
(23) T. Horn, N. Kizsenick & B. Post, J. Appl. Crystallogr. **8**, 457 (1975).
(24) D. D. Wagman, W. H. Evans, V. B. Parker, I. Hallow, S. M. Baily & R. M. Schumm, *NBS Technical Notes* 270-3 (1968).
(25) G. Dolling, *Inelastic Scattering of Neutrons in Solids and Liquids*, Vol.II., p.37. IAEA, Vienna (1963).
(26) C. Sebenne, D. Bolmont, G. Guichar & M. Balkanski, Phys. Rev. B **12**, 3280 (1975).
(27) R. R. Zucca, J. P. Walter, Y. R. Shen & M. L. Cohen, Solid State Commun. **8**, 627 (1970).

(28) R. Fletcher & M. J. D. Powell, Comput. J. **6**, 163 (1963).

(29) M. C. Zerner & M. Hehenberg, Chem. Phys. Lett. **62**, 550 (1979).

(30) L. C. Snyder, J. W. Moskowitz & S. Topiol, Phys. Rev. B **26**, 6727 (1982).

(31) J. M. Ricart & F. Illas, J. MOl. Struct. **120**, 309 (1985).

(32) N. N. Gerasimenko, M. Rolle, L. J. Cheng, Y. H. Lee, J. C. Corelli &J. W. Corbett, Phys. Stat. Sol. **B90**, 689 (1978).

(33) B. N. Mukashev, K. N. Nussupov & M. F. Tamendarov, Phys. Lett. **72A**, 381 (1979).

(34) H. J. Stein, J. Electron. Mater. **4**, 159 (1975).

(35) B. B. Nielsen *MRS Symposia Proceedings, Vol 59.*, p.487 (1986).

SPECTROSCOPIC STUDIES OF POINT DEFECTS IN SILICON AND GERMANIUM

H. G. Grimmeiss, M. Kleverman, K. Bergman, and L. Montelius
Department of Solid State Physics, Lund University,
P.O.Box 118, S-221 00 Lund, Sweden

ABSTRACT

The electronic structure of point defects in silicon and germanium is discussed. After some general remarks on defects generated by doping with elements from group III and V in the Periodic Table, special attention is given to defects with binding energies of the ground state much larger than those predicted by the effective mass theory (EMT) such as double donors (group VI) and double acceptors (group II) as well as transition metals. All these point defects exhibit sharp line spectra in absorption and photoconductivity which allow detailed studies of their electronic structure including states which are not seen in optical dipole transitions. As an example, detailed results are given on spin triplet states and avoided crossings of double donors in silicon. It is shown that the excited Coulomb states of these point defects are well described by the EMT.

1.INTRODUCTION

The properties of most semiconductor devices are determined by defects in general and quite often by impurities in particular. Many of these impurities form point defects. Without the proper exploitation of such defects the development of microelectronics would have been unthinkable. However, point defects are not always useful while manufacturing semiconductor devices. Some of them have the ability to affect carrier lifetimes and, hence, influence the dynamic properties of almost any semiconductor device. This change in carrier lifetime may imply considerable improvement of the electronic properties of some devices while for other applications any change in carrier lifetimes may be unwanted. Considering the important role defects play in most currently used devices it is quite obvious that detailed knowledge concerning the electronic properties of defects is needed not only for academic reasons but also for the improvement of semiconductor devices. It is therefore not surprising that a great number of different experimental methods have been developed in the past with the aim of quantitatively determining electronic parameters and to study the chemical identity and nature of the local environment of such defects. Methods often used are ESR, photoconductivity, luminescence and junction-space charge (JSC) techniques. Any of these methods may give unique information on certain properties of the defects and some of them have been used for decades with great success. It is, however, important to realize that the information provided by one of these methods alone in general is not sufficient to uniquely identify the defect with respect to its electronic properties and its chemical identity. One of the purposes of this paper is to show how defect characterization and identification can be improved considerably by combining several different experimental methods. In particular the usefulness of high resolution Fourier transform spectroscopy (FTS) in conjunction with uniaxial stress will be demonstrated.

For the sake of simplicity the discussion of point defects in this paper is restricted to localized states in silicon and germanium. The simplest point defects are obtained in these materials by doping with elements from groups III and V in the periodic table (Fig.1). Most of these dopants have binding energies of the ground state which are small and close to the value which is expected from the effective mass theory (EMT). They are therefore often called shallow impurities. In addition to the ground state, the electronic structure of such impurities exhibits excited states. The approximate magnitude of the binding energy of the ground state and the excited states can be estimated from simple consideration of an atom-like model. If, for example, a host atom in a silicon crystal is replaced by a phosphorous atom, a localized defect results with an electronic structure which differs from other constituents of the lattice by one extra nuclear charge and one extra valence electron (Fig. 2). In a first approximation such a defect can therefore be considered as an electron moving around a positive charge in a quasi neutral surrounding, very much like a hydrogen atom.

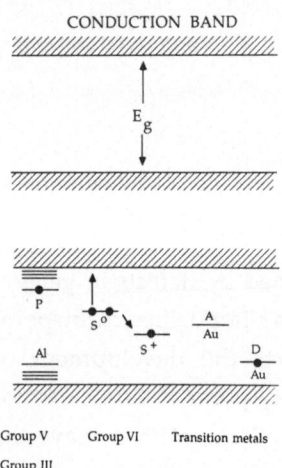

Fig.1 Typical examples of point defects in silicon.

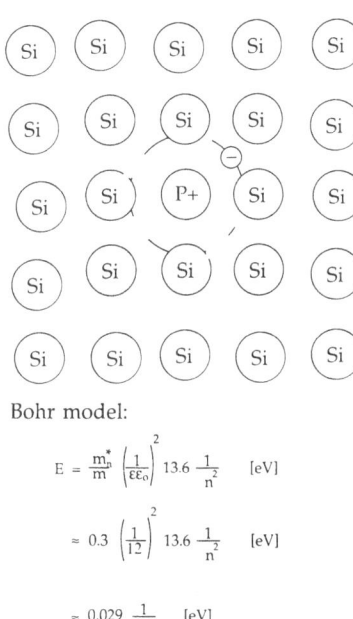

Bohr model:

$$E = \frac{m^*_n}{m} \left(\frac{1}{\varepsilon\varepsilon_0}\right)^2 13.6 \frac{1}{n^2} \quad [eV]$$

$$\approx 0.3 \left(\frac{1}{12}\right)^2 13.6 \frac{1}{n^2} \quad [eV]$$

$$\approx 0.029 \frac{1}{n^2} \quad [eV]$$

Fig. 2 Bohr's model applied to substitutional phosphorus atoms in silicon.

Considering that the mass of an electron in a crystal is not the free electron mass m_n but the effective mass m^*_n and that the electron is not moving in free space but in a medium of dielectric constant ε, the allowed energies E(n) of such a defect atom can be estimated by applying Bohr's model. Since the binding energy of the ground state of a hydrogen atom is 13.6 eV, we obtain

$$E(n) = \frac{m^*_n}{m_n} \cdot \frac{1}{\varepsilon^2} \cdot \frac{13.6}{n^2} \quad (eV) \qquad (1)$$

Assuming an isotropic effective mass $m^*_n = 0.3\, m_n$ and a dielectric constant $\varepsilon = 12$, a value of 29 meV is calculated for the ground state of the defect (n=1) using eq. 1. In a similar way the binding energies of the first (n=2) and the second excited state (n=3) are obtained as E(2)=7.2 meV and E(3)=3.2 meV. It will be seen later that these values, in spite of the simplicity of the model, are rather close to the actual binding energies of a shallow donor as obtained from spectroscopic studies.

Binding energies of excited states are normally calculated using the EMT. In these calculations the effective mass of the electron in silicon is no longer considered as isotropic. One of the consequences of introducing an anisotropic effective mass is that the excited states (with the exception of s-states) split as indicated schematically in Fig.3.

Taking into account the selection rules for (electrical dipole) transitions from the ground state into excited states which are readily obtained from group theory, it is fairly easy to predict the overall features of an absorption spectrum expected for a shallow donor as illustrated by arrows in Fig. 3. According to the EMT, the absorption line with the smallest energy should originate from transitions from the ground state into the $2p_0$ excited state followed by the next line due to transitions from the ground state into the $2p_\pm$ state, and so on. Comparing these predictions with measured absorption spectra of, for example, phosphorus in silicon[1] (Fig. 4), it is readily seen that the agreement between theoretical considerations and experiment is good. A more detailed comparison shows that the excited states of such defects are very well described by the EMT, whereas the agreement obtained for ground states is not as good. This implies that the energy difference between absorption lines are well accounted for by theory but that the energy position of absorption lines cannot be predicted with the same accuracy. However, by combining experimental data with the EMT, very accurate information on the binding energy of the ground state of a donor can nevertheless be obtained. If, for example, the energy position of the absorption line due to the transition from the ground state into the $3p_\pm$ excited state has been measured with high accuracy, the binding

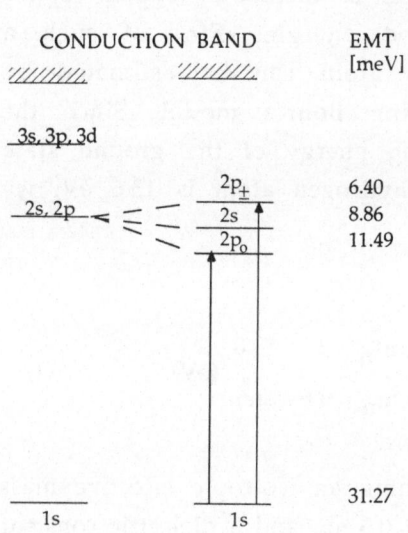

Fig. 3. Binding energies of the lowest energy states of an electron bound to a donor as described by the effective mass theory.

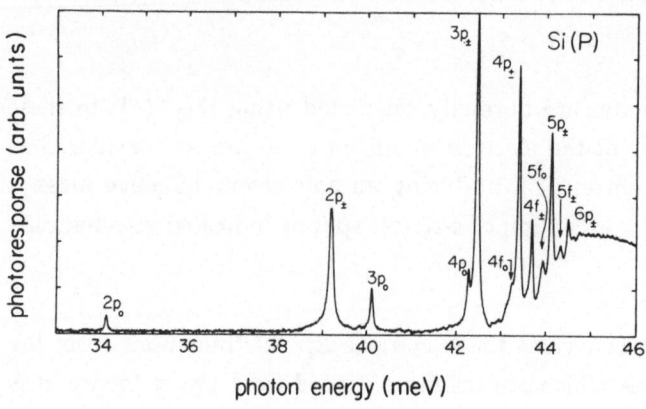

Fig. 4. Absorption spectrum of phosphorus in silicon (from Ref. 1).

energy of the ground state is readily obtained by adding the measured transition energy of the absorption line to the calculated binding energy of the $3p_{\pm}$ state. In this way the entire electronic structure of such defects can be studied in detail including excited states up to $6 p_{\pm}$ and even higher.

2. LOCALIZED STATES WITH BINDING ENERGIES OF THE GROUND STATE MUCH LARGER THAN THOSE PREDICTED BY THE EMT.

Point defects are generated in silicon and germanium not only by impurities from group III and V of the Periodic Table but also from other groups. The properties of these defects differ, however, at least in one respect from those described earlier. If, for example, silicon is doped with one of the chalcogens S, Se or Te, and one of the host atoms is replaced by a dopant atom, it is important to realize that the sulfur impurity has two electrons more than the host atoms and that therefore a neutral chalcogen donor can emit two electrons instead of one as in the case of phosphorus. Point defects binding two excess electrons are called double donors. Similar arguments are valid for some of the group II dopants in silicon and germanium which form double acceptors.

Intuitively it is not surprising that the binding energies of the ground state for double donors and acceptors are much larger than the value predicted by the EMT, as confirmed by experiments. It has often been argued whether or not the electronic structure of such deep centers should differ from group III and V impurities and in particular whether or not these deep defects should have excited states. Considering the fact that excited Coulomb states of substitutional sulfur[2] were already observed in 1962, it is difficult to understand why it has been assumed that defects with larger binding energy of the ground state than that predicted by the EMT can only bind charge carriers through the strong localized central cell potential and why the possibility of generating excited states by the tail of the coulombic potential has been neglected. Since excited states originating from the coulombic potential are not expected to show any chemical shift (as in the case for the central cell potential which is responsible for the large increase of the binding energy of the lower s-states, in particular the ground state) there is no reason to assume that centers where the ground state energy is determined by the central cell potential should not have excited states which are well described by the EMT. On the other hand it seems quite reasonable to expect that the ground state of such deep centers may show new, interesting features which are not observed for EMT-like states.

3. DOUBLE DONORS

To illustrate how well the electronic structure of excited states for centers with binding energies of the ground states much larger than those expected from the EMT is described by effective mass theory we will focus our interest in this chapter on group VI donors in silicon, in particular on S, Se and Te. Since the central cell potential and, hence, the binding energy of the ground state is much larger for these impurities than for group-V donors, the valley-orbit splitting of s-states has to be taken into consideration when discussing the electronic structure of such centers. As in the case of shallow centers, s-states of chalcogens in silicon are not well described by the EMT. Silicon has six equivalent conduction band minima and *all* states are therefore at least six-fold degenerate. Due to valley-orbit interaction this degeneracy is partly lifted, resulting in a splitting of states. Since the probability of finding a bound electron close to the impurity is higher for s-states, the splitting is largest for these states, in particular for 1s-states. This is also valid for shallow donors. But the splitting is much less pronounced for shallow centers than for deep centers and can,

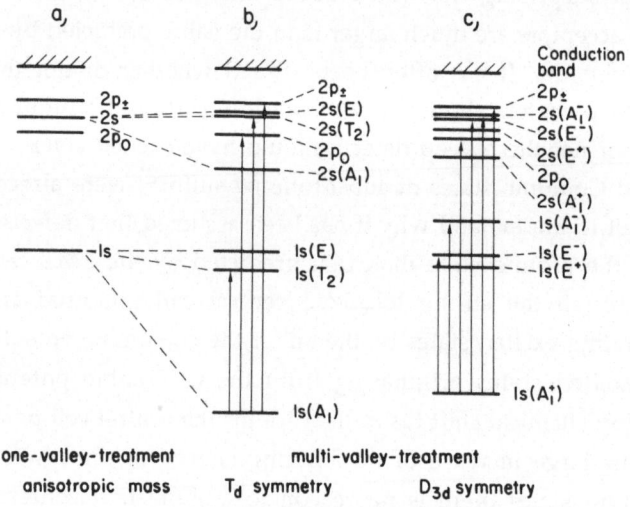

Fig. 5. Lowest part of the energy spectrum of donors with different symmetry in silicon. Symmetry allowed transitions are marked with arrows.

therefore, hardly be observed for several of the group V donors in low-temperature absorption measurements.

The splitting of s-states due to valley-orbit interaction depends on the symmetry of the center and is presented for a single substitutional impurity in silicon in Fig. 5. In tetrahedral (T_d) symmetry the six-fold degenerate 1s state splits in a (singlet) $1s(A_1)$ ground state, a (triplet) $1s(T_2)$ and a (doublet) 1s(E) state[3]. Similar but much smaller splittings are expected for s-states of smaller binding energies. Including selection rules as in the case of shallow centers (see arrows in Fig. 5) the absorption spectrum of a deep donor is expected to be somewhat different from that of a shallow donor, since in T_d symmetry the absorption line with the lowest energy no longer originates from the transition from the $1s(A_1)$ ground state to the $2p_0$ state but to the 1s split state $1s(T_2)$. Apart from the $ns(T_2)$ states, the rest of the spectrum is expected to be similar to the spectra observed for shallow donors. These considerations are in perfect agreement with experimental results as shown in Fig. 6, where the absorption spectrum of the neutral selenium donor in silicon[4] is presented. Comparing this spectrum with a spectrum obtained for arsenic[1] (Fig. 7) and normalizing the spectra with respect to the $2p_0$ lines, the good agreement between the excited states of shallow and deep centers is easily seen. Similar good agreement of excited states is observed for the first and second electron of a double donor as shown in Fig. 8. Here the energy scale of the Se^+ spectrum has been divided by four since the charge affecting an electron bound at a Se^+ center is twice as large as for a neutral Se^0 center and the energy spacing between the excited states is, therefore, four times larger. For comparison, Fig. 5 also shows the electronic structure for a center with D_{3d} symmetry which, for example, is expected for a Se pair. The arrows indicate the expected absorption lines when selection rules are

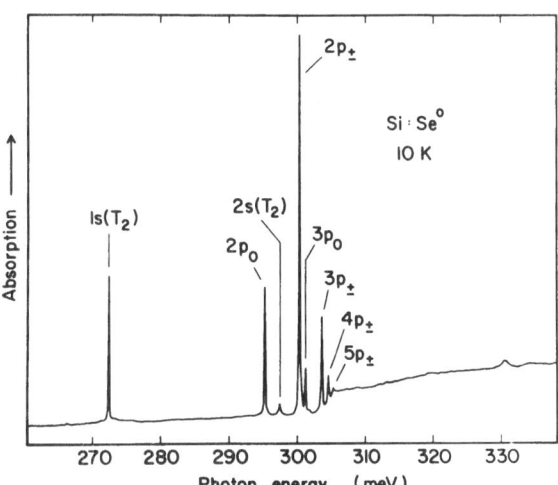

Fig. 6. Absorption spectrum of neutral selenium in silicon

included. Due to the different splitting of the s-states it is quite obvious that absorption spectra provide information not only on binding energies of different states but also on the local symmetry of a center.

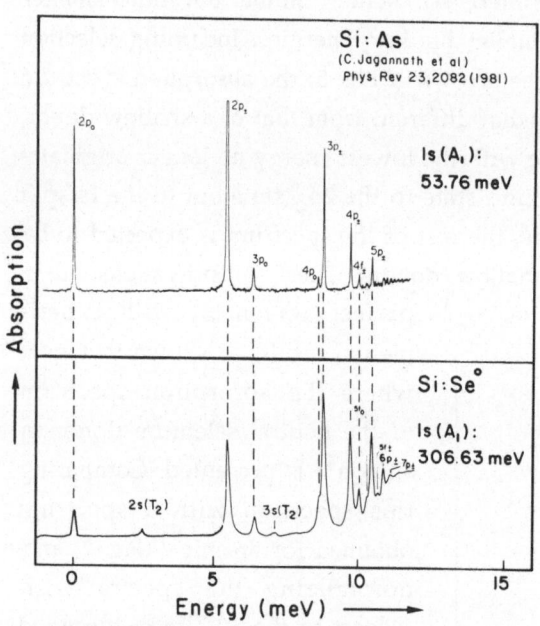

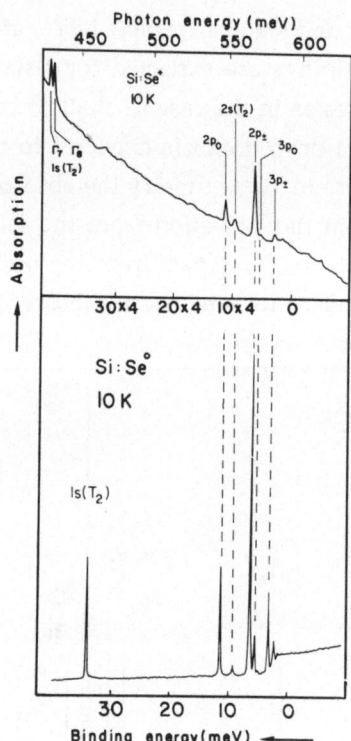

Fig. 7. Comparison of line spectra obtained for arsenic and neutral selenium in silicon.

Fig. 8. Comparision of line spectra obtained for neutral and charged selenium in silicon

Because of the selection rules normal absorption measurements provide only information on energy levels which can be excited from the ground state by allowed transitions. Such studies hardly give any information on states such as 1s(E) or 2s(A$_1$). These states can nevertheless be studied by using higher order absorption processes such as Fano resonances.

The two interacting processes in the case of double donors are a no-phonon excitation from the ground state to the continuum and a transition from the ground state to a bound excited state accompanied by emission of a phonon. The structure observed in absorption or photoconductivity is therefore seen above the ionization limit (C.B. in Fig. 9).

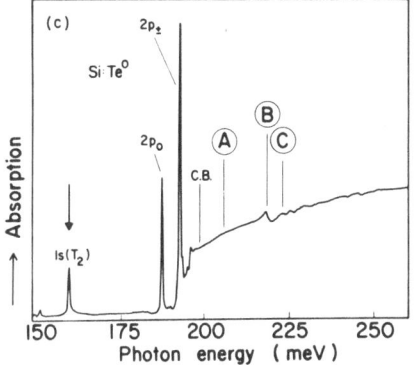

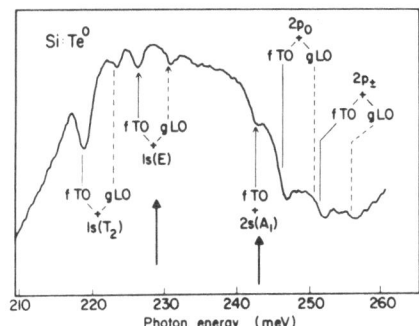

Fig. 9. Photoexcitation spectra of Te⁰ centers in silicon. A, B and C mark the excitation energy of the 1s(T₂) state plus the energies of three different intervalley phonons (from Ref. 5)

Fig. 10. Photoconductivity continua for Te⁰ in silicon (see also Fig. 9)

It has previously been shown that in silicon, in zeroth order approximation, only intervalley phonons, in particular the g LO(Δ'_2) phonon of 63.9 meV and the fTO(S_1) phonon of 59.2 meV, are allowed in intraband electron-phonon interactions. A closer inspection of Fig. 10 reveals that this is indeed observed. These structures give information not only on states such as $1s(T_2)$ and p states which can also be studied in no-phonon transitions and therefore make the assignment of the resonances observed fairly straightforward, but also on states such as $1s(E)$ and $2s(A_1)$ which cannot be studied in direct optical transitions (see arrows in Fig.10). When binding energies of $1s(E)$ and $2s(A_1)$ states have been determined in this indirect manner, those of higher $ns(E)$ and $ns(A_1)$ states can be inferred by a scaling procedure[4].

By comparing and combining data deduced from various measuring techniques it is possible to obtain very detailed information on the energy position of ground states, multi-valley split levels and Rydberg states as well as on the symmetry and charge state of point defects. This is shown in Fig. 11 using the Se⁰ center as an example. Similar detailed results have been obtained for some of the group II impurities in silicon showing excited states which are well described by the EMT and higher order excitation processes such as Fano resonances[6,7].

4. UNIAXIAL STRESS EXPERIMENTS

Symmetry properties of a center are particularly easy to study using uniaxial stress. Since the wave functions of donor states which are well described by the EMT are closely related to the conduction band minima, the stress dependence of bound electron states can be accounted for by the deformation potential approximation (DPA) which describes the behaviour of the conduction band minima under stress. When uniaxial stress is applied to a silicon crystal the conduction band minima shift in energy, which is shown in Fig. 12, with respect to their centre of gravity for directions of stress of high symmetry.

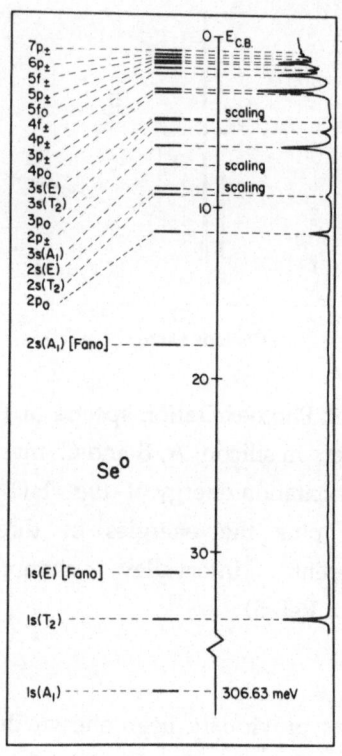

Fig. 11. Energy states of the first electron bound to a selenium donor in silicon as deduced from absorption spectra, Fano resonances and scaling.

States which are well described by single valley EMT (e.g p-states) follow their respective valley. This implies, for example, that the splitting of the $2p_\pm$ states is 3Δ for $\overline{F} \| [001]$ and $3\Delta/2$ for $\overline{F} \| [110]$ where Δ can be taken as a convenient measure of the applied stress. Other states such as the 1s(E) state which are affected by the valley-orbit interaction, do not normally split like the valleys and therefore show a different behaviour. Since the $1s(T_2)$ state is rather p-like it is not surprising that this state follows the valleys and splits like the conduction band minima. This is seen in Fig. 13 (a), where the stress behaviour of the $1s(T_2)$, $2p_0$ and $2p_\pm$ lines is shown in the [001] direction for the Se^0 center[8]. At first sight, the spectra seem to confirm the expected linear splitting of the lines. However, it is interesting to note that at a certain value of the stress ($\Delta = 16.1$ cm^{-1}), an additional weak absorption line

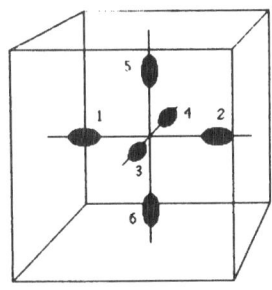

appears below the lower energy component of the $1s(T_2)$ line. Figures 13(b) and 13(c) show a magnification of this stress region for stress applied along the [001] and [110] axes respectively. It is clearly seen that the low-energy branch of the $1s(T_2)$ line in this stress region interferes with a new line in both directions. A similar interference is observed for the higher-energy component for $\overline{F} \| [110]$ but not for $\overline{F} \| [001]$. The interference can be explained by additional states that cross the $1s(T_2)$ components under stress, but which are optically forbidden and only become visible at near-resonance. The interference observed for the higher-energy component is easy to explain when it is realised that not only the $1s(T_2)$ but also the $1s(E)$ state splits under uniaxial stress. Fig. 14

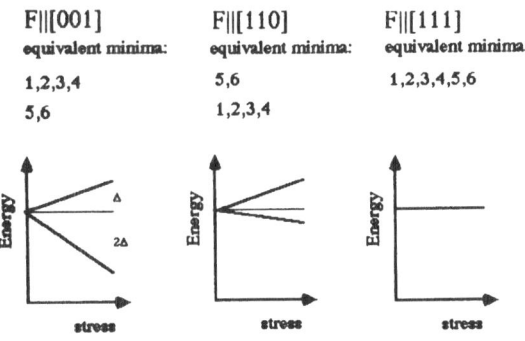

Shift in energy of minima for different directions of stress

Fig. 12. Schematic splitting diagram of the conduction band valleys under uniaxial stress.

shows schematically the splitting behaviour of these states. Since only components of the same symmetry can mix, it is quite obvious that a crossing of a $1s(E)$ component with a $1s(T_2)$ component should only be seen for the higher-energy branches in the [110] direction, in agreement with the experimental results (Fig. 13).

The interference observed for the lower-energy components of the $1s(T_2)$ line cannot be explained in a similar straightforward manner as the crossing with the $1s(E)$ state, since no hydrogenic state exists between the $1s(T_2)$ and the ground state which could be pushed upwards in energy by the applied stress. To explain this observation, spin effects have to be included, in analogy with the helium atom, since a double donor is a two-electron system. In addition to the spin-singlet states we, therefore, expect every configuration to have spin-triplet states as well, except where prohibited by

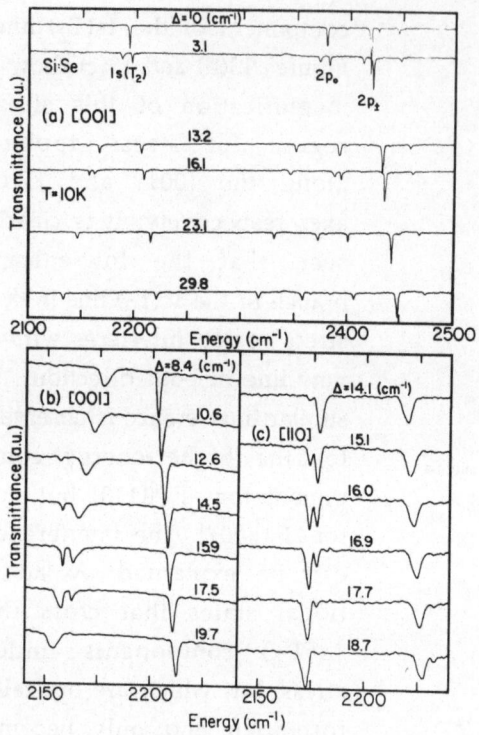

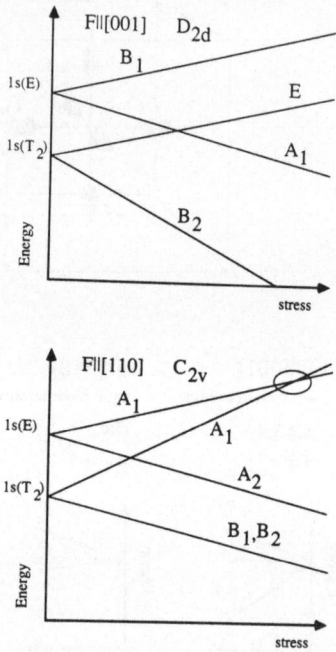

Fig. 13. (a) Overall view of the splitting of Si:Se^0 excited under uniaxial stress in the [001] direction. (b) and (c) Magnified parts of the singlet-triplet crossing regions for stress applied along [001] and [110] directions respectively.

Fig. 14. Schematic energy level diagram showing singlet E and T$_2$ crossing for different stress directions.

the Pauli's principle, as is the case for the ground state. According to Hund's rule, the spin-triplet terms are expected to have higher binding energies than their spin-singlet counterparts due to the exchange interaction. The s-o interaction splits the spin triplet term into a multiplet, but may also mix equal-symmetry states of different spins if they are sufficiently close in energy. The s-o interaction strenght is proportional to the gradient of the electronic potential and thus stronger for orbitals where the electron penetrates the central cell, such as the 1s orbitals, than in more extended orbitals, such as the 2p orbitals. (The contribution to the s-o interaction from the lighter silicon atoms may be neglected compared to the contribution

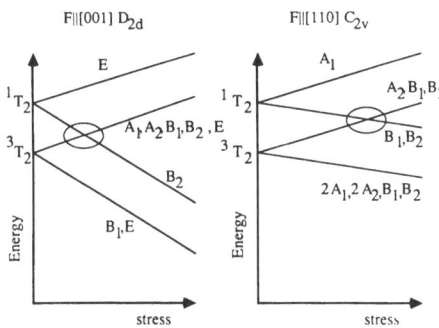

Fig.15 Schematic energy level diagram showing singlet and triplet T_2 crossing for different stress directions.

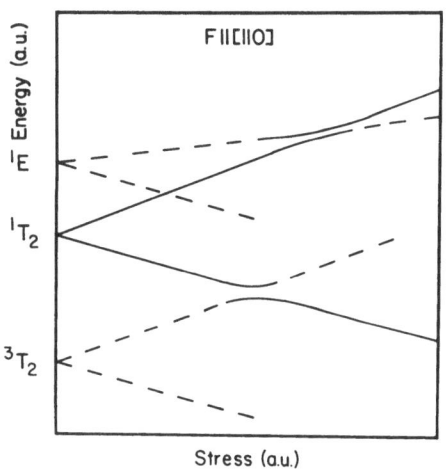

Fig. 16 Schematic energy level diagram showing singlet E and T_2 crossing on the upper branches

from heavier atoms such as selenium or tellurium.) Thus the $1s(T_2)$ orbital, which is the most penetrating orbital of all the excited states, is the most likely candidate for showing effects of mixing of spin-singlet and spin-triplet states. Because transitions to pure spin-triplet states are optically forbidden, it is the mixing with the singlet states that makes the triplet visible in the absorption experiments. Fig. 15 describes schematically the stress behaviour of the spin-singlet state and the spin-triplet state.

Since both states belong to the same configuration the behaviour of the spin-triplet state under stress is expected to follow closely that of the singlet. It is then quite clear from Fig.15 that for both stress directions the lower spin singlet term of the $1s(T_2)$ state crosses a triplet state which has been pushed upwards in energy by the uniaxial stress and which leads to the avoided crossing behaviour discussed above. Fig. 16 summarizes schematically the uniaxial stress results. Solid lines are experimental results, whereas the dotted lines represent extrapolated data. These data clearly show that uniaxial stress experiments not only provide information on the symmetry of a center but can also be used to study particular features of the electronic structure. It should be mentioned here that the value obtained for the energy position of the 1s(E) state from stress experiments was in excellent agreement with the value obtained from studies of Fano resonances as discussed in section 3.

5. DOUBLE DONORS IN GERMANIUM

Whereas acceptor states in germanium have been studied in detail, very little has been known about deep donor states until recently. Data which have been published on the chalcogen impurities S, Se and Te in germanium are summarized in Fig. 17.

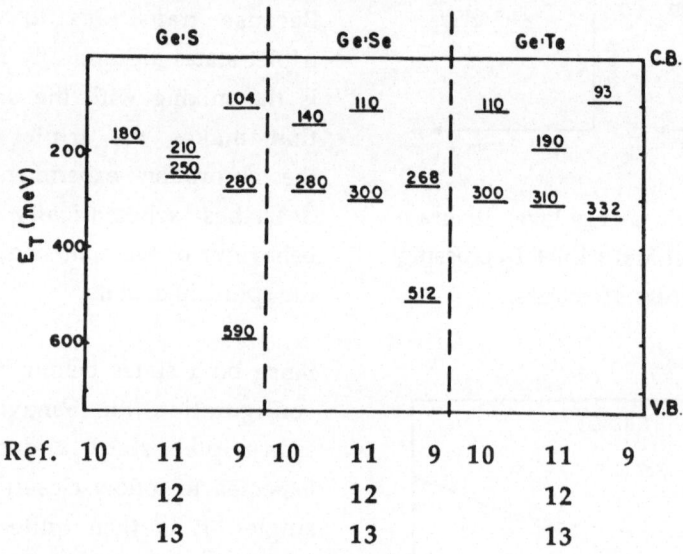

Fig. 17. Survey of published data on the energy positions of centers observed in sulfur-, selenium- and tellurium-doped germanium. E_T is the binding energy of a center.

In most of these studies two energy levels have been observed in the upper half of the band gap for different chalcogens. Since chalcogens on single substitutional sites are supposed to form double donors also in germanium, the levels with smaller energy have normally been attributed[10] to neutral centers whereas the levels with higher energy have been considered to be due to charged centers. Most of these data have been revealed by junction space-charge techniques, and none of the previous results have been obtained by spectroscopic methods. Since the charge state of a center is deduced in a less straightforward manner from junction space charge techniques such as DLTS,than from optical experiments such as photoconductivity or absorption measurements, as shown in Fig. 8, high resolution photoconductivity studies on S, Se and Te doped germanium were performed in order to reveal the charge states of the different levels. Fig. 18 shows such a measurement for the

E_c-0.268 eV donor level in selenium-doped germanium[9]. A comparison with the EMT clearly shows that the charge state of this center is neutral and that the center therefore cannot be coupled with the center about 0.1 eV below the conduction band as previously suggested in the literature[11].

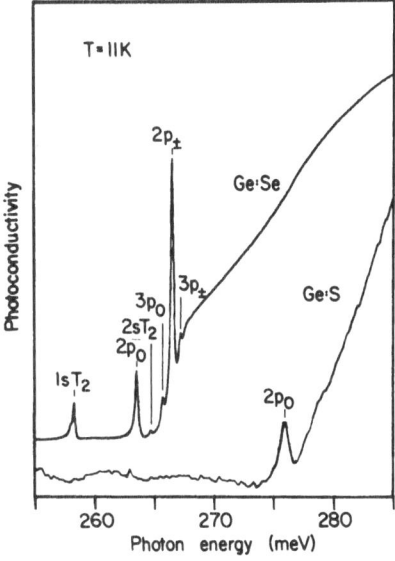

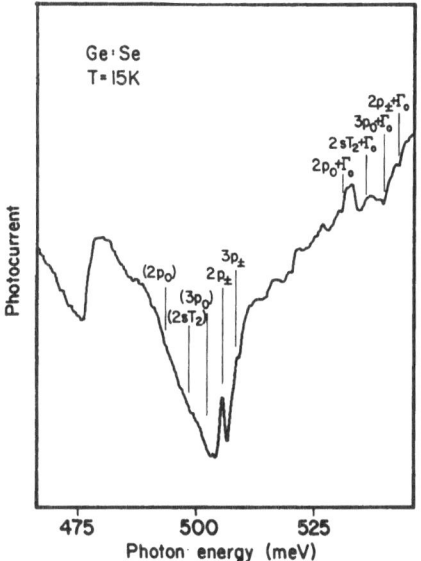

Fig. 18. Photoconductivity spectra of S and Se doped Ge.

Fig. 19. Photocurrent spectrum of Se doped Ge at 15 K showing the line spectrum and Fano resonances of the E_c-0.51 eV center and the sensitivity edge for the hole excitation of the E_c-0.268 eV center.

If the chalcogens have similar electronic properties in germanium as they have in silicon then one may expect to find the single positively charged state of the center in the lower half of the bandgap where hitherto no chalcogen related centers have been observed.

Fig. 19 shows that at almost exactly twice the binding energy of the E_c-0.268 eV level below the conduction band a charged center has been detected[9] which is only seen in selenium doped germanium and which has a binding energy of 512 meV. This is a typical example for demonstrating the usefulness of Fourier spectroscopy.

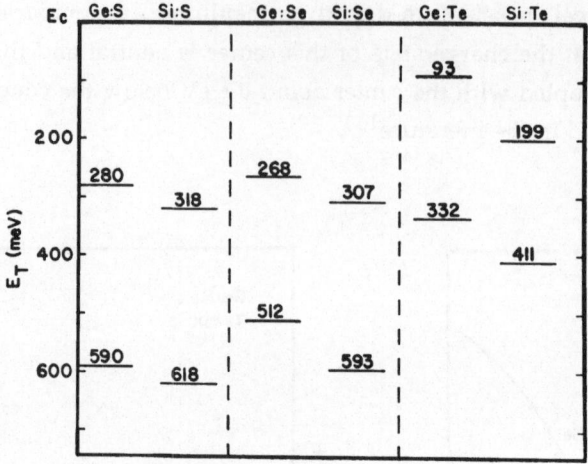

Fig. 20. Comparison of energy positions for isolated chalcogen centers in germanium and silicon (from Ref. 9)

Summarizing all the data which have been obtained for chalcogen-doped germanium and comparing them with the results found for silicon (Fig. 20) it becomes obvious that chalcogens in both semiconductors have similar properties and that the binding energies of sulfur and selenium differ only by about 10% in germanium and silicon. Furthermore, in both materials deep donors have similar electronic structures featuring sharp line spectra and higher order excitation processes such as Fano resonances.

6. TRANSITION METALS IN SILICON

Having discussed some properties of group II, III, V and VI impurities in silicon we will now focus our interest on another group of impurities which are of particular importance for both basic research and technical applications, namely transition metals. Some of the transition metals are known to be fast diffucers in silicon. They are widely used for interconnections, ohmic contacts and Schottky barriers in the production of integrated circuits. In addition, these impurities act as efficient lifetime controllers in devices. It is therefore of great interest to have good insight into their capture and recombination mechanisms as well as their diffusion and complexing properties. None of these insights can be gained properly without detailed information on their electronic structure.

Until recently very little has been known about the electronic structure of transition metals in silicon. One of the reasons for this was that only broad and smooth optical spectra have been available. A typical spectrum of the photoionization cross section for the gold acceptor in silicon which has been obtained by junction space charge techniques is shown in Fig. 21. No structure has been observed, and no theoretical model could be used to deduce binding energies from such spectra with reasonable accuracy. Furthermore, spectra such as in Fig. 21 never revealed the charge state of the center studied. The situation improved considerably when Fourier spectroscopy was used for investigating transition metal impurities[15,16]. Fig. 22 shows an overall spectrum of gold doped silicon clearly revealing the manifold of sharp lines detectable in such a sample by high resolution spectroscopy[16]. The spectrum reveals lines and Fano resonances from

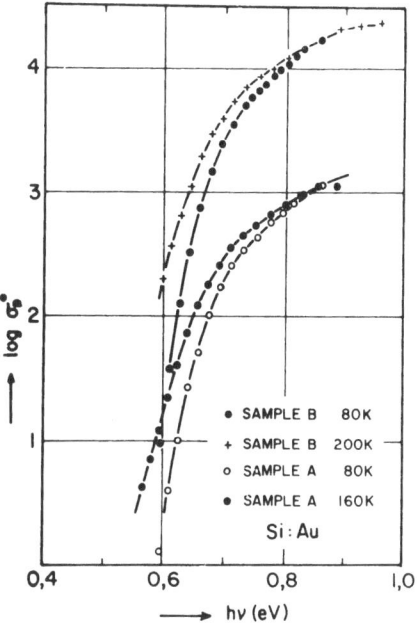

Fig. 21. Photoionization cross section spectra for the gold acceptor in silicon measured with junction space charge techniques (from Ref.14)

several gold related centers. A magnification of the part of the spectrum close to about 0.6 eV (4950 cm^{-1}) shows a set of lines (Fig. 23) which, by comparision with the corresponding spectra for shallow acceptors, could be identified as $P_{3/2}$ acceptor lines. By adding the corresponding EMT binding energy to the measured absorption line, of e.g. I_2, the binding energy of the ground state of the center could be determined with high accuracy giving a value of 622 meV. The energy position of the ground state of the well known acceptor in gold doped silicon is about 615meV above the top of the valence band as deduced from DLTS measurements[17]. The $P_{3/2}$ line series is therefore identified as originating from the deep gold acceptor.

The intensity of the line structure at about 0.68 eV (5300 cm^{-1}) which is only weakly observed in absorption is considerably increased by using photoconductivity measurements. A comparision with spectra obtained for group III acceptors in silicon (Fig.23) identifies these as $P_{1/2}$ transition lines together with strong Fano

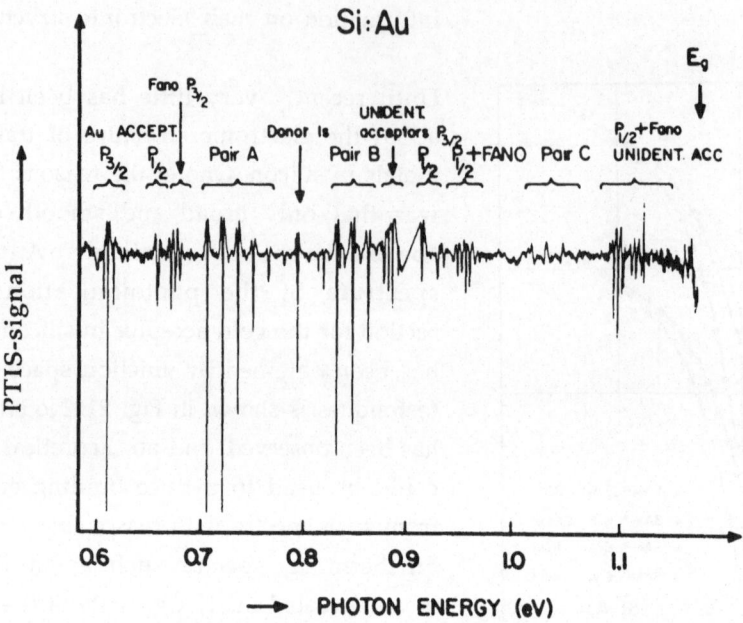

Fig. 22. High resolution photoconductivity spectrum of gold doped silicon.

resonances involving the $P_{3/2}$ state. These results demonstrate that Fano-resonances play an important role also in optical processes involving deep acceptor states. Considering that part of the gold spectrum which concerns the gold acceptor (Fig.23) it is quite clear that the overall spectral features of both the $P_{3/2}$ and the $P_{1/2}$ states are well described by the EMT and are in good agreement with spectra obtained for group III impurities in silicon such as indium.

Similar high resolution spectra have been obtained for platinum in silicon[15,18] (Fig. 24). However, when the series of $P_{3/2}$ states originating from bound-to-bound hole transitions of the deep platinum acceptor is compared with the corresponding series for the deep gold acceptor (Fig. 25) it is readily seen that the platinum spectrum contains many more lines than the gold spectrum. A closer inspection of the platinum spectrum shows that it consists of a no-phonon $P_{3/2}$ spectrum and two sets of phonon replicas to the no-phonon $P_{3/2}$ series. Furthermore, phonon replicas to the 2p′ line are also observed. The pseudo-localized phonon participating in this process has an energy of about 7.2 meV. The width of the first phonon replica is only about 50% larger than the one for the no-phonon line showing that the

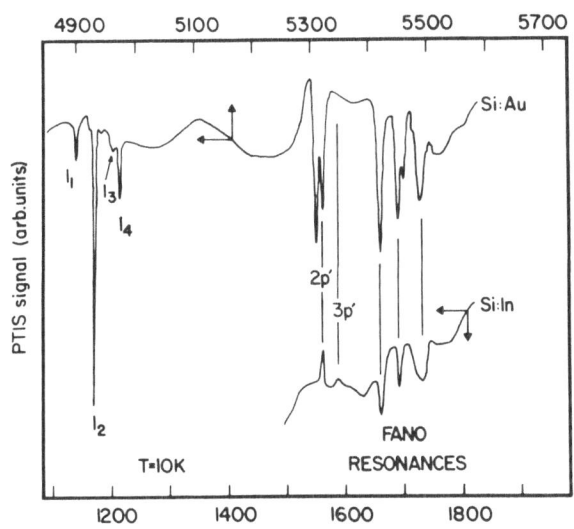

Fig. 23. Photothermal ionization spectra of Si:Au showing a series of excited $P_{3/2}$ states at about 4950 cm^{-1} originating from the gold acceptor. At higher energies the $P_{1/2}$ and Fano resonances of the gold acceptor are compared with those of Si:In (from Ref. 6).

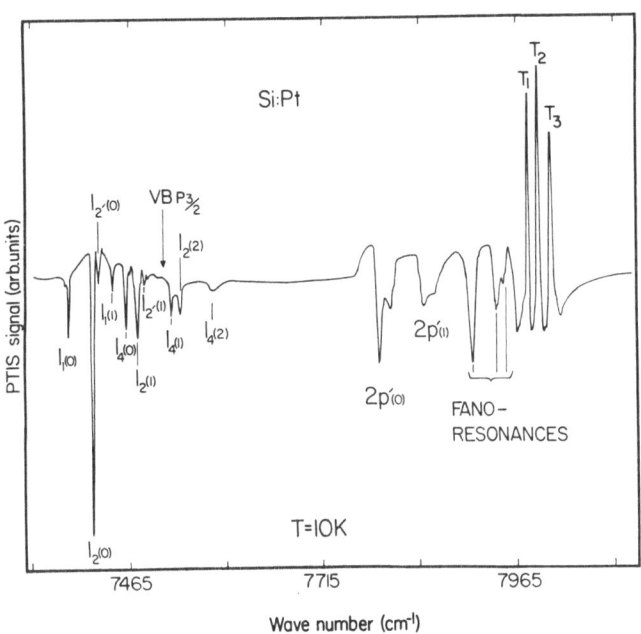

Fig. 24. High resolution photothermal ionization spectrum of platinum-doped silicon.

interaction with the acoustic band phonons is small. To the best of our knowledge this is the first time that phonon replicas of a whole Rydberg series have been observed for a deep acceptor in silicon.

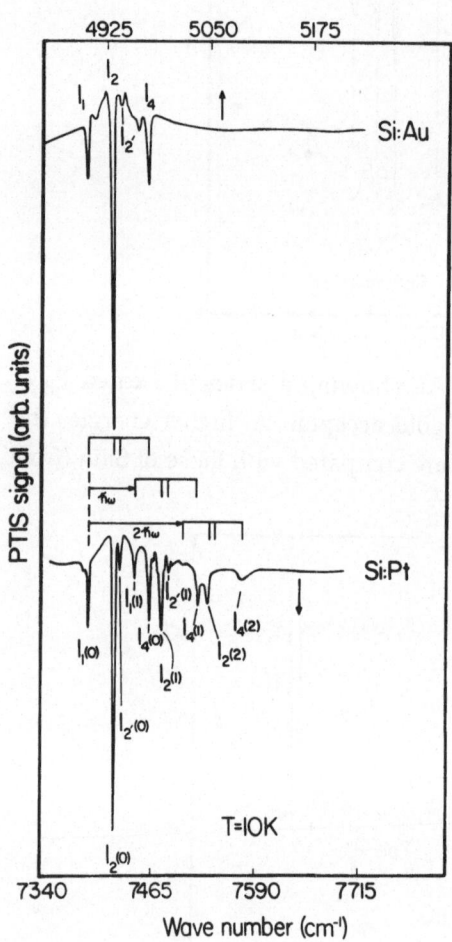

Fig. 25. Comparision of the $P_{3/2}$ lines in Si:Au with those obtained in Si:Pt.

The central-cell correction for the excited p-states of the Au and Pt acceptors is expected to be small since the relative intensities and the energy spacings show only weak impurity dependence. It is, therfore, interesting to note that weak high-energy partners ($I_{2'}$ in Fig. 25) to the strongest line I_2 are observed for both the Au and Pt acceptors and that the energy difference between the corresponding partners is impurity dependent. A closer inspection shows that the energy splitting between the 2p′ partners (2p′(+) and 2p′(-)) is equally large than the corresponding splitting between the $I_{2'}$ and I_2 lines for both Au and Pt. This strongly suggests that the splitting of the 2p′ and I_2 lines has the same origin. In principle, the doublet 2p′ line of the Au and Pt acceptor could be caused by a point group symmetry lower than T_d. Since the final state in the 2p′ transition is delocalized it is expected that such a state is well described by the EMT and transforms according to the orbital-singlet Γ_6 irreducible representation of T_d implying that a lowering of the point group symmetry will not cause a split of the final state. It is therefore belived that the ground state of the Au and Pt acceptor has an energy partner of about 1meV smaller binding energy. The temperature dependence of the relative intensities of 2p′(+) and 2p′(-) is small indicating that the relative hole occupancies of the two ground state partners are not determined by a Boltzmann factor. Instead the occupancies are caused by the experimental conditions such as the broad spectral band width of the light from the Fourier spectrometer which allows for a continuous two-step excitation via the ground state balanced by recapture. This explanation is supported by uniaxial stress measurements[19].

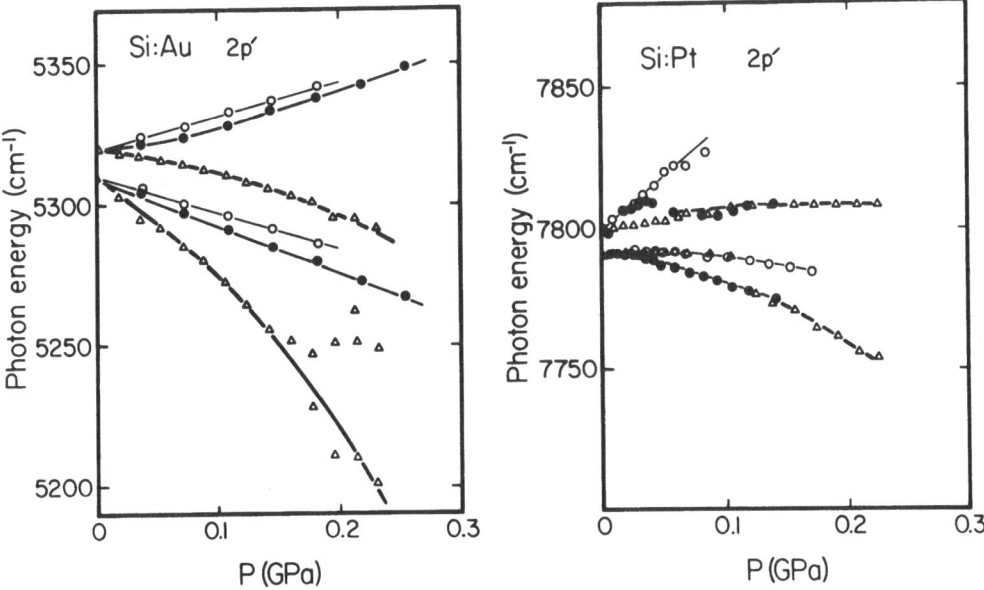

Fig. 26 The stress response of the 2p´of Au. Δ [100], ● [110], and o [111].

Fig. 27. The stress response of the 2p´ lines of Pt. Δ [100], ● [110], and o [111].

The result of uniaxial stress on the 2p´ line of shallow acceptors is a splitting of the Γ_8 ground state. In the case of Au and Pt no such splitting of the 2p´ line into further components is seen for moderate stresses parallel to the high symmetry directions [100], [110], and [111]. For stresses in the [111] and [110] directions the energy position of the 2p´(+) and 2p´(-) line is observed for both Au and Pt whereas for stresses in the [100] direction the behaviour of Au is very different (Fig. 26) from that of Pt (Fig. 27). It should be noted that for Au the energy difference Δ2p´ (between the 2p´ partners), as a function of stress is, for *all* directions (neglecing the zero stress splitting), close to the splitting observed for group III acceptors (Fig. 28). This suggests that under uniaxial stress the two ground state components behave similarly to the ones generating the Γ_8 ground state of group III acceptor. However, when the

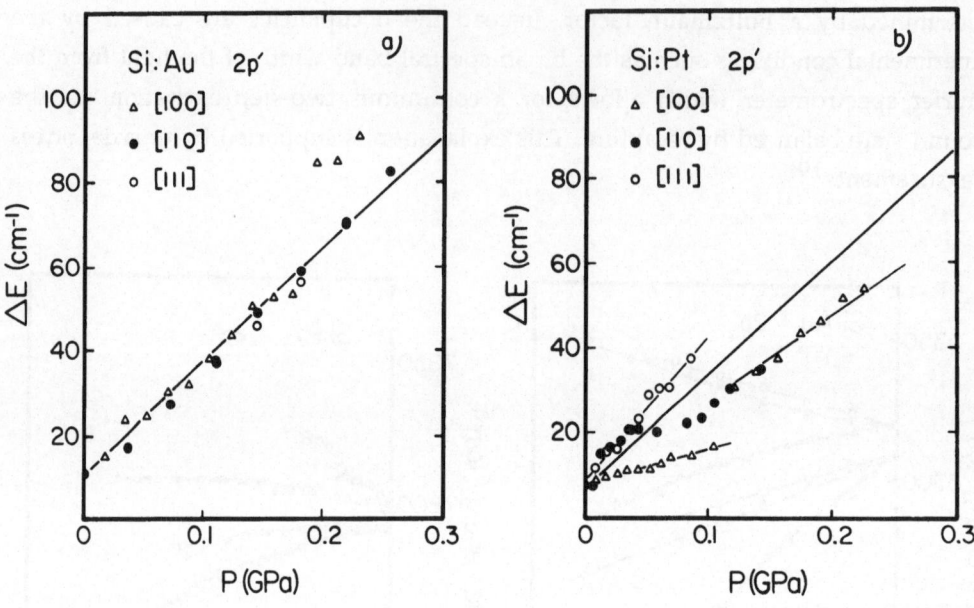

Fig. 28. Comparison between the energy splitting of the 2p′ line of B (straight line) and the corresponding splitting of a) Au and b) Pt.

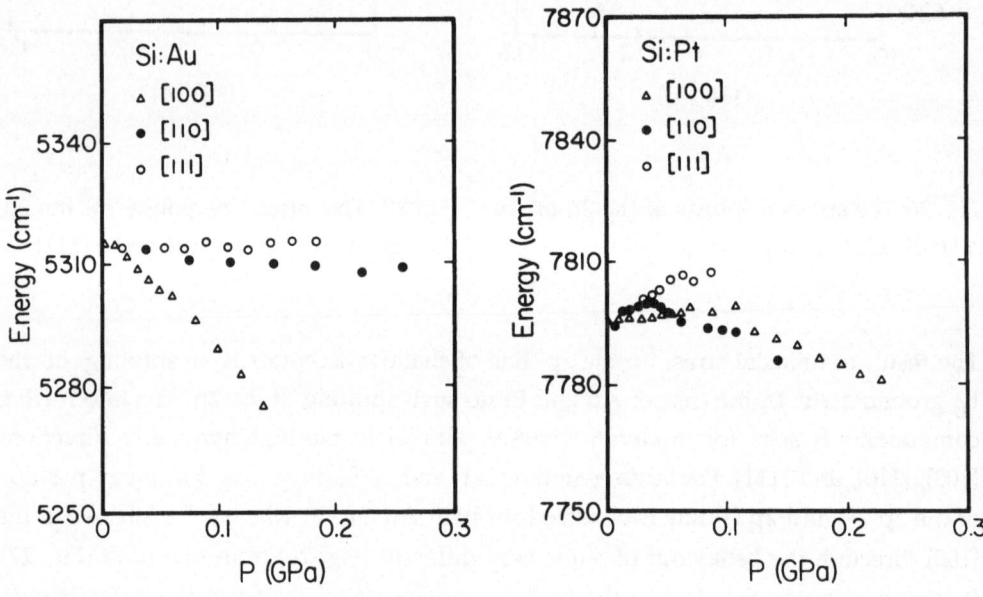

Fig.29. The center of gravity shift for the 2p′line of Au and Pt as a function of uniaxial stress.

centers of gravity of the 2p′ partners for the Au and Pt acceptors are plotted as a function of stress it is readily seen (Fig. 29) that the ground state of the Au acceptor is particularly sensitive to stress in the [100] direction. Together with the results presented in Fig.26 this may indicate that the Au acceptor is off-center[20].

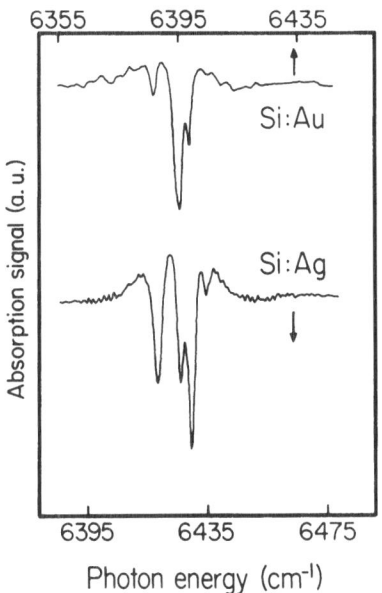

Fig.30 1s(E+T$_2$) complex of Au doped silicon in comparision with similar structure observed in Ag doped silicon.

A line structure previously observed in luminescence in Au doped silicon has been assumed to originate from intra 1s transitions of the Au donor[21]. These transitions are clearly observed as four sharp lines in absorption and photoconductivity measurements also (Fig. 30), and are believed to be caused by transtions from the 1s(A$_1$) ground state to the 1s(E+T$_2$) excited 1s states. Transitions to excited p-states have not been observed so far for the Au donor which is surprising since, in the case of chalcogen doped silicon, transitions from the ground state to p-states show the highest intensities. The difference may be explained by recalling that the deep potential of the chalcogen impurities[22] pulls an anti-bonding A$_1$ state from the conduction band into the band gap whereas for transition metals[23] bonding states from the valence band are pushed into the band gap.

Since wavefunctions of excited p-states in the case of donors are derived only from a narrow region close to the bottom of the conduction band, the overlap in $\bar{k}$-space between the ground state of the Au and Pt donor and the p-states is negligible which may explain why such transitions have so far not been observed.

Fig.30 also shows part of an absorption spectrum[24] obtained from a Ag doped Si sample. It is interesting to note that the structure at about 6400 cm^{-1} is very similar to the intra 1s lines observed for Au suggesting the same origin of the lines in both cases. A closer inspection of the Ag spectrum, however, shows a line at about 100 cm^{-1} below the 1s lines which seems to be associated with the Ag structure. This suggests a more complex electronic structure than in the case of Au and Pt.

Since sharp line spectra are now available for transition metals in silicon, similar investigations can be undertaken on these impurities as were previously done on group VI impurities. Although more work has to be performed, it is quite obvious that part of the electronic structures of transition metals in silicon is well described by the EMT.

7. CONCLUSIONS

It should be clear from the results presented in this paper that, in general, a combination of different experimental methods is needed to characterize and identify point defects in semiconductors unambiguously. Although EPR and junction space charge techniques are important experimental methods for such investigations, it has become more and more evident that high resolution spectroscopy can provide valuable additional information. Furthermore, previously discussed dissimilarities between shallow and deep centers, such as certain properties caused by excited Coulomb states, should be reconsidered in view of the fact that point defects in silicon, so far studied in detail, show well resolved and sharp line spectra, which, at least partly, are well understood in terms of EMT.

REFERENCES

1. C. Jagannath, Z.W. Grabowski, and A.K.Ramdas, Phys. Rev. B23, 2082 (1981)

2. W.E. Krag and H.J. Zeiger, Phys. Rev. Lett. 8, 485 (1962)

3. W. Kohn and J.M. Luttinger, Phys. Rev. 98, 915 (1955)

4. E. Janze´n, R. Stedman, G. Grossmann, and H.G. Grimmeiss, Phys. Rev. B29, 1907 (1984)

5. E. Janze´n, G. Grossmann, R. Stedman, and H.G. Grimmeiss, Phys. Rev. B31, 8000 (1985)

6. M. Kleverman, K. Bergman, and H. G. Grimmeiss, Semicond. Sci. Technol. 1, 49 (1986)

7. M. Kleverman and H. G. Grimmeiss, Semicond. Sci. Technol. 1, 54 (1986)

8. K. Bergman, G. Grossmann, H.G. Grimmeiss, and M. Stavola, Phys. Rev. Lett. 56, 2827 (1986)

9. H. G. Grimmeiss, L. Montelius, and K. Larsson, to be published

10. W.W. Tyler, J. Phys. Chem: Solids 8, 59 (1959)

11. S.J. Pearton, Solid State Electronics 25, 499 (1981)

12. S.J. Pearton and A.J.Tavendale, J. Appl. Phys. 54, 820 (1983)

13. S.J. Pearton, Aust. J. Phys. 35, 53 (1982)

14. S. Braun and H. G. Grimmeiss, J. Appl. Phys. 45, 2658 (1974)

15. G. Armelles, J. Barrau, M. Brousseau, B. Pajot, and C. Naud, Solid State Commun. 56, 303 (1985)

16. M. Kleverman, J. Olajos, H.G. Grimmeiss, Phys. Rev. B35, 4093 (1987)

17. S. D. Brotherton and J. Bicknell, J. Appl. Phys. 49, 667 (1978)

18. M. Kleverman, J. Olajos, H.G. Grimmeiss, unpublished

19. M.Kleverman, J.Olajos, and H.G.Grimmeiss, to be published

20. The stress result for 2p´(-) of Au and Pt for small stresses in the [100] direction vary with experimental conditions caused, probably, by inhomogenous stress and/or electrical field. The discussion in this article should not be affected by this.

21. D. Thebault, J. Barrau, G. Armelles, N. Launet, and J.P. Noguier, Phys. Stat. Sol. (b) 125, 357 (1984)

22. V. A. Singh, U. Lindefelt, and A. Zunger, Phys. Rev. B27, 4909 (1983)

23. J. L. A. Alves and J. R. Leite, Phys. Rev. B34, 7174 (1986)

24. J. Olajos, M.Kleverman, and H.G.Grimmeiss, to be published

Deep Levels in Cz-Si due to Heat Treatment at 600...900 °C

K.Schmalz and F.-G. Kirscht
Institute for Physics of Semiconductors
PSF 409, DDR-1200-Frankfurt (Oder)

The oxygen behavior in Cz silicon due to heat treatment (ht.) is of great interest for device technology regarding to the fact that device yield can be influenced significantly due to effects related to oxygen precipitation /1,2/.

The intensive research using different spectroscopic methods in the field of the 450 °C-thermal donors has given a better knowledge concerning the structure of these donors, but the electrical properties of the defects (new donors, recombination/generation centers) formed due to ht. at temperatures $\gtrsim$ 600 °C are less known. For oxygen-rich Cz-Si heat treated at this temperature range the formation of extended defects inhomogeneously distributed in the volume seems to be characteristic of. In this case, spectroscopic methods usually fail to give information concerning the microscopic structure of the defects.

It has been observed that the recombination lifetime of minority charge carriers in Cz-Si of n- and p-type, resp., decrease strongly due to longer ht. at 600...900 °C. This fact has been explained in the model of electrically active oxygen clusters /6, 7, 10/ similar as for as grown Cz-Si /11/, or a correlation with secondary defects has been assumed /12/.

The present work will focus on the formation of dl defects due to oxygen precipitation at temperatures $\gtrsim$ 600 °C summarizing results for Cz-Si /13, 14, 15/. The experimental results are obtained by means of TEM- and DLTS-investigations comparing the formation of the dl defects with the volume defect formation due to oxygen precipitation, and investigating the carbon influence on these processes.

From the experimental observations the conclusion has been drawn that there are dl defects induced by oxygen precipitation, which are closely related to a large supersaturation of Si-interstitials, achieved by the growth of oxygen precipitates /14/.

This stage of oxygen precipitation can be achieved in quite different way in dependence on the material parameter of Cz-Si and the ht.-procedure employed.

1. On the oxygen precipitation due to heat treatment

The ht. of Cz-Si with high oxygen content leads at 600...900 °C primarily to the formation of SiO_x precipitates being at lower temperatures still spherically shaped / e. g. 16, 17/ at least for short duration of ht.. With increasing temperature of ht. plate-like precipitates with a thickness of a few nm develop nearly independent on T and the duration of the ht. /18/. The structure of plate-like precipitates was shown to be amorphous.

An increase in ht. temperature leads to (i) an increase of precipitate size, (ii) a decrease of precipitate density and (iii) reaching earlier the equilibrium for precipitation. Differences for different kinds of Cz-Si with nearly the same oxygen and carbon concentrations $([O_i]_o, [C_s]_o)$ could be mainly attributed to an essential changed as-grown nuclei spectrum (s. Tab. 1).

Table 1: Typical values characterizing defect formation due to different ht. (HVTEM investigations)

ht. conditions	sample type $([O_i]_o/10^{17} cm^{-3}/$ $[C_s]_o/10^{16} cm^{-3})$	N_p/cm^{-3}	$L_p//um$	N_d/cm^{-3}
64 h 750 °C, N_2		7×10^{11}	0.11	increasing
64 h 800 °C, N_2	A (9,0/<1)	2×10^{11}	0.20	*)
64 h 850 °C, N_2		1×10^{11}	0.60	
16 h 900 °C, N_2	c (14,7/10,5)	3×10^{12}	0,03	2×10^{11}
	f (9,1/8,1)	7×10^{11}	0,06	5×10^{11}
	i (6,2/35,0)	1×10^{11}	0,04	–
64 h 900 °C, N_2	d (9,6/1)	7×10^{9}	0,75	7×10^{9}
	c	4×10^{12}	0,03	3×10^{11}
	i	2×10^{9}	0.38	–
240 h 600 °C, N_2	d	4×10^{12}	0,03	2×10^{12}
+64 h 900 °C, N_2	c	9×10^{11}	0,07	2×10^{11}
240 h 800 °C, O_2	d	4×10^{11}	0,17	1×10^{8}
	i	2×10^{11}	0,05	–

N_p: average precipitate density
N_d: avergage density of dislocation loops
L_p: mean precipitate size
*) in the range $10^9 ... 10^{10}$ cm^{-3}

Simultaneously to oxygen precipitation there is a production of silicon-interstitials (Si_i) which tend to aggregate independently of oxygen. There is a quantitative agreement between the emitted Si_i, as deduced from HREM observation of the residual strain around SiO_x precipitates and the agglomerated Si_i in different forms /19/. Carbon has a great influence on the Si_i emission as well Si_i clustering and thus, indirectly, on the shape of amorphous precipitates, and controls partly the nucleation of the oxygen precipitation.
After long ht. at higher temperatures the resulting misfit strain at the precipitate interface increasingly causes dislocation formation. At temperatures < 700 °C the formation of rod-like defects dominates. Differences in defect formation due to ht. at higher temperatures

(800...900 °C) can be interpreted as being caused by the interaction of Si_i with C_s. In carbon-rich material this could lead to a lowering of the critical size of nuclei for precipitation or compensation of the so called exigent volume at precipitation, resp. /2C/. As a consequence in oxygen-rich Cz-Si extra-ordinary high precipitate density in conjunction with strong secondary defect formation is observed. In carbon-lean material precipitation at about 900 °C is clearly dominated by growing-induced microprecipitates as indicated by low precipitate density.

Extremely carbon-rich material ($[C_s]_o \approx 4 \times 10^{17} cm^{-3}$) with low oxygen concentration shows also a retardation of nucleation. Thus, a reduced fraction of precipitated volume in conjunction with a suppression of dislocation loops is observed for ht. at 800...900 °C.

2. Deep levels due to heat treatment at 600...900 °C

The electrical activity of oxygen precipitation-induced defects depends on the as-grown parameters of the Cz-Si, especially on $[O_i]_o$ and $[C_s]_o$. Some characteristic features of DLTS-spectra reflect certain stages of defect formation achieved in dependence on the as grown Cz-Si due to different ht. at 600...900 °C with quenching rate in the order of 1K/s /14, 15/.

2.1. DLTS-results for Cz p-Si

In Fig. 1 characteristic DLTS-spectra are presented for Cz-Si (slice E: $[O_i]_o = 9 \times 10^{17} cm^{-3}$, $[C_s]_o = 3 \times 10^{16} cm^{-3}$) heat treated at 700...850 °C for duration up to 240 h ($2N\Delta C/C$-effective dl concentration) /15/.

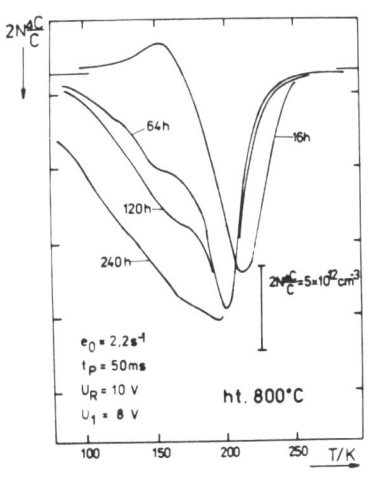

Fig. 1
DLTS spectra at 800 °C ($e_o = 2.2 \ s^{-1}$, $t_p = 50$ ms, $U_R = 10$ V, $U_1 = 8$ V)

After short ht. at 750...800 °C a DLTS-peak appears the shape of which is characteristic of a point-like defect. With longer ht. the peak concentration increases and the peak is broadened. Furthermore, for very long ht. the "broadened" peak transforms into a "broad" band-like spectrum with superimposed peak. For the ht. at higher T = 800 °C the concentration is reduced compared to the ht. at 750 °C. It should be noted that qualitatively different DLTS-spectra were measured after 850 °C-ht.. The peak concentration is considerably smaller and the broad band-like spectrum is not formed. In difference to ht. at 750...850 °C in the case of 700 °C-ht. it is not possible to separate the spectrum in a peak-like part and a "broad" band.

The shape of the peak and the formation of the band-like spectrum are found to be strongly dependent on $[C_s]_0$ and $[O_i]_0$.

Table 2a: DLTS-spectrum parameters obtained for A-samples
($[O_i]_0$ = 9,0x10^{17}cm^{-3}, $[C_s]_0$ < 1x10^{16}cm^{-3}) after ht. at 800 ^oC (e_0 = 220 s^{-1}. t_p = 50 /us)

t/h	16	32	64	120
2N$\triangle$C/C/10^{11}cm^{-3}	9	7	6	15
$\triangle$T/T$_M$		0,17	0,15	0,18	0,28

In Tab. 2a for carbon-lean Cz-Si the 2N$\triangle$C/C and the normalized peak half-width $\triangle$T/T$_M$ are given. Only for the longest ht. the $\triangle$T/T$_M$ increases significantly. Compared to oxygen-rich Cz-Si with low $[C_s]_0$ for oxygen and carbon rich Cz-Si the formation of the dl defects is yet possible at 900 ^oC. It is however essential to note that for Cz-Si with very high carbon but low oxygen content ($[O_i]_0$ = 6x10^{17}cm^{-3}, $[C_s]_0$ = 4x10^{17}cm^{-3}) after long ht at 800...900 ^oC there were not introduced dl (2N$\triangle$C/C < 10^{11}cm^{-3}). For oxygen and carbon-rich Cz-Si ($[O_i]_0$ = 9x10^{17}cm^{-3}, $[C_s]_0$ 1x10^{17}cm^{-3}) after ht. at 900 ^oC DLTS-spectra with strongly broadened peak are measured. The $\triangle$T/T$_M$ decreases for long 900 ^oC-ht. without remarkable change in 2N$\triangle$C/C (s. Tab. 2b).

Table 2b: DLTS-spectrum parameters obtained for f-samples
($[O_i]_0$ = 9,1x10^{17}cm^{-3}, $[C_s]_0$ = 8,1x10^{16}cm^{-3}) after ht. at 900 ^oC (e_0 = 220 s^{-1}, t_p = 50 /us)

t/h	8	16	32	64
2N$\triangle$C/C/10^{12}cm^{-3}	3,2	2,6	2,3	2,2
$\triangle$T/T$_M$	0,36	0,40	0,36	0,27

This is a further argument for assuming that the properties of these dl defects can be gradually changed by ht.

The dl defects are also formed due to ht. at T = 600...650 ^oC, but the detection is difficult regarding to the enhanced formation of oxygen related donors.

The investigation of Cz-Si with long ht. at 600 ^oC and a post-ht. at 900 ^oC have given evidence that for such ht.-procedures DLTS-spectra consist of a broad band-like spectrum with the broadened characteristic peak. If 2N$\triangle$C/C at the peak temperature T_M is taken as a measure for the dl concentration, values ranging from 10^{13}...10^{14}cm^{-3} are estimated with the higher ones for Cz-Si with higher $[O_i]_0$ and $[C_s]_0$.

Analyzing the results obtained for Cz p-Si heat treated it is possible to distinguish qualitatively different DLTS-spectra types as follows:

I: a sharp peak the peak temperature T_M of which is somewhat shifted in dependence on the parameters of the Cz-Si ($[O_i]_o$, $[C_s]_o$..) and the ht.-procedure

II: a peak at T_M strongly broadened toward lower temperatures in the spectrum, the peak half-width is significantly changed due to ht.

III: going out from the peak a broad band is formed toward lower temperatures in the spectrum

For these spectrum types the dependences on the filling time t_p are qualitatively different. For the type I in dependence on ht. there is needed a t_p in the range 50 /us...5ms to reach saturation in the DLTS-signal. For types II and III, resp., the t_p-dependence was nearly a logarithmic ones (s. Fig. 2) being characteristic of barrier-limited capture process /20, 21/.

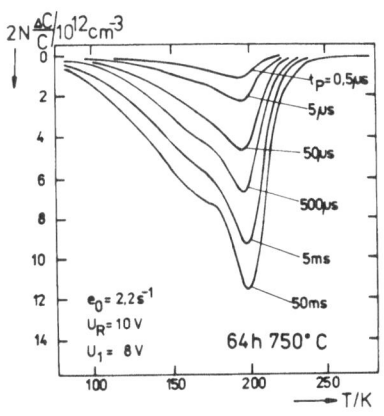

Fig. 2
DLTS spectrum of Cz p-Si (slice E) with ht. at 750 °C in dependence on t_p ($e_o = 2,2$ s^{-1}, $U_R = 10$ V, $U_1 = 8$ V)

The defects leading to the spectrum type I should be the initial stage for further formation of dl defects, as the results from the kinetics have been suggested.

From the Arrhenius-plot of the normalized emission rate e_p/T^2 vs. $1/T$ the activation energy E_a and the effective capture cross section were obtained. There were found values in the range 0.4...0.55 eV for E_a.

It has to be noted that for shorter ht. leading to effective dl concentration in the range of few 10^{11} cm^{-3} to about 10^{12} cm^{-3} an activation energy E_a of about 0.4 eV has been found.

But, for longer ht., especially at T = 750...800 °C, leading to dl concentration in the range of few 10^{12} cm^{-3} to about 10^{13} cm^{-3} an E_a in the range 0.5...0.55 eV has been measured. Therefore, it can not be excluded that there are even two different groups of ht.-induced dl corresponding to E_T-E_V of about 0.4 eV and E_T-E_V of 0.5..0.55 eV, respectively.

More precise values for deep levels with $E_a = 0.5...0.55$ eV could not be estimated due to a certain dependence on the dl occupancy and on the electric field. These results has lead to the suggestion of ht.-induced dl. clusters.

An further argument for dl clusters is the broadening of the DLTS-peak for longer ht.. In the simplest case peak broadening can be explained assuming a Gaussian distribution of dl energy in the band gap /22/. However, in this way the peak shape can be fitted satisfactorily only for shorter ht.. After longer ht. an asymmetric shape is characteristic of /14/. The capture cross section σ_p was obtained from t_p-dependence. The theoretical analysis taking into consideration a barrier-limited capture process /21/ gives for σ_p a value of 10^{-17} ... 10^{-16} cm^2, and due to hole capture a higher potential barrier in the case of longer ht..

2.2. DLTS-results for Cz-n-Si

Fig. 3 ($[O_i]_o$ = 8,1x10^{17}cm^{-3}, $[C_s]_o$<1x10^{16}cm^{-3}) shows characteristic DLTS-spectra obtained for Cz n-Si with ht. at 750...850 $^{\circ}$C /15/. In difference to the DLTS-spectra obtained for heat treated Cz p-Si there was not found a pronounced DLTS-peak, but only "broad" spectra.

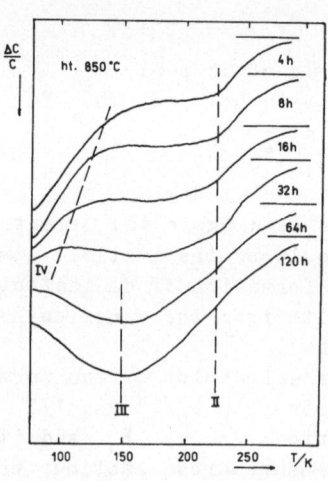

Fig. 3
DLTS spectra of Cz n-Si with ht. at 850 $^{\circ}$C (e_o = 2,2 s^{-1}. t_p = 50 ms, U_R = 10 V, U_1 = 9 V)

However, for these "broad" spectra it was possible to observe three characteristic components (II, III, and IV) depending on the Cz-Si parameters as well as on the temperature and duration of the ht..

The component IV has been found to correlate with the concentration of the oxygen related donors. The weights of the components II and III in the spectra were estimated by the concentrations 2NΔC/C at the respective characteristic temperatures T_{II} and T_{III}, resp.

The relation of these concentrations 2NΔC/C$_{II}$ and 2NΔC/C$_{III}$ depend on the conditions of the ht. (temperature, duration, ambient) and on the Cz-Si parameters.

A dominating formation of the component III was observed due to longer ht. at 800...850 $^{\circ}$C.

For oxygen and carbon lean Cz n-Si the concentrations 2NΔC/C$_{II}$ and 2NΔC/C$_{III}$ are lower than 10^{11}cm^{-3} due to ht. at 600...900 $^{\circ}$C.

In the case of carbon lean Cz-Si with $[O_i]_o$ = (7...10)x10^{17}cm^{-3}, heat treated at 800 $^{\circ}$C, the relation of 2NΔC/C$_{II}$ and 2NΔC/C$_{III}$ depends strongly on the secondary defect formation.

Thereby, the component III is observed only for a concentration of 2NΔC/C$_{II}$ $\gtrsim$ 5x10^{12}cm^{-3}, and is related to a strong formation of dipole-like perfect dislocation loops and isolated dislocation loops.

For oxygen precipitation not leading to the formation of dislocation loops, only the component II was measured with increasing concentration with ht. duration (s. Fig. 4; $[O_i]_o$ = 9x10^{17}cm^{-3}, $[C_s]_o$<1x10^{16}cm^{-3}).

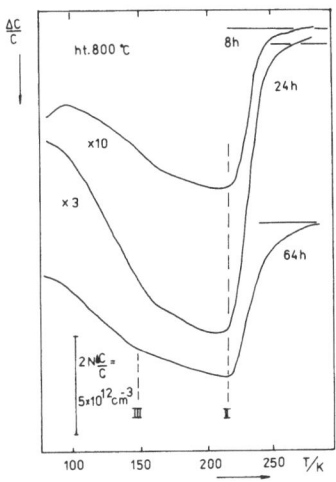

Fig. 4
DLTS spectra of Cz n-Si with ht. at 800 °C (e_o = 2,2 s^{-1}, t_p = 50 ms, U_R = 10 V, U_1 = 9 V)

It has to be noted that for low concentrations (few $10^{11} cm^{-3}$) the component II shows a peak shape without broadening. By means of Arrhenius-plot an activation energy E_a of about 0.4 eV was measured. Obviously, the dl with E_C-E_T of 0.4 eV corresponds to the initial stage of defect formation in a similar way as the dl with E_T-E_V of about 0.4 eV. In difference to the dl with E_T-E_V = 0.4...0.55 eV observed for Cz p-Si a band-like energetic distribution of deep levels was found in the case of n-Si for significantly smaller dl concentrations.

Besides this, even after long ht. no evidence has been found for a strongly pronounced barrier limited capture process in contrast to the results obtained for the heat treated Cz p-Si. Filling pulse times tp in the order of 1 µs were sufficient to reach about the half of the DLTS-saturation signal for the components II and III, resp..

For Cz n-Si with 600 °C-ht. two durations of ht. are of especial interest: (i) short ht. not leading to the formation of oxygen related donors, (ii) the saturation concentration of the od 's has been reached and does not change due to longer ht..

For (i) a broad band in the DLTS-spectrum has been measured after short 600 °C-ht. (about 5...20 h).

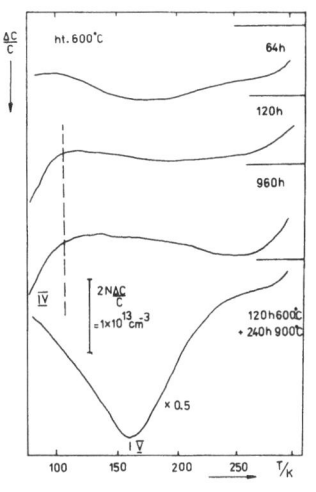

Fig. 5
DLTS spectra of Cz n-Si with ht. at 600 °C (e_o = 2,2 s^{-1}, t_p = 50 ms, U_R = 10 V, U_1 = 9 V)

The effective dl concentration $2N\Delta C/C$ of this band increases up to about $10^{13} cm^{-3}$ and does not change significantly even due to longer 600 °C-ht. But, a long 600 °C-ht. of 100...1000 h (case ii), leads to a transformation in the shape of the DLTS-band (s. Fig. 5, $[O_i]_o$ = 8,9x$10^{17} cm^{-3}$, $[C_s]_o$ = 6x$10^{16} cm^{-3}$). The weight of the broad band is shifted to higher temperatures in the DLTS-spectrum, especially for carbon-lean Cz-Si, indicating that a continious transformation of

electrically active defects occurs due to 600 °C-ht., presumably associated with the transformation stages of rod-like defects. A qualitative different defect transformation has been detected for Cz-Si with
long 600 °C-ht. (100...300 h) due to post-ht. at 900 °C. In this case,
the appearance of a high density of dislocation loops is correlated
with a strongly broadened DLTS-peak (corresponding to dl with E_C-E_T of
about 0.3 eV) with an effective dl concentration $2N\Delta C/C$ up to about
$10^{14} cm^{-3}$.

The experimental results have shown that the formation of the broad
band due to 600 °C-ht. and of the spectrum components II and III could
not be related to the oxygen-related donors. Therefore, there is only
a correlation between the od's and the effective dl concentration of
the component IV. Regarding to this, an evaluation of electrical
properties of oxygen-related donors from the whole DLTS-spectrum as in
/23/, not taking into acount the different origin of the spectrum
components, is not correct.

3. On the relationship of deep level defects with the oxygen
   precipitation at 600...900 °C

The occurence of dislocation loops as secondary defects is not a
prerequisite for the generation of the dl defects. On the other hand,
the formation of oxygen precipitates is not necessarily related to dl
defect generation.
The occurence of the initial dl defects (E_T-E_V = 0.4...0.55 eV, E_C-E_T=
0.4 eV) is associated with an early stage in the generation of dislocation loops and/or oxygen precipitate transformation which is
established under definite conditions of ht. and parameters of the Cz-
Si. A common characteristic seems to be a large supersaturation of
silicon interstitials Si_i in the surrounding of a growing precipitate.
The dl defects related to the Si_i-supersaturation are assumed
tentatively to be point-like defects, with properties depending
strongly on the strain field near the precipitates.
The formation of these dl defects could be supported by back stress
due to piled-up Si_i (in addition to stress arising from misfit strain)
at oxygen precipitate interfaces. Thus, stress relief due to the
generation of inner Si_i sinks (secondary defects) would cause dl
defects to transform. The properties of the dl defects assumed would
also depend on other Si_i sinks in the Si matrix as e. g. substitutional carbon atoms.
Analysis of the dl defects with E_T-E_V of about 0.5 eV has revealed
that the properties of these dl defects depend in certain limits on
the actual stage of volume defect formation. A longer ht. would result
in stronger strain in the surrounding of the precipitates with the
begin of secondary defect formation, leading to the broadening of the
dl energetic distribution and also to a locally increased dl concentration causing mainly barrier-limited capture of holes. Larger potential barriers in the surrounding of volume defects for hole capture in
p-Si as compared to electron capture in n-Si should be the main reason

for the observed great differences in minority carrier lifetime degradation due to ht. /9/ between p- and n-Si.
A stronger secondary defect formation (mainly dislocation loops) leads to a band-like energetic distribution for the dl in the upper and in the lower half of the gap, probably due to the interaction of dangling bond defects. For a high dislocation loop density deep levels with $E_C - E_T$ of about 0.3 eV are characteristic of.
It is necessary to consider that in the case of Cz-Si heat treated the dislocation-induced dl observed for plastically deformed silicon /24/ has not been detected. This might be a further argument for the proposed origin of the dl defects.

Acknowledgements

The authors want to thank to Dr. H. Richter for valuable advice.
The IFE, Halle is highly acknowledged for giving the opportunity of using the HVTEM equipment. We would like to thank Mr. K. Tittelbach-Helmrich for taking part in the DLTS-investigations.

References

/1/ H. Richter
    Proc. 1st Internat. Autumn School GADEST 1985, p.1,
    GDR, Ed. H. Richter
/2/ H.R. Huff and F. Shimura
    Solid State Technology/March 1985 p.103
/3/ P. Gaworzewski and K. Schmalz
    phys. stat. sol.(a) 55,699 (1979), and ref. therein
/4/ A. Kanamori and M. Kanamori
    J. appl. Phys. 50, 8095 (1979)
/5/ P. Gaworzewski and K. Schmalz
    phys. stat. sol (a) 77, 571 (1983)
/6/ M. Miyagi, K. Wada, J. Osaka, and N. Inoue
    Appl. Phys. Lett. 40, (8), 719 (1982)
/7/ R. Ohtaki, Y. Matsushita, M. Tajima
    Defects in Semicond. 1984, Proc. p. 571
/8/ J.Whitfield, C.J. Varker, S.S. Chan, S.R. Wilson,
    R.W. Carpenter, S.I. Krause, E.R. Weber
    Proc. SPIE, Int. Soc., Opt. Eng., 623, 83 (1986)
/9/ J.M. Hwang and D.K. Schroder
    J. appl. Phys. 59, (7), 2476 (1986)
/10/ K.D. Glinchuk, N.M. Litovchenko, Z.A. Salnic, and S.I. Skryl
    Cryst. Res. and Technol., 20 (4), 485 (1985)
/11/ K. Nauka, H.C. Gatos, and J. Lagowski
    Appl. Phys. Lett. 43 (3), 241 (1983)
/12/ K. Ikuta, Y. Matsouka, and H. Takaoka
    Jap. J. appl. Phys. 19, Suppl. 19-1, 621 (1980)
/13/ K. Schmalz, P. Gaworzewski, and F.-G. Kirscht
    phys. stat. sol. (a) 81, K 165 (1984)

/14/ K. Schmalz, F.-G. Kirscht, H. Klose, H. Richter,
     and K. Tittelbach-Helmrich, phys. stat. sol. (a), 100, 567 (1987)
/15/ K. Schmalz, F.-G. Kirscht, H. Richter,
     and K. Tittelbach-Helmrich
     phys. stat. sol. (a), in preparation
/16/ P. Gaworzewski, E. Hild, F.-G. Kirscht, and L. Vecsernyes
     phys. stat. sol (a) 85, 133 (1984)
/17/ W. Bergholz, J.L. Hutchinson, and G.R. Booker
     Semicond. Silicon 1986, Proc. p. 874
/18/ K. Wada, N. Inoue, K. Kohra
     J. Cryst. Growth, 49, 749 (1980)
/19/ A. Bourret
     Semicond. Silicon 1986, Proc. and ref. therein
/20/ T.Y. Tan and C.Y. Kung
     J. appl. Phys. 59, 917, (1986)
/21/ K. Schmalz, K. Tittelbach-Helmrich, and H. Richter
     phys. stat. sol. (a) 99, K 97 (1987)
/22/ P. Omling, L. Samuelson, and H.G. Grimmeiss
     J. appl. Phys. 54, 5117 (1983)
/23/ K. Hölzlein, G. Pensl, and M. Schulz
     Appl. Phys. A 34, 155 (1984)
/24/ V.V. Kveder, Yu. Ossipyan, W. Schröter, and G. Zoth
     phys. stat. sol. (a) 72, 701 (1982)

INTERPRETATION OF THE ELECTRIC
FIELD DEPENDENT THERMAL
EMISSION DATA OF DEEP TRAPS

T.Pavelka and G.Ferenczi
Research Institute for Technical Physics of the HAS
1325 Budapest, P.O.B.76, Hungary

Abstract

The paper considers the consequencies of the simultaneous thermal
emission and recapture of carriers in the edge layer of a reverse
biased junction on the interpretation of the thermal emission probability measurements. It is concluded that the contribution of the
capture process if properly taken into account proves the validity
of the Poole-Frenkel model of the electric field dependence.

It is well established that the thermal emission probability from
deep levels are influenced by the local electric field within the
space charge layer of a semiconductor junction. The theoretical interpretation of the electric field dependence of the thermal emission
probability may help in understanding the lattice coupling properties
of deep levels.

In the present work we have studied in detail the low electric
field behaviour of the thermal emission probability as extracted from
DLTS measurements.

The measurements were carried out on nitrogen doped n-type GaP
Schottky barriers using the Differential DLTS mode [1] of a Semi Trap
DLS-82E Deep Level Spectrometer. The thermal emission of a nitrogen
related deep electron trap [1] was studied in detail. The characteristic dependence of the thermal emission probability (e_n) on the junction's electric field is seen on Fig.1/a. e_n at high electric fields
closely follows the functional dependence predicted by the barrier
lowering model of Poole-Frenkel [2], however, at low fields the experimental data are significantly larger than the theoretical prediction.

This deviation can be explained by the fact that the low field
data are measured in the vicinity of the cross-over of the Fermi-level

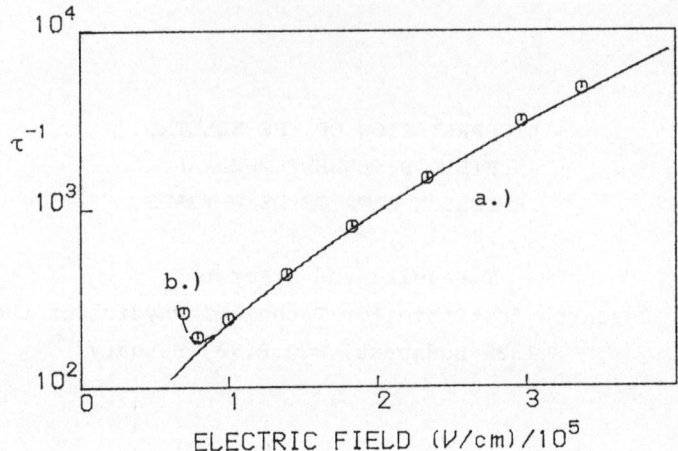

F i g . 1.
Electric field dependent thermal emission probability measured
at 200 K on an n-type GaP Schottky diode (N_D=1.2x10^{17}cm-3)
using the Differential DLTS approach. The squares are the
data points.
a.) Full line: Poole-Frenkel model
b.) Broken line: Taking into account the recapture process
    the theory fits the measured data.

and the deep level i.e. around the edge layer. In this part of the
space charge region the influence of the free electron Debye-tail can
not be neglected [3], hence the rate equation for emptying an electron
trap is:

$$\frac{dN}{dt} = (N_T-N) c_n n(x) - N e_n(x) \qquad /1/$$

Where N_T is the total concentration of the deep traps, e_n and
c_n is the emission and capture probabilities and

$$n(x) = n_0 \exp[-\frac{N_D q^2}{2 \epsilon kT} x^2] \qquad /2/$$

is the local concentration of free electrons [4]. As it is seen
from eqn /1/ the time constant for emptying a trap is

$$\tau^{-1} = [e_n + c_n n(x)] \qquad /3/$$

From eqn/3/ follows, that the increase of the number of free electrons will increase the rate of return to the thermal equilibrium occupancy, hence the apparent thermal emission probability extracted naively from DLTS data is larger than it should be due to the carrier recapture around the cross-over point. Using eqns/1-3/ the measured data follows closely the Poole-Frenkel model as seen on Fig.1/b.
This result allows the following conclusions:

- The zero field activation energy can be determined from the temperature shift of the Poole-Frenkel fitted emission probability data. In case of the nitrogen related defect, the true thermal ionisation energy is 0.455 eV, as compared to 0.401 eV evaluated from the low electric field data without taking the recapture process into account.

- It is generally valid for all measurements of the electric field dependent thermal emission, that the low electric field data do not represent the true thermal emission probability but rather the time constant of eqn/3/. Consequently fitting theoretical models to uncorrected data (see e.g. [5-9]) is to be questioned. To prove this point, we have successfully refitted the published data of [5] with the Poole-Frenkel model taking into account the recapture mechanism. (See Fig.2.) Similarly the Poole-Frenkel model seems to be valid for GaAlAs using data points of Ref.[6]. An independent proof of this

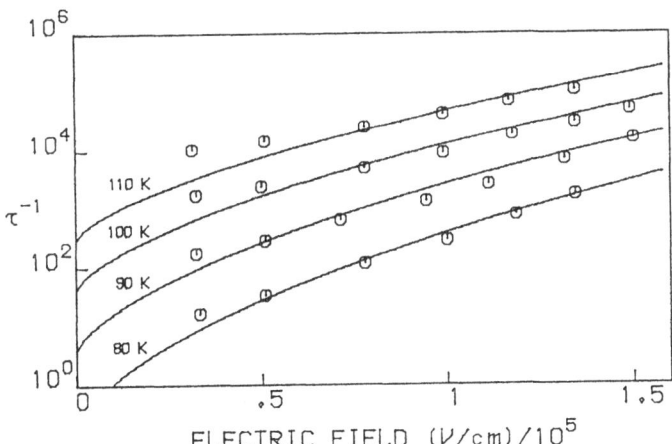

Fig. 2.
Refitted data for the A center in Si with the Poole-Frenkel model. Data points are from ref.[5].

simple picture is demonstrated on Fig.3. We have plotted the electric field dependent thermal emission data measured on the same nitrogen

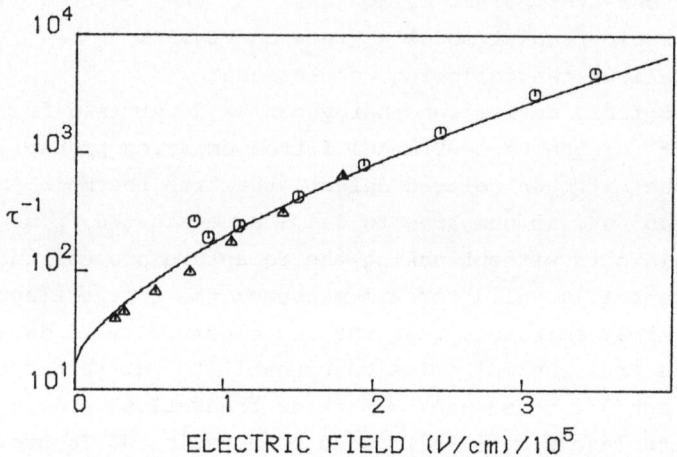

F i g . 3.
Thermal emission data measured on GaP Schottky barrier with shallow doping level at $1.2 \times 10^{17} cm^{-3}$ (squares) and on an n-type GaP Schottky barrier with shallow doping level at $1 \times 10^{16} cm^{-3}$ (triangles).

related trap in GaP in two different samples having different shallow doping concentration. As it is seen on the Fig.3 the deviation from the Poole-Frenkel fit is only due to the recapture process.

Conclusions

We have demonstrated, that the free carrier recapture in the edge layer of semiconductor junctions seriously influence the value of the thermal emission probability measured at low electric field. Accounting for the carrier recapture, the measured data are surprisingly well fitted with the simple Poole-Frenkel model. This observation questions the use of more elaborate models.

References

[1] G.Ferenczi, P.Krispin and M.Somogyi
J.Appl.Phys., 54, 3902 (1983)

[2] J.Frenkel
Phys.Rev. 54, 647 (1983)

[3] J.M.Noras
Solid St.Commun. 39, 1225 (1981)

[4] E.H.Rhoderick
Metal-semiconductor Contacts, Clarendon Press, Oxford (1980)

[5] K.Irmscher, H.Klose and K.Mass
Phys.Stat.Sol.(a) 75, K25 (1983)

[6] J.R.Morante, P.Roura, A.Perez, J.Samitiers, A.Cornet and A.Herms
Materials Science Forum 10-12, 533 (1986)

[7] S.Makram-Ebeid
Appl.Phys.Lett. 37, 464 (1980)

[8] G.Vincent, A.Chantre and D.Bois
J.Appl.Phys. 50, 5484 (1979)

[9] T.T.Nguyen, K.L.Wang and G.P.Li
Appl.Phys.Lett. 44, 211 (1984)

Electrochemical Characterization of GaAs and Its Multilayer Structure Materials

Peng Rui-wu, Luo Mao-min, Wang Zhou-cheng,
Ding Yong-qing, Chen Zi-yao
Shanghai Institute of Metallurgy,
Chinese Academy of Sciences
Shanghai, China

1. Introduction

GaAs and its multilayer structure materials are particularly interesting for the fabrication of modern semiconductor devices and integrated circuits, but in some cases it is difficult to characterize these materials by conventional solid state techniques. The semiconductor/electrolyte (SE/EL) junctions were thought to offer many advantages compared with solid state junctions [1] and remained very attractive in the simplicity of their fabrication. Up to now, the electrochemical method based on the liquid junctions and anodic dissolutions has been believed to be a sensitive, flexible and versitale technique for characterizing the compound semiconductor materials. Using this technique many physical parameters were successfully determined even for the GaAs multilayer structure materials [2]. This paper presents a review of our recent work on studying the GaAs/electrolyte junction, characterizing the bulk and surface properties of GaAs compounds and profiling several GaAs multilayer structures.

2. Experimental

The experimental arrangement for the electrochemical study has previously been described [3]. In the photoelectrochemical study a monochromatic light is emploied to directly illuminate the SE/EL interface through a quartz window and the photocurrent produced by the generation of electron and hole pairs is carefully detected by a lock-in amplifier as shown in Fig.1.

The measurements of deep level transient spectrum (DLTS) and surface photovoltages (SPV) were performed with conventional apparatus used in our recent work [4,5].

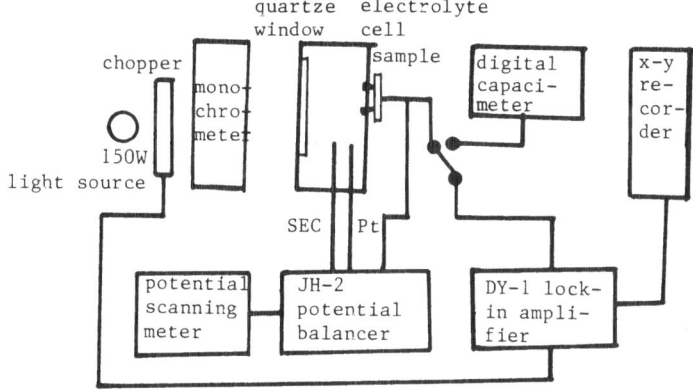

Figure 1: Setup for electrochemical measurements.

3. Results and Discussion

3.1 SE/EL junction

3.1.1 I-V characteristics

The basic characteristics of GaAs/electrolyte junction can be first achieved from the current versus potential curves in the dark and under illumination. Fig.2 shows the typical I-V behaviors of n-GaAs/ 1N KOH and n-GaAs/1:1 HCl junctions in the dark, which is similar to the published data [6]. But it is not consistent with the earlier prediction by Myannlin and Pleskov [7], who proposed the following relationship:

for large current density, $V \propto i^2$ (1)

for small current density, $V \propto \ln i$ (2)

The potential range in Fig.2 where no detectable current crosses the interface is not so large and depends not only on the free carrier concentration in the semiconductors, but also on the bulk and surface properties at both sides of liquid junction. It is our opinion that unlike the Tranchant et al.'s work [8], I-V characteristics seem not suitable for the determination of carrier concentrations [9].

In general, I-V curves may provide a useful means of identifying the conductivity of semiconductor materials as described below.

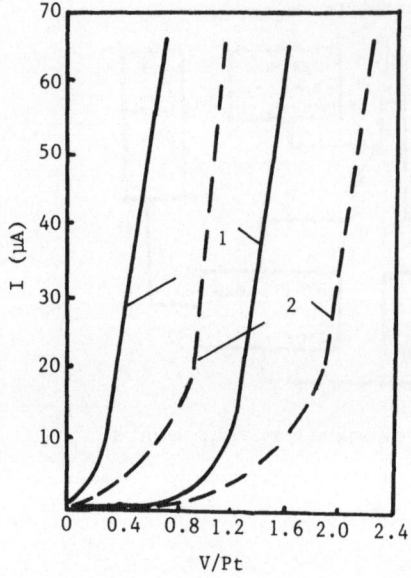

Figure 2: I-V plots for n-GaAs
with different concentration:
(1) in 1 N NaOH; (2) in
1:1 HCl:H_2O in the dark.

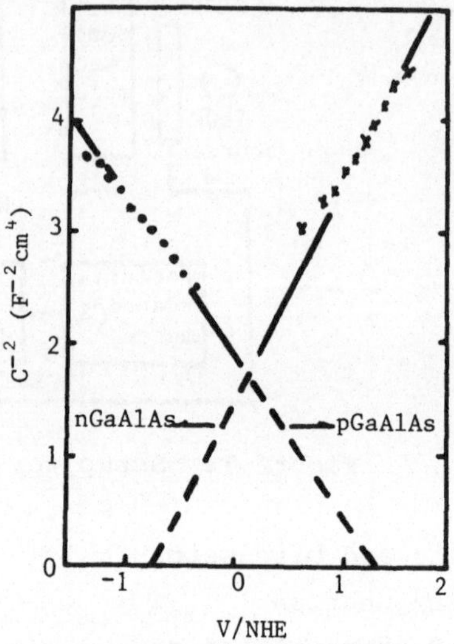

Figure 3: C^{-2} versus V plots
for n- and p-GaAlAs electrodes
in HCl (pH=1) [2].

3.1.2 C-V characteristics

If we assume that no electrochemical reaction takes place at the
SE/EL interface, then a simple conclusion on the potential drop in
concentrated electrolytes (C>0.1 M L^{-1}) mainly on the semiconductor
side of its interface may be drawn. In such case the space charge
capacitance (C_{sc}) can be written as follows:

$$1/C_{sc}^2 = 2/e\,\varepsilon\varepsilon_o N\,(U_{sc} - kT/e) \qquad (3)$$

where ε_o is the semiconductor dielectric constant, e is the electron
charge, N is the density of ionized donors and U_{sc} is the band bending.

Fig.3 illustrates the typical C-V curves of p- and n-GaAlAs electrodes in HCl solution [2]. The dependence of C_{sc}-V becomes a
straight line in coordination (C_{sc}^{-2}, V), which confirms the eq.(3) and
gives a so called Mott-Schottky plot.

It is worth to note that the frequency dependence effects were minimized in this study by using a high frequency of 100 KHz. However, a divergence from the Mott-Schottky relationship at low potentials and an increase in capacitance after the sample had been driven to high potentials were observed. It was also shown in C^{-2}-V curve of GaAs either for a liquid junction or for a Schottky barrier (Fig.4) [10]. This can be attributed to the presence of deep levels in GaAs as suggested by Hesse et al. [11].

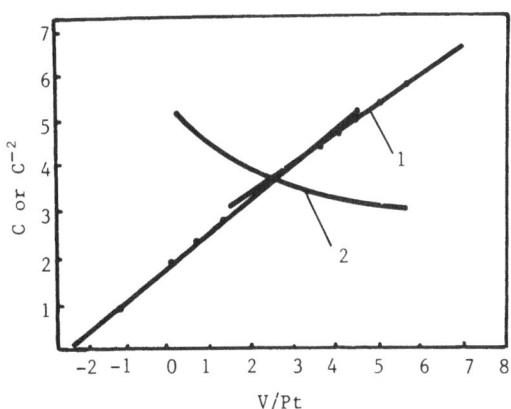

Figure 4: Capacitance-voltage characteristics
for n-GaAs/2M KCl contact: (1) C^{-2}-V; (2) C-V.

According to eq.(3), if U_{sc} changes only with the electrode potential, V, the Mott-Schottky plot should yield a slop, proportional to 1/N, which makes it possible to determine the carrier concentrations of donors or acceptors, N_D or N_A and the intersection point of this straight line with potential axis U_{sc} gives the flat band potential, V_{fb}, according to following diffination

$$U_{sc} = V - V_{fb} \qquad (4)$$

The knowledge of V_{fb} is necessary to calculate the position of valence and conduction bands at surface on the electrochemical potential scale and to establish a relationship between the energy levels in the semiconductor and the electrolyte, by which the possible charges transfer across the interface can then be analysed in terms of electrochemical reactions. The surface energy diagram based on the values of V_{fb} for several GaAs compounds are illustrated in Fig.5 [12] together with other published data [13].

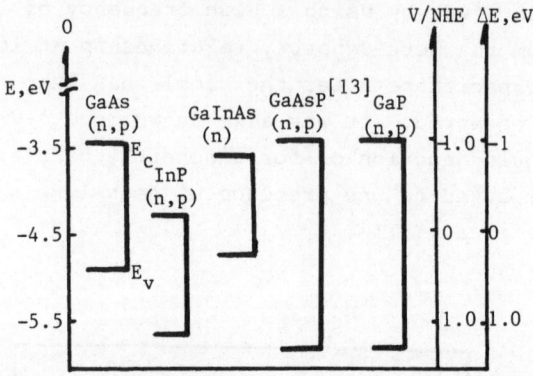

Figure 5: Surface energy diagram for several
GaAs related compounds in pH=1 solution [12].

3.1.3 I-t characteristics

The variation of anodizing current with time, I-t curve, for n-GaAs
sample in the dark is shown in Fig.6 [9]. It is observed that the
current starts at an initially high value and charges up the semiconductor at a rate which gives [14].

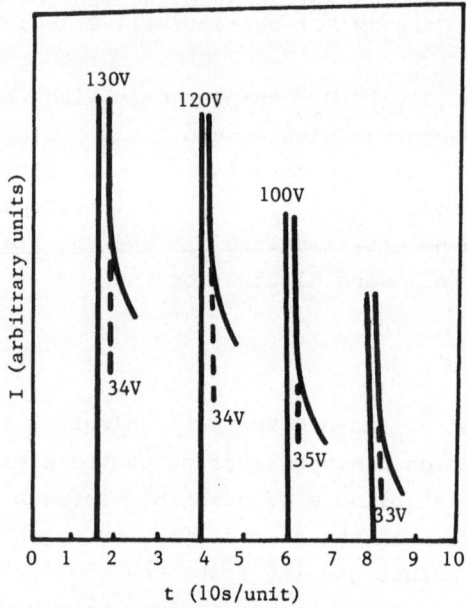

Figure 6: I-t curves for n-GaAs epitaxial
layer (n=1.2x10^{16}/cm^3) [9].

$$I = I_s e^{t/\tau} \text{ with } \tau = C_D R_S \tag{5}$$

Where R_S is the series resistance and C_D is the depletion layer capacitance which is, of cause, voltage dependent. Typically τ is the order of milli-seconds and thus the build-up of the space charge layer usually takes place before the active-passive transition occurs, this is, the surface breakdown voltage, V_B, is practically always reached before the sample begins to be oxidized. Therefore, the initial current drop due to eq.(5) is very steep and it occurs usually well before the current decay due to anodization whose behavior is then given by the following expression

$$i = \frac{V - V_B}{R_S} e^{-t/\tau} \text{ with } \tau = R_S/E_{ox} \cdot A \tag{6}$$

where E_{ox} is the field in oxide, A is the oxide formation rate and V and V_B are the applied bias voltage and surface breakdown voltage respectively. According to eq.(5) and (6), the surface breakdown potential can be taken at the break of the I-t curves from Fig.6 as discussed below. Beside this, we have used the i-t curve to measure the residual current during the current decay for further characterization of the GaAs materials [9].

3.1.4 i_{ph}-V characteristics

The photocurrent versus potential, i_{ph}-V curve, at fixed photon energy can usually be described by the Gartner equation [15]

$$i_{ph} = -q\phi[1 - \frac{\exp(-\alpha W_D)}{1 + \alpha L_p}] - qp_o \frac{D_p}{L_p} \tag{7}$$

where ϕ is the photon flux and p_o is the concentration of holes at equilibrium, A is the illuminated semiconductor surface area, α is the optical absorption coefficient and W is the space charge region width.

Fig.7 shows the I-V and I_{ph}-V curves of GaAs epilayers. In the case of n-GaAs the anodic currents are limited to a very low value, then increase rapidly at a given electrode potential, while the cathode currents increase with potential both under illumination and in the dark. In the case of p-GaAs the situation is different. These results are similar to the published data [16].

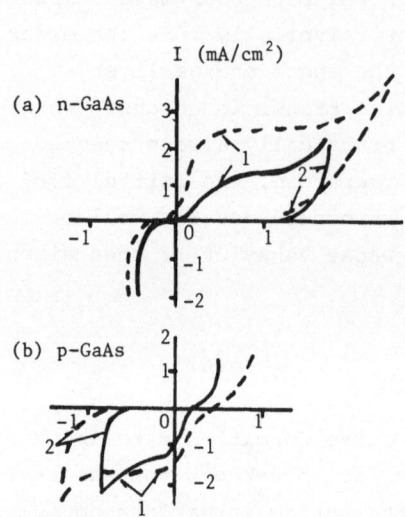

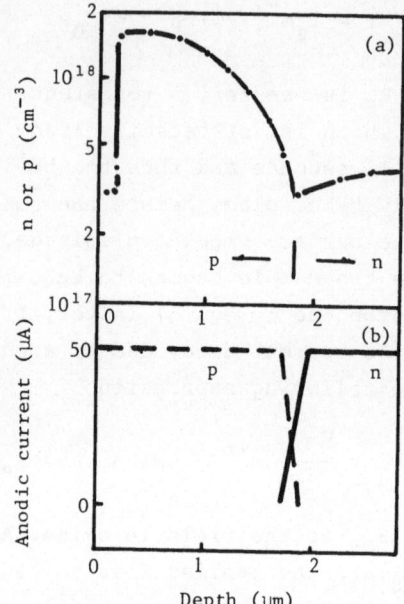

Figure 7: I-V plots for n- and p-GaAs in 1N KOH solution (converted to pH=1): (----) after Hollan et al.[16]; (1) under illumination; (2) in the dark [2].

Figure 8: Continuous carrier concentration profile of p-GaAlAs/p-GaAs/n-GaAs multilayer (a) concentration profile; (b) conductivity type and corresponding dissolution current: (——) under illumination; (----) in the dark [2].

According to the different photoeffect in n- and p-GaAs, the conductivity type and then the p-n interface can be identified as shown in Fig.8. Fig.8a shows the concentration of the p- and n-GaAlAs layers and their interface positions, while Fig.8b shows the correlation between the change in dissolution current and change in conductivity type under the controlled potential.

In addition, using the I_{ph}-V curves or Gartner equation, the information needed for computation of minority carrier diffusion length can be obtained.

3.2 Bulk properties
3.2.1 Diffusion length

The importance of minority carrier diffusion length, L, in GaAs is well known for modern opto-electronic devices, especially for photocathods and solar cells.

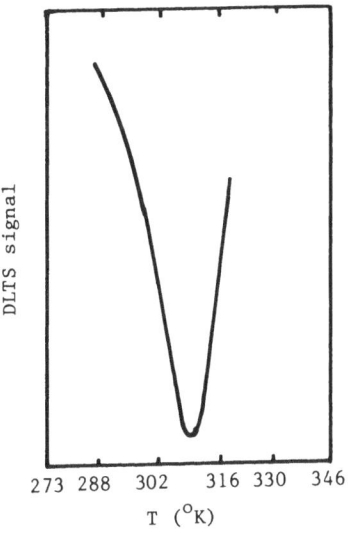

DLTS signal

273 288 302   316 330 346

T (^{o}K)

Figure 9: Liquid junction
DLTS spectra for n-GaAs/2M KCl
contact: Reverse Bia=4V, Rate
window=173ms:86ms; Pulse hight
=2.2V.

Fig.9 illustrates the typical LJDLTS spectra with a peak which may
be related to the break of C^{-2}-V curve (Fig.4) and thought to be associated with the presence of deep levels. The deep level concentration of $1.7x10^{14}cm^{-3}$ with capture cross section of $1.5x10^{13}cm^{2}$ located
at E_{na}=0.823eV was calculated. Because the same position and same
concentrations of this peak were obtained not only by the conventional
DLTS for one sample, but also by LJDLTS for various n-GaAs sample,
this deep level may be considered as an electron trap called EL-2.

Although only one of the electron trap, El-2, was detected, the
LJDLTS can still be recoginized as an important promising technique
for rapid determination of EL-2 in GaAs, which is common in vapor
phase epitaxial (VPE) and liquid encapsulated Czochrolski (LEC) or
Horizonal Bridgman (HB) grown GaAs and plays an important role in
compensating undoped semiconducting materials used in GaAs large scale
integrated circuits.

3.3 Surface qualities
3.3.1 Near surface defects

Even though the GaAs samples have similar surface morphology, the
different defect structures near or at the surface have been revealed
by the electrochemical etching under controlled condition as shown in
Fig.10 [17]. Ion microprobe analysis indicates that two samples (see
Fig. 10, b and c) have more defects and more lower breakdown voltages

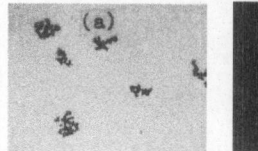 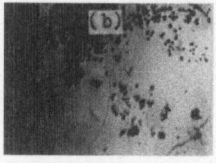 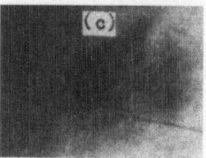

Figure 10:   Near surface defects of S-doped GaAs
epitaxial layers (a) (110), n=1.2x10^{16}cm^{-3}, V$_B$=34V;
(b) (110), n=1.2x10^{16}cm^{-3}, V$_B$=17V; (c) (100),
n=5x10^{16}cm^{-3}, V$_B$=14V [17].

near surface than other one.   These defects seem to be more like precipitates or inclusions   rather than dislocations.   An improvement in
the growth procedures allows one to prepare a high quality epilayers
with excellent surface morphology and without any structural defects
near the surface as pictured in Fig.10, a [18].

3.3.2 Surface breakdown voltages

The correlation between the surface quality and the electrical
properties of epilayers can be studied by the electrochemical I-t
method as shown in Fig.11.   It is evident that there is a parallel
relation between the surface defects revealed by the electrochemical
etching and the surface breakdown voltage taken at the break of I-t

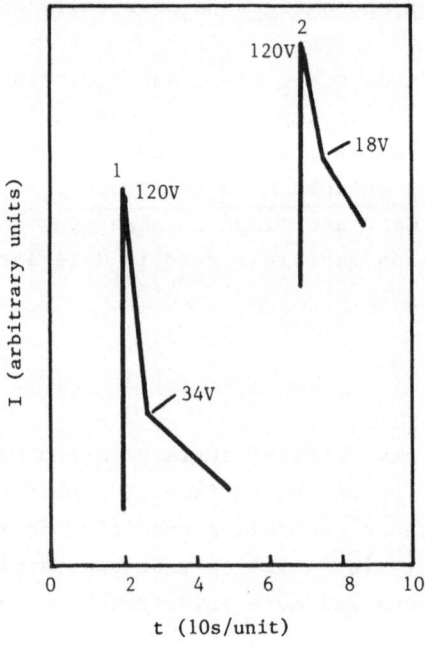

Figure 11:   I-t curves for
n-GaAs epitaxial layers:
(1) D-22-24; (2) D-22-28
[9].

curves. The surface breakdown voltage (18V) for sample D-22-28 with more defects appears to be smaller a factor of two than that for sample D-22-24 (34V). This fact may be explained by the local avalanche breakdown or by the tunnelling process on the epilayer surface [16].

Fig.12 shows the dependence of the surface breakdown voltages, V_B, obtained by the i-t curves on the carrier concentrations.

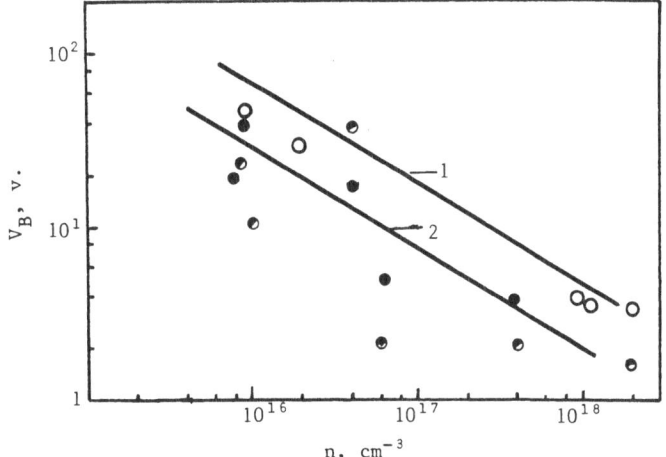

Figure 12: Variation of the breakdown voltage with carrier concentration.
(1) theoretical curve [19];
(2) average value measured by conventional C-V method in 1974
●,◉ - measured by i-t curve and C-V method respectively, Peng Rui-wu [9]
○ - measured by i-t curve, Colguhoum [14].

It is obvious that the V_B taken from i-t curves shows a strong dependence on the carrier concentrations and somewhat coincides not only with the theoretical curve, but also with the experimental data from our published results [9] and Colguhoum's work [14] using conventional C-V and electrochemical i-t method respectively. However, the V_B obtained by i-t method are higher and in fact much more reproducible than those obtained by the conventional one.

3.4 Continuous profiling and interface characteristics
3.4.1 Continuous carrier concentration profiling

The primary method for determining the carrier concentration is based on capacitance data and Mott-Schottky plots as described above. The combination of C-V measurement and anodic dissolution provides a useful tool for continuous profiling the carrier concentration.

Fig.13 shows the typical carrier concentration profiles of single, double and hi-lo-hi structures with S-doped submicrometer active layers [18]. The hi-lo-hi structure exhibits more abrupt transitions in the contact layer-active layer, active layer-buffer layer and buffer-substrate interface, but it still has some impurity tails caused by outdiffusion as mentioned in previous work [17].

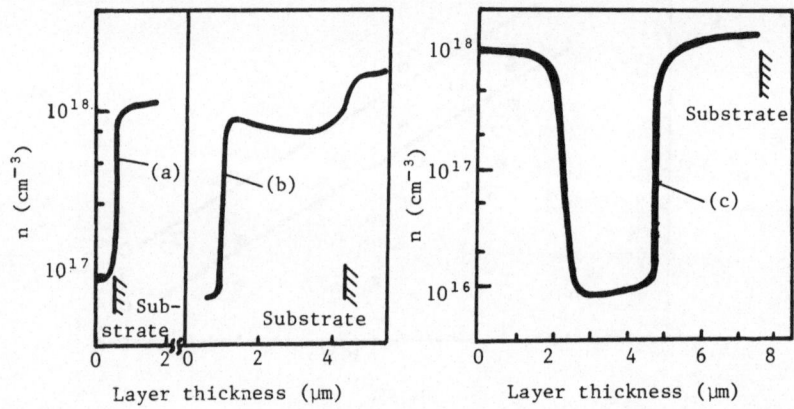

Figure 13: Carrier concentration profiles for S-doped GaAs epilayers: (a) simple layer; (b) double layer; (c) hi-lo-hi structure [18].

3.4.2 Interface carrier concentration profiling

The determination of the interface carrier concentration is also based on the combination of the C-V measurement and anodic dissolution. It has been observed that the interface carrier concentration profile of undoped layers on semi-insulating Cr-doped GaAs are strongly dependent on the quality of the substrates.

The interface carrier concentration profiles for low temperature epilayers measured by both the electrochemical and the conventional C-V methods are in good agreement each other as shown in Fig.14.

If auto-doping of Cr from the substrate is not a serious problem, the bad profile shape in the interface region may be attributed to outdiffusion effects only [17].

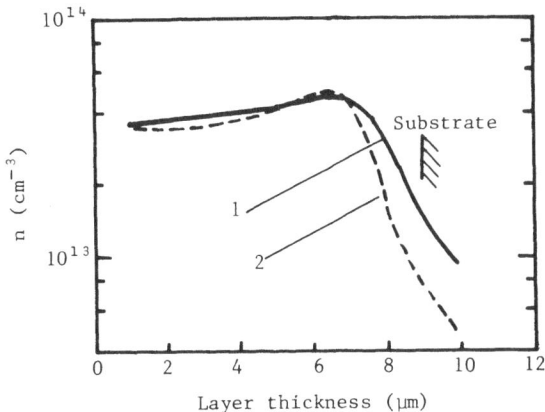

Figure 14: Interface carrier concentration profiles
measured by: (1) conventional; (2) electrochemical
method [17].

3.4.3 Continuous compositional profiling

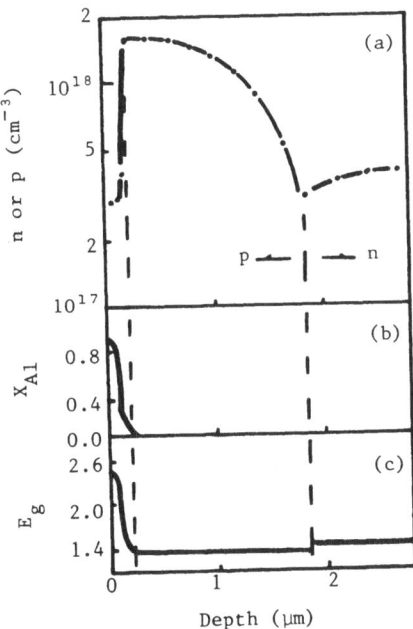

Figure 15: Full characterization of p-GaAlAs/p-GaAs/n-GaAs
multilayer material: (a) carrier concentration and p-n
junction position; (b) Al-com-
position; (c) bandgap [2].

The compositional analysis rests on determining the bandgap
through measurement of the dependence of the photocurrent of the

semiconductor/electrolyte barrier upon wavelength of illumination at fixed electrode potential by photoelectrochemistry [13]. Knowing the bandgap, the Al composition of GaAlAs can be deduced. Fig.15 illustrates the full profile of a p-GaAlAs/p-GaAs/n-GaAs multilayer structure which indicates the variation of the carrier concentration, Al composition bandgap and depth as well as conductivity type of all the layers [2].

3.4.4 P-n junction profiling

The combination of C-V and I_{ph}-V measurements with anodic dissolution has provided an efficient tool for profiling p-n junction structures. The typical concentration profiles for p-n and n-p-p$^+$ multiepilayers prepared by Cd- and S doping technique in-situ [20] are measured by electrochemical method as shown in Fig.16 [20]. The p-n or n-p-p$^+$ multilayer structures have a good concentration profile in epilayers and a more abrupt transition in the interface region. It is evident that the electrochemical method may be an unique technique for profiling the p-n junction structure, which will be very important for modern devices.

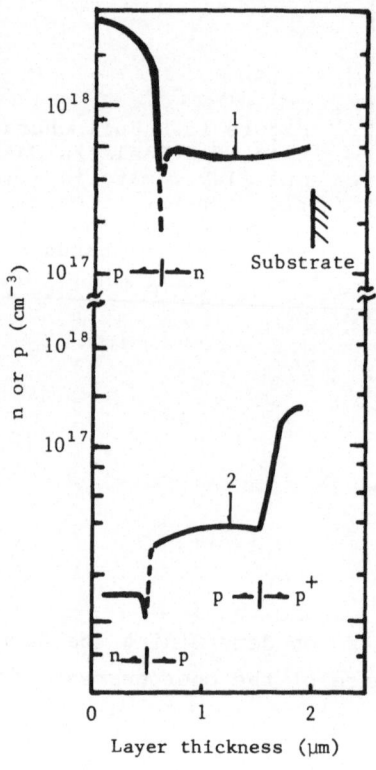

Figure 16: Carrier concentration profiles for S and Cd doped GaAs epilayers: (1) p-n; (2) n-p-p$^+$ junction structures [20].

3.4.5 Heterostructure profiling

Using the difference behaviors in I-V and C-V curves for Ge and GaAs materials and combining the C-V measurement with anodic dissolution, the concentration profile of GaAs/Ge/GaAs heterostructure grown by a halid vapor phase epitaxy in situ [21] has been plotted in Fig. 17. The measured layer thicknesses are 1 and 2μm for GaAs and Ge respectively. Although the carrier concentrations in both layers are up to $10^{18}/cm^3$, the interface between them is conderablely sharp.

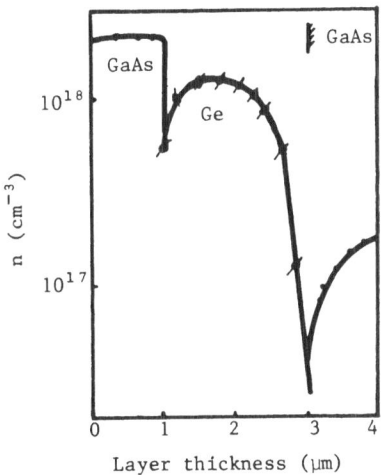

Figure 17: Carrier concentration profile for GaAs/Ge/GaAs multilayer structure [21].

4. Conclusions

Using the electrochemical and photoelectrochemical measurements the important bulk and surface properties of GaAs, such as diffusion length, deep levels, near surface defects and breakdown voltages were determined with reasonable success on the basis of discussion of SE/EL interface behaviors. Combining the above measurements with anodic dissolution, the various profiles of GaAs multilayer structures including the carrier concentration, composition, bandgap, p-n junction and heterostructure as well as the conductivity type were simultaneously achieved for the first time.

References

[1] Yu.Ya. Gurevich and Yu.A. Pleskov: Photoelectrokhimiya Poluprovodnikov (Nauk, MOCKOW 1983)
[2] Peng Rui-wu, Chen Zi-yao and Shao Yong-fu, J. Crystal Growth 69, 469 (1984)

One of the most attractive electrochemical method for their determination is based on the i_{ph}-V behavior by means of Gartner equation (eq.6). In practice, this method has met with moderate success [4], probably due to the variation of surface conditions and involvement of many adjustable parameters and numerical solution. For this reason, a conventional surface photovoltage (SPV) - like technique was suggested by author et al. [5] in a first attempt to determine the minority carrier diffusion length, L, which we called liquid junction surface photovoltage (LJSPV). In table 1 the values for L determined by LJSPV are listed together with those determined by conventional SPV only for n-GaAs, since the later technique is neither sensitive nor suitable for p-GaAs due to the low and undetectable voltage signal.

Table 1   Values of L_n or L_p [5]

Sample No.	Conductivity type	Carrier concentration (cm^{-3})	Diffusion length	
			LJSPV	Conven. SPV
D-05	n	~10^{18}	2.67±0.24	2.69±0.21
D-49-18	n(epilayer)	~10^{17}	5.06±0.36	4.91±0.36
A-02	p	~10^{19}	4.81±0.39	-
A-03	p	~10^{19}	4.63±0.29	-

It can be seen from table 1 that the L_p for n-GaAs obtained by both methods are in good agreement with each other and the L_n for p-GaAs obtained by LJSPV is very reproducible. This supports that the LJSPV is a more versitale and more simple method for determination of minority carrier diffusion length in both type of GaAs than conventional one.

3.2.2 Deep levels

The deep level traps in GaAs have been widely studied by the deep level transient spectroscopy (DLTS), but the analogous on electrochemical DLTS, which we have first suggested and called liquid function DLTS (LJDLTS), has not yet been received any attention.

The effectiveness of LJDLTS for deep level study rests on the fact that liquid junctions are similar to solid Schottky barriers. It exhibits a high response rates for voltage signal and a considerable temperature range for electrolytes as well as a high stability for junction parameters [4].

[3] Chen Zi-yao, Shao Yong-fu and Peng Rui-wu, J. of Science (Kexue Tongbao) 27, 1057 (1982) (in English)

[4] Luo Mao-min and Peng Rui-wu, in: Proc. 1985 Natl. Symp. on GaAs and Related Compounds (Chinese Soc. of Metals, 1987)

[5] Wang Zhou-cheng and Peng Rui-wu, to be published in 1987 Natl. Symp. on GaAs and Related Compounds, Beijing, 1987.

[6] M.M. Faktor, T. Ambridge, C.R. Elliott and J.C. Regnault, in: Current Topics in Materials Sciences, Vol.6, Ed. E. Kaldis (North-Holland, Amsterdam, 1980) p.1

[7] V.A. Myannlin and Yu.A. Pleskov, Electrokhimiya Poluprovodnikov (Nauk, MOCKOW, 1965)

[8] J.C. Tranchart and L. Hallon, J. Electrochem. Soc. 125 (1978) 1185

[9] Peng Rui-wu and Ding Yong-qing, in: Proc. 1981 Natl. Symp. on GaAs and Related Compounds (Chinese Society of Metals, 1983) p.203

[10] Luo Mao-min and Peng Rui-wu, in: Proc. 1985 Natl. Symp. on GaAs and Related Compounds (Chinese Soc. of Metals, 1987)

[11] K. Hesse and H. Strack, Solid State Electronics 15, 767 (1972)

[12] Chen Zi-yao, Shao Yong-fu and Peng Rui-wu, in: Proc. 1981 Natl. Symp. GaAs and Related Compounds (Chinese Society of Metals, 1983) p.209

[13] A.M. Redon, J. Vigneron and J. Chevalber, J. Electrochem. Soc., 127, 613 (1980)

[14] A. Colguhoum, Solid State Electronics 19, 819 (1976)

[15] W.W. Gartner, Phys. Rev. 116, 84 (1959)

[16] L. Hollan, J.C. Tranchart and R. Memming., J. Electrochem. Soc., 126, 857 (1979)

[17] Peng Rui-wu, J. Crystal Growth 56, 350 (1982)

[18] Peng Rui-wu, Sun Chang-zhu and Shen Song-hua, J. of Electronics (China) 1, 53 (1984)

[19] S.M. Sze, Physics of Semiconductor Devices, John Wiley and Sons Inc., 1969

[20] Peng Rui-wu, Xu Chen-mei, Li Chui-yun and Wang Bo-hong, J. of Electronics (China) 3, 71 (1986)

[21] Zhan Xia-fang and Peng Rui-wu, Presented at 4th Natl. Symp. on Semiconductor Microwave Devices, Photoelectronic Devices and Compound Materials, Guangzhou, China, 1986.

POSITRON STUDY OF DEFECTS IN GaAs

M. Misheva, G. Pasajov, G. Tubmev*, R. Yakimova

Sofia University, Faculty of Physics,

5 A. Ivanov Blvd, 1126 Sofia, Bulgaria

* Bulgarian Academy of Sciences, Institute for Nuclear Research
and Nuclear Energy, 72 Lenin Blvd, 1784 Sofia, Bulgaria

1.   INTRODUCTION

Study of defect properties in III-V semiconductors is important
from the fundamental and the technological point of view. Positron
annihilation [1] is sensitive to neutral and negatively charged vacancy-type defects, defect complexes of vacancies, impurities atoms
and antisites. During the last five years an increased interest has
appeared in using positron annihilation for study of lattice defects
in semiconductor compounds, especially in III-V.

In the present work positron lifetimes as well as the probability
for three photon annihilation have been measured.

2.   POSITRON ANNIHILATION BACKGROUND

In a perfect crystal positrons annihilate mainly with the valence
electrons of the solid. The annihilation rate (the inverse of the
mean positron lifetime) depends on the overlap of positron and electron densities. The vacancy-type defects, mentioned above, are capa-
ble to trap thermalized positrons which results in changes of the
annihilation characteristics - the positron mean lifetime increases
comparable with that of the "free" positrons. The lifetime spectra
measured can be analysed by the multy-exponential function fitting
programme.

According to the trapping model [2] in case that n-kinds of positron traps exist, and that the positrons captured can not escape, the
specimens characteristics are as follows:

τ_b and τ_{dj} - lifetimes of positrons annihilated in perfect region and trapped states, respectively ($j=1,\ldots,n$);

k_{dj} - trapping rate of various traps;

c_{dj} - concentration of a one type traps;

μ_{dj} - specific trapping rate of the same traps.

These characteristics are related with the observed lifetimes τ_i and their intensities I_i ($i=1,\ldots,n+1$) by the equations:

$$\tau_b = \frac{1}{\sum\limits_i \dfrac{I_i}{\tau_i}}, \quad \tau_{dj} = \tau_{j+1}, \quad k_{dj} = I_{j+1}\left[\frac{1}{\tau_1} - \frac{1}{\tau_{j+1}}\right], \quad c_{dj} = \frac{k_{dj}}{\mu_{dj}}.$$

The mean lifetime is:

$$\tau_m = \sum_i \tau_i I_i.$$

When there is only one defect type to trap positrons, the lifetime spectrum is two-exponential.

Three-gamma annihilation of a free positron is realised with a probability of 0.27%. When positrons and electrons are combined to form positronium-like system, the probability for three-photon annihilation increases. A necessary condition for existence of such a system (qPs-states) in the solid is the availability of vacant volumes in the crystal lattice. Vacancy type defects (clusters, agglomerates) can respond to this requirement. The formation of the qPs-system is characterized by the ratio $P/P_{Al} = \text{Eff}$ [3]. P is the probability of tree-gamma annihilation, which is assumed to be equal to the integrals ration I_{341}/I_{511} in the energy spectrum regions of 341 keV and 511 keV. P_{Al} is same probability in aluminum (assuming that no qPs-states are forming there). Increasing Eff indicates increased probability of positronium-like system formation.

The behaviour of positrons in GaAs is studied by a number of authors [4-9], but the knowlege of defect properties is still far from the complete. The aim of the present work is to characterize point defects in GaAs single crystals, used for epitaxial growth substrates, by means of positron annihilation techniques. In order to complete the information about the positron annihilation in III-V semiconductors InAs crystals are studied as well.

3.   *EXPERIMENTAL*

All samples studied were single crystals grown either by Czochralski or Bridgman method, undoped and doped n-type, and semiinsulating. Their typical characteristics are shown in Table 1. From (100) oriented wafers (for exeption sample 8) we cut samples of 10x10x0.7 mm^3, which were mechanically polished and chemically etched.

Three kinds of additional treatement was applied to the samples doped with Te, Cr and In, respectively:

i) ^{60}Co gamma-irradiation up to 57 MR at room temperature;

ii) Annealing (I) under epitaxial growth conditions (750°C, 1 hour in H$_2$);

iii) Annealing (II) to create As-vacancies (900°C, 2 hours in Ar).

Commonly used ^{22}NaCl positron source, sealed between two thin mylar foils, was sandwiched by the samples.

The lifetime spectrometer used was of a fast-fast type with a time resolution 260 ps (FWHM). During 2 hours about 1.1x10^6 counts were accumulated for each spectrum. The samples were measured at least three times.

Lifetime spectra were analysed by means of POSITRONFIT-EXENDED programme [10], which was able to decompose up to four lifetimes.

We used three-gamma spectrometer [11] with a resolution of coincidence circuit 20 ns. During 17 hours 3x10^4 events were registered.

All measurements were carried out at room temperature.

4.   *RESULTS AND DISCUSION*

4.1.   Lifetime measurements

4.1.1.   Gallium Arsenide

Two-, three- and four-component analysis of the lifetime spectra were attempted. The best fitting of the experimental data was obtained by three-exponential decomposition. The results are summarized in Table 2. The third component (τ_3 = 1710±4 ps, I$_3$ = 3.72±0.05 %) was practically constant and was attributed to the positron source.

The annihilation characteristic values calculated according to the trapping model (assuming one type positron traps) are presented in Table 3. The results were obtained by averaging of a large number of measurements.

Table 1. Crystal parameters

Sample	Dopant	Growth method	N_n, cm^{-3}	N_d, cm^{-2}	Orientation
1	Te	Czochralski	2×10^{18}	2.0×10^4	(100)
2	Cr	Czochralski	SI	2.3×10^4	(100)
3	In	Czochralski	SI	1.0×10^4	(100)
4	Sn	Czochralski	4×10^{17}	2.3×10^3	(100)
5	Undoped	Czochralski	SI(EL2)	5.0×10^3	(100)
6	Sn	Bridgman	5×10^{17}		(100)
7	Sn	Bridgman	1×10^{18}		(100)
8	Undoped	Bridgman	4×10^{16}(Si)	1.5×10^5	(111)

N_n - free electron concentracion, N_d - dislocation density

Our experimental results indicate that the positron lifetimes in Te-doped and undoped n-type as-grown crystals have comparable values. Similar data are observed in [4]. However the intensity of the second component and the trapping rate in sample 8 are over of all values measured, which may be due to the high dislocation density in this material. In SI as-grown crystals the lifetimes are slightly lower (about 5 ps) than those in n-type GaAs [8,9], independently of the compensating mechanism in the crystals. It is important to notice that in all cases studied we observed two lifetime components, while most of the authors [6-8] found only one component in semiinsulating material.

Table 2. Positron lifetime results for GaAs

Sample	τ_1, ps	I_1, %	τ_2, ps	I_2, %
1	219±3	72±3	389±17	25±2
3	224±3	82±2	430±15	15±1
5	224±4	80±2	421±12	16±2
8	214±2	69±2	377±10	27±2
1^I	223±2	73±2	386± 7	23±2
1^{II}	228±4	75±3	396±16	22±3
2^{II}	228±2	81±2	426±12	15±2
3^{II}	231±1	85±1	460±16	11±1

I, II indicate the annealing regime

Gamma-irradiation doesn't change the annihilation characteristics, which can be related to a possible selfannealing of the defects at room temperature. Although one could not expect a creation of vacancy type defects in the substrates during annealing I, some redistribution of point defects is possible. Lifetime measurements seem to be insensitive to such kind of changes, independently of dopants. The annealing II affects differently lifetime values in Te-doped and SI gallium arsenide (see Table 3). In the first case the positron mean lifetimes

Table 3. Mean τ_m, bulk τ_b positron lifetimes and trapping rate k_d

Sample	τ_m, ps	τ_b, ps	k_d, ns^{-1}
1	263±1	247±2	0.52±0.06
3	256±1	242±2	0.33±0.03
5	257±1	243±2	0.35±0.04
8	261±1	244±2	0.57±0.04
1^I	262±1	248±2	0.45±0.02
1^{II}	267±1	253±3	0.43±0.06
2^{II}	259±1	246±1	0.32±0.03
3^{II}	258±1	245±1	0.26±0.02

I, II indicate the annealing regime

increase, when in SI-samples they tend to keep the initial values. This result can be explained taking into account the widely accepted suggestion that As-vacancies form donor-type levels in the energy gap of GaAs [12]. Assuming that they are deep donors [8] the ability to capture positrons is depending on the Fermi-energy. In n-GaAs(Te) with electron concentration 2×10^{18} cm^{-3} the Fermi-level lies very close to the conduction band and deep donor levels are un-ionised, thus they can trap positrons. In SI-GaAs these donors can be easily compensated and no trapping of the positrons occurs. As is described in [8] the ratio τ_d/τ_b gives information about the structure of defects which trap positrons. For n-GaAs both doped and undoped we obtain value 1.55±0.05 that corresponds to divacancy type of defects. In SI-GaAs this ratio is 1.77±.04 and can be connected with the result obtained from X-ray topography of similar samples, where large regions (comparable with the lattice parameter) of structural imperfections were observed. Assuming the specific trapping rate for divacancies equal to 4×10^{-9} cm^{3}s^{-1} [8] we have found that their concentration is in the range of $1.1 \div 1.4 \times 10^{17}$ cm^{-3} for n-type GaAs.

4.1.2. Indium Arsenide

The samples studied were grown by the Czochralski method in a sealed ampule and were cut in 100-orientation. They were undoped n-type with the free electron concentration 5×10^{16} cm^{-3}, and dislocation density 1×10^4 cm^{-2}. The annihilation spectra are analyzed supposing that the annihilation component of the positron source is the same as in GaAs measurements. Then four-component decomposition fits the experi-

Table 4. Positron lifetimes for InAs

τ_1, ps	I_1, %	τ_2, ps	I_2, %	τ_3, ps	I_3, %
239±3	70±2	416±7	24±2	2270±33	1.9±0.1

mental data well ($\tau_{source}=1710$ ps, $I_{source}=3.72$ %), which are shown in Table 4.

The calculated results by the trapping-model, in assuming of two defect types, are as follows:

$\tau_m=324\pm1$ ps, $\tau_b=274\pm1$ ps, $k_{d1}=0.45\pm0.03$ ns^{-1}, $k_{d2}=0.07\pm0.03$ ns^{-1}.

As is seen the value of 274 ps for the bulk lifetime in InAs is with 30 ps higher then that in GaAs, which is due to the higher lattice distances in the first crystals. Tha ratio $\tau_{d1}/\tau_b=1.52\pm0.03$ points to divacancies. The similar ratio for the second type of defects is much higher, which suggests that the formation of qPs-states is possible.

4.2.   Three-gamma annihilation in GaAs

We measured GaAs-crystals doped with Cr, Te, Sn (grown by Czochralski method) and doped with Sn but grown by horizontal Bridgman technique. If we consider that $P_{Al}=2.65\pm0.03$ the values of Eff=P/P_{Al} can be estimated, which characterizes a possible formation of qPs in our samples. The data obtained are shown in Table 5 (sample numbers correspond to Table 1) and they show that in Sn-doped Bridgman grown samples qPs-states can be formed. It seems to be connected with the crystal growth conditions under which more complicated defect complexes appear. In the Czochralski grown material no evidence for positronium-like system is obtained, independently of the dopant used.

Table 5. Tree-gamma annihilation data

Sample	Eff
1	1.07±0.04
2	1.00±0.04
4	1.06±0.04
6	1.19±0.04
7	1.26±0.04

5.   *CONCLUSIONS*

   Positron lifetime measurements and three-gamma annihilation have
been applied to study point defects both in undoped and doped GaAs.
The main results are summarized bellow:
   i) in GaAs(Te) and undoped n-type GaAs the most effective positron
      traps are divacancies;
   ii) semiinsulating GaAs-crystals exhibit two-component lifetime spec-
       tra, which is related to some clustering of defects presented;
   iii) in Bridgman grown GaAs doped with Sn data about qPs-states forma-
        tion are obtained.
   In addition InAs crystals have been measured. In positron lifetime
spectra taken from these samples three components have been observed,
the longest one being related to positronium-like system formation.

REFERENCES

1. P. Hautojärvi (ed.) - Positrons in Solids, Springer Verlag, Berlin,
   Heidelberg, 1979
2. R. N. West - Adv. Phys. , 22, 263(1973)
3. A. Bisi, C. Bussolati, S. Cova, L. Zappa - Phys. Rev. , 141, 348(1966)
4. O. Takai, Y. Hisamatsu, N. Owada, H. Ishimura, K. Hinode, S. Tanigawa,
   M. Doyama - Phys. Lett. A, 76, 157(1980)
5. S. Dannefaer - J. Phys. C, 15, 599(1982)
6. S. Dannefaer, B. Hogg, D. P. Kerr - Phys. Rev. B, 30, 3355(1984)
7. Xiong Xing-min - private communication
8. G. Dlubek, O. Brummer, F. Plazaola, P. Hautojärvi - J. Phys. C, 331(1986)
9. M. Stuky, C. Corbel, B. Gefroy, P. Moser, P. Hautojärvi -
   Materials Science Forum, 10-12, 265(1986)
10. P. Kirkegaard, M. Eldrup - Comp. Phys. Comm. , 7, 401(1974)
11. A. Georgiev, M. Misheva, P. Mishev, G. Tumbev - Proceedings of
    the First European Meeting on Positron Annihilation,
    Wernigerode, GDR, 23-27. 03. 1987
12. F. A. Kröger - The Chemistry of Imperfect Crystals, vol. 2,
    Amsterdam, 1974

DEEP LEVEL PROFILING TECHNIQUE IN THE SEMICONDUCTOR
OF MIS STRUCTURE

V.S.Lysenko,   A.N.Nazarov   and   T.E.Rudenko
Institute of Semiconductors, Academy of Sciences of the Ukr.SSR
252028, Kiev-28, Prospect Nauki II5, USSR

When studying the processes of defect annealing and impurity acti-vation in the surface layer of the semiconductor being a part of the MIS structure, the spacial distribution of residual radiation damage must be established.

This problem is quite complicated because some energy levels are usually present in the semiconductor, the distribution of the implan-ted impurity in the semiconductor surface region is not uniform and, after all, the effect of surface electronic states (SES) must be taken into account.

To establish the deep level distribution in the semiconductor sur-face layer, we have proposed to use the method of thermally stimula-ted current release (TSCR) combined with measurements of non-equilib-rium quasistatic and high-frequency C-V dependences.

The main point of the method for determining the defect profile may be explained in the following way. The temperature dependence of the current measured by TSCR method under the linear regime in a de-finite temperature range, reflects the energy distribution of deep levels in the band gap irrespective of their spatial distribution /I/.

The energy distribution of SES at the interface between the semi-conductor and the insulating layer is usually determined by means of equilibrium low-frequency and quasistatic C-V characteristics /2/. Clearly, in the case when the deep centres are present in the semicon-ductor volume, the quasistatic C-V characteristics are determined both by energy and spacial distribution of these centres. In the case of local surface states the TSCR spectra must correspond to the quasi-static C-V characteristics.

If the centres are distributed in the semiconductor volume then the current maximum must be observed in the TSCR spectrum but such a maximum is absent in the dependence of the low-frequency capacitance C_{lf} on the potential φ at the corresponding potential value φ_s. At the same time, the dependence $C_{lf} - C_{hf} = (\varphi_s)$ is determined by the distribution of thes centres across the semiconductor depth.

In the case of quasi-continuous level spectrum in the band gap, the smooth change of current with temperature is observed in the TSCR the temperature interval with an upper limit easily determined from the TSCR spectrum. The current maximum due to thermal generation of electron-hole pairs is clearly defined in the TSCR spectrum in the regime of non-equilibrium depletion.

3.Under high-frequency conditions the deep level is not filled during the half-period time. Under the low-frequency conditions the equilibrium filling of the deep level is established during the half-period time due to majority carrier emission into the allowed energy band.

4.The deep level in the quasi-neutral semiconductor volume is fully occupied at the temperatures under investigation.

5.If some distributed deep levels are present in the upper half of the band gap, then we consider that their energy separation is sufficiently large (above 2kT). In this case we may choose the temperature region so that $e_{n-I}(T) \gg e_n(T) \gg e_{n+I}(T)$ (here e_n, e_{n-I} and e_{n+I} are the rates of thermal emission from the n-th, (n-I)-th and (n+I)-th levels, respectively.

If the density $N_{tI}(x)$ of distributed centres with a single energy level exceeds appreciably the SES density then the current peak in the TSCR spectrum will be observed at a temperature (T_{max}) determined by the energy separation E_{tI} from the conduction band edge. The position of this peak in a temperature scale slighly depends on the heating rate.

If the temperature is sufficiently low ($T_o < T_{max}$) and the sweep rate of the gate voltage is sufficiently high then the deep level has no time to change its charge. In this case the low-frequency and high-frequency C-V characteristics are similar. The spacial distribution of the free carriers is determined from C-V characteristics as follows

$$n(W) = \frac{\varepsilon_o \varepsilon_s}{q} \frac{dV_s}{W dW} = -\frac{C^3}{q \varepsilon_o \varepsilon_s} \frac{dV_g}{dC} \qquad (I)$$

where W is the width of the depletion layer. In the case under consideration this distribution coincides with the doping profile $N_d(x)$ (in the case of donor levels) or with the distribution of the value $N_d(x) - N_{tI}(x)$ (in the case of acceptor levels). Let us suppose that we have for deep donor levels: $N_t(x) \ll N_d(x)$. If the gate voltage is changed then the deep level considered changes its charge at T_I T_{max} as a result of electron emission. This process gives its contribution into the low-frequency (quasistatic) capacitance and leads to distortion of the C-V characteristics.

spectra. Because the time constant of carrier emission τ depends
exponentially on the level energy depth E_t then at low temperatures
only the shallow levels (for which $w_{lf}\tau \lesssim 1$) give their contribution
into the low-frequency capacitance C_{lf}. If we suppose that the cross
section of traps is independent of energy, then we may introduce the
dividing level E_{td}, which divides the re-charging levels. It is
obvious that this level moves to the band gap middle with increasing

temperature.

Thus, at those ψ_s when the Fermi level on the semiconductor
surface intersects the level E_{td}, the $C_{lf}(\psi_s)$-dependence will have
a sharp bend at low temperatures. This is the case for SES with a
quasi-continuous energy spectrum.

The above-mentioned considerations show that the combined analisis
of TSCR spectra and quasistatic C-V characteristics at different temperatures makes is possible to distinguish SES at the semiconductor-
insulator interface from the levels in the semiconductor volume. In
the last case the spatial distribution of deep centres may be determined from the quasistatic and high-frequency C-V dependences measured in the temperature range established from the TSCR spectrum. The
distribution of centres with a single energy level may be evaluated
by choosing the temperature interval in which the quasistatic and
high-frequency C-V characteristics are measured. This choose is made
by means the TSCR spectra as follows: the distribution of levels at
E_{tI} is determined at T_I (where $T_{maxI} < T_I < T_{max2}$), the distribution
of level at E_{t2} is determined at T_2 (where $T_{max2} < T_2 < T_{max3}$), ets.
The T_{maxI}, T_{max2} and T_{max3} are temperatures of current peaks in the
TSCR spectra due to emission of carriers from deep levels with energy
depths at E_{tI}, E_{t2} and E_{t3} respectively.

Let us make some suppositions for calculating the deep level concentration in the semiconductor volume.

1. We will investigate a MIS structure containing an n-type semiconductor. Some centres are distributed in the semiconductor volume and
their energy level is located in the upper half of the band gap. The
rate of electron emission from this level into the conduction band
(e_{ni}) exceeds considerably the rate of hole emission into the valence
band (e_{pi}).

2. During the measuring time of C-V dependences the minority carrier
generation does not occur, i.e. the C-V characteristics are of a non-
equilibrium nature relating to minority carrier density. In the case
of the level under consideration ($e_n \gg e_p$) this condition is valid in

A following relation between the low-frequency and high-frequency
C-V dependences and the spacial distribution of the deep level density may be easily obtained:

$$\frac{C_{lf}}{C_{hf}} - 1 = \frac{N_{t_1}(W_1)(dW_1/dW)(1-W_1/W)}{N_{t_1}(W_1)(dW_1/dW)(W_1/W)+N_d(W)} . \qquad (2)$$

As seen, the numerator in the right-hand term of eq.(2) corresponds
to the deep level density distribution $N_t(W_I)$ and differs from this
density by $(dW_I/dW)(I - W_I/W)$ where W_I is the intercepting point
between the level considered and Fermi level. Let us call this numerator the effective profile of the deep level density. It can be
easily shown that the denominator in the right-hand term of eq.(2)
is equal $(\varepsilon_o \varepsilon_s/q)(dV/WdW)$ where V is the voltage drop across the semiconductor. Then the effective profile of the deep level density may
be described as follows:

$$N_{t_1 eff} = \frac{C_{hf}^3(T_1)}{q \varepsilon_o \varepsilon_s \frac{dC_{hf}}{dV}(T_1)} \cdot \frac{\left[\frac{C_{lf}(T_1)}{C_{hf}(T_1)} - 1\right]}{\left[1 - \frac{C_{hf}(T_1)}{C_d}\right]} , \qquad (3)$$

where $C_{hf}(T_I)$ is the specific high-frequency capacitance of the MIS
structure measured at temperature $T_I < T_{max}$. Indeed, the distribution
of deep level density $N_{tI}(W)$ may be obtained as effective distribution divided by the correcting factor (dW_I/dW) $(I - W_I/W)$.

The value of this factor may be obtained from the dependence of
the intercepting point between the deep level energy E_{tI} and the Fermi
level energy E_{fn} on the width of the depleting layer $W_I(W)$. According
to the previous paper /3/ it may be obtained by the following numerical integration:

$$\frac{\varepsilon_o \varepsilon_s}{q}\left[E_{fn}(N_d(x),T_1) - E_{t_1}\right] = \int_{W_1}^{W}(x-W_1)N_d(x)dx, \qquad (4)$$

where E_{fn} is the Fermi energy at temperature T_I. By numerical integration of equation (4) we may obtain $W_I(W)$, $dW_I/dW(W)$ and the correcting factor (dW_I/dW) $(I - W_I/W)$. In this integration we must use the
dependence $N_d(x)$ obtained from the C-V dependence at $T_o < T_{max}$ and
the value E_{tI} obtained from the TSCR spectrum.

If some centres with single energy levels in the band gap are
present then TSCR spectrum consists of the current peaks which do not
overlapp one with the other. In this case we may distinquish the
contributions of different levels by choosing the temperature of
measurements. The figure illustrates the calculated distribution of
deep level density as well as the free carrier density in the surface

region of silicon in the Al-SiO$_2$-Si structure at SiO$_2$ thickness of 850Å. The structure was previously implanted by P$^+$ ions at an energy of II0keV and implantation dose of I,25 I0^{I2}cm^{-2} and then was thermally annealed at 550°C for I5 min.

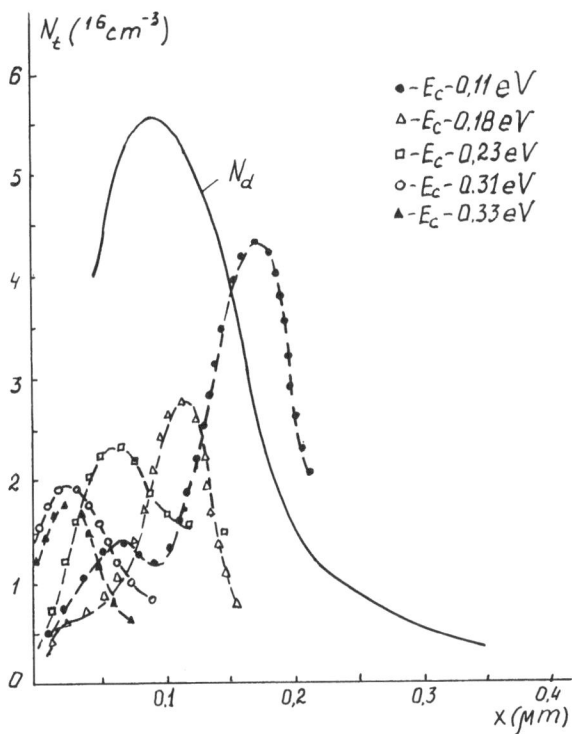

N_t (^{16}cm^{-3})

- •$-E_c-0,11$ eV
- △$-E_c-0,18$ eV
- □$-E_c-0,23$ eV
- ○$-E_c-0,31$ eV
- ▲$-E_c-0,33$ eV

N_d

x (µm)

Figure. Distribution of deep level density in the Si near-surface layer for P$^+$ ion implanted and then thermally annealed Al-SiO$_2$-Si structure.

REFERENCES

I. Lysenko V.S. and Nazarov A.N. Radiation damage in the near-surface layer of oxidized silicon ion-implanted by different impurities.-Poverhnost.I982,N5,8I-88 (in Russian)

2. Kuhn M.: A quasi-static technique for MOS CV and surface state measurment.-Sol.St.Electr.I970,I3,N6,873-885.

3. Qin G.G. and Sach C.T.: Theory of concentration profiling for semiconductor with many deep levels.-Sol.St.Electr.I982,25,NI0,I045-I053.

TRANSITION METAL IMPURITIES IN SILICON

C.A.J. Ammerlaan and T. Gregorkiewicz

Natuurkundig Laboratorium, Universiteit van Amsterdam

Valckenierstraat 65, 1018 XE Amsterdam, The Netherlands

Abstract

Magnetic resonance has made significant contributions to the charac-
terisation of point defects and small aggregates in semiconductors. A
particularly clear demonstration of the potential of this technique
is provided by research on transition metals in silicon. Basic atomic
and electronic structural information of these centres was revealed
by magnetic resonance. Two aspects of such studies, both dealing with
the degree of covalency of the 3d transition metal impurity iron in
silicon, will be discussed in the present paper. The fine structure in
the electron paramagnetic resonance (EPR) spectra is analysed by
taking into account the crystal field of the relevant symmetry and
spin-orbit interaction. Agreement with experimental data requires significant covalent delocalisation of impurity electrons. This is then
directly verified and confirmed by electron nuclear double resonance
(ENDOR). From the hyperfine interactions with the ligand 29-silicon
nuclei the spin density around the impurity is mapped in detail. Consistent with the observed delocalisation the central impurity hyperfine coupling is reduced considerably when compared with the free-ion
value.

1. Introduction

Magnetic resonance can provide detailed information on the atomic and
electronic structure of paramagnetic impurities in a diamagnetic host
crystal. By applying a magnetic field energy levels are split into
magnetic sublevels, an interaction known as the Zeeman effect. The
splitting is quantitatively described by the spectroscopic splitting
tensor, more commonly called the g-tensor. The tensor is usually measured by electron paramagnetic resonance (EPR). The structure of the
tensor directly provides a classification of the symmetry of the
defect in terms of a crystallographic system. The principal values of
the g-tensor will deviate from the free-electron value $g_e=2.0023$, for
a spin $S=1/2$ system, if orbital contributions to the magnetism are
present. For higher values of the spin, effects of orbital momentum

manifest themselves through crystal fields based on higher-order
interactions. In these cases the g-values bear information on the
orbital structure of the centre. This information is, however, usually
difficult to extract, as knowledge on wave functions and energy levels
of ground and excited states is required [1]. The use of the g-tensor
to gain a detailed insight into defect structure is therefore rare.
The present paper will attempt to carry out such an analysis for the
specific case of positively charged iron in silicon.

As opposed to the fine structure with its related g-tensor, the
interpretation of hyperfine interactions is much more straightforward.
These interactions, specified by the A-tensors, can be measured with
great precision and over a wide range of coupling strengths, by the
method of electron nuclear double resonance (ENDOR). From these hyperfine data the distribution of spin and charge around the impurity can
be mapped in great detail [2,3,4,5]. Section 3 of this paper will
present such an analysis, also for positive iron in silicon, and the
consistency with the results from the g-tensor data will be checked.

Besides the information based on experimental research methods, in
recent years the theoretical description of impurities in semiconductor hosts has made considerable progress. Recent calculations have
employed Green's function methods to solve the Schrödinger equation
[6], or the multiple scattering Xα method in a molecular cluster model
[7], all in a self-consistent manner. The detailed discussion of these
results is beyond the scope of the present paper.

2. Fine structure

2.1 Axial crystal field

Iron as an impurity in silicon will occupy interstitial lattice sites.
According to the Ludwig and Woodbury model all valence electrons are
transferred to the 3d-shell [8]. In the positive charge state the iron
ion has configuration $3d^7$. With parallel exchange coupling of the
three holes in the d-shell the spin of the centre is S=3/2. The triplet ground state has effective orbital momentum l'=1. Iron in this
isolated form has been observed by EPR [9,10]. Besides, EPR observations have been made for complexes where iron has formed an impurity
pair with an acceptor on a substitutional lattice site [10,11,12,13].
In table I the g-tensors of these centres are given. In addition, two
other centres, A27 and A28, which appear to have a related structure,
are mentioned [14]. In all cases the analysis of the EPR spectra has
used effective spin J=1/2. As table I shows an interesting set of g-

values, ranging from as low as 0.59 to as high as 6.389 is found in the experiments. Their interpretation is a challenge for the theoretical analysis.

Table I. Spectroscopic data for positive interstitial iron and related complexes in silicon.

Centre	Symmetry	g-Values			α	Δ_{ax} (meV)	Δ_{rh} (meV)
		g_z	g_x	g_y			
Fe	Cubic	3.524	3.524	3.524	-0.286	0	0
FeB	Trigonal	2.0676	4.0904	4.0904	-0.256	-16	0
FeAl(1)	Trigonal	6.389	1.138	1.138	-0.346	+43	0
(2)	Orthorhombic-I	5.885	1.236	1.612	-0.3	+66	13
(3)	Orthorhombic-I	1.73	2.51	5.36			
FeGa(1)	Trigonal	5.087	2.530	2.530	-0.284	+11	0
(2)	Orthorhombic-I	6.19	0.59	0.69			
(3)	Orthorhombic-I	2.02	3.37	4.65			
FeIn	Orthorhombic-I	2.070	3.78	4.40	-0.268	-16	1.3
A27	Monoclinic-I	1.96	3.24	4.78	-0.188	-13	2.6
A28	Monoclinic-I	2.15	4.10	4.20	-0.360	-18	0.3

Table II. Spectroscopic data for positive interstitial iron and related complexes in silicon, in axial approximation.

Centre	Symmetry	g-Values		α	x	$\Delta_{ax}/\alpha\lambda$	Δ_{ax} (meV)
		$g_\parallel$	$g_\perp$				
Fe	Cubic	3.524	3.524	-0.286	2	0	0
FeB	Trigonal	2.0676	4.0904	-0.256	7.513	-4.44	-16
FeAl(1)	Trigonal	6.389	1.138	-0.346	0.318	+8.71	+43
(2)	Orthorhombic-I	5.885	1.424	-0.165	0.420	+7.08	+17
(3)	Orthorhombic-I	1.73	3.935				
FeGa(1)	Trigonal	5.087	2.530	-0.284	0.933	+2.61	+11
(2)	Orthorhombic-I	6.19	0.64	-0.140	0.170	+17.91	+36
(3)	Orthorhombic-I	2.02	4.01				
FeIn	Orthorhombic-I	2.070	4.09	-0.256	7.444	-4.40	-16
A27	Monoclinic-I	1.96	4.01	-	-	-	-
A28	Monoclinic-I	2.15	4.15	-0.360	6.029	-3.52	-18

Although some of the centres actually have lower symmetry they will all be treated as axial in a first analysis. This assumption will allow an analytical more transparent treatment, which is presented first. Where necessary the tensors of the lower symmetry centres will be forced to appear as axial by averaging the two g-values which are already closely equal in several cases. An approximate perpendicular

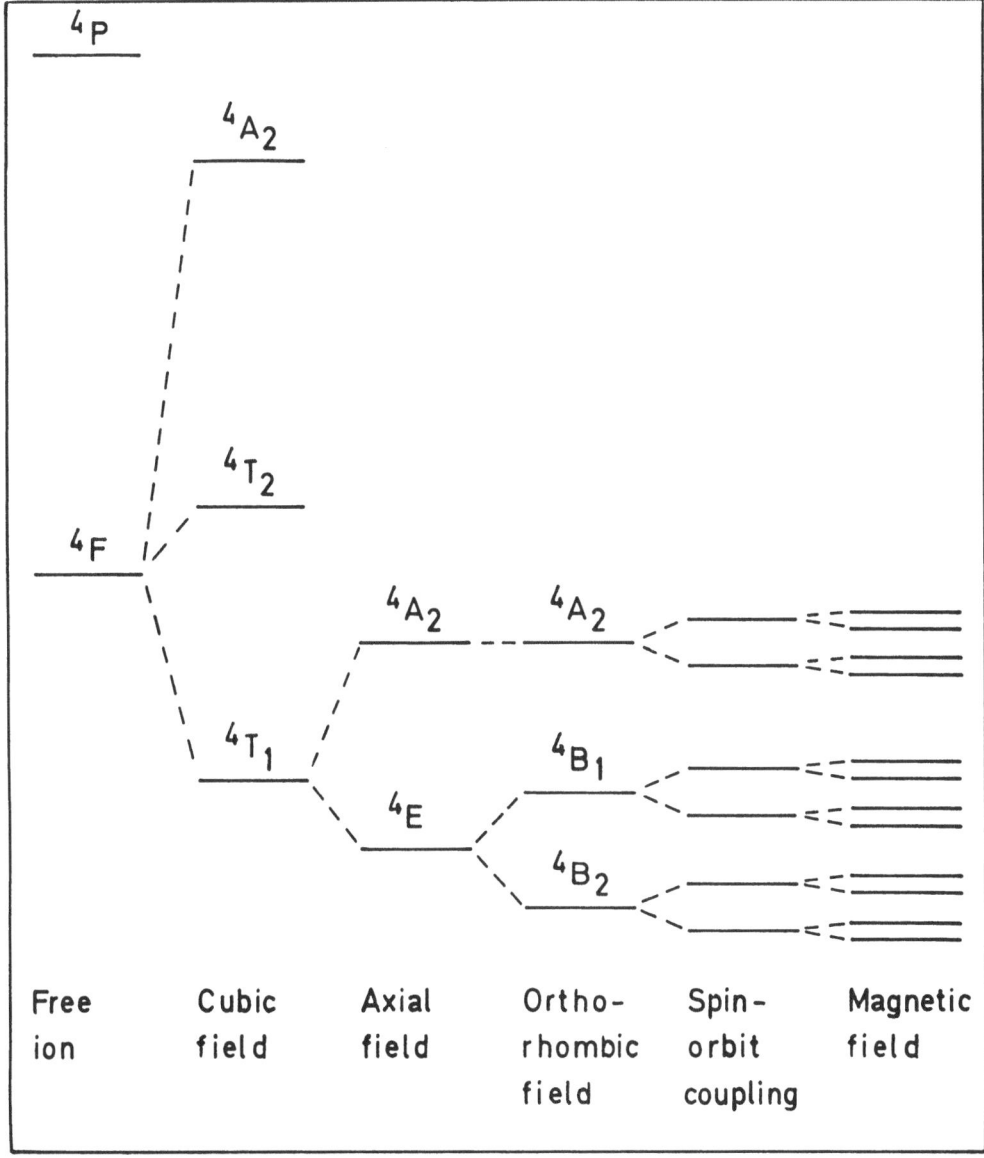

Fig. 1   Energy level diagram for the Fe^{+}-ion in electronic configuration $3d^{7}$.

g-value $g_\perp$ is defined as $g_\perp = (g_x + g_y)/2$. The resulting data are found in table II. The more general case taking account of the orthorhombic or monoclinic symmetry, which requires computer methods to be solved, will be discussed more briefly in paragraph 2.2.

Figure 1 presents a schematic energy level diagram of positive iron, electronic configuration $3d^7$, on an interstitial site in a silicon crystal. The free-ion ground state ^4F is split by the cubic crystal field (cf) leaving a 12-fold degenerate $^4T_1$ state lowest. In complexes of lower than cubic symmetry the crystal field, of trigonal or orthorhombic symmetry, will further split this level. Spin-orbit (so) interaction will finally lift the four-fold spin degeneracy, resulting in the splitting of the $^4T_1$ level into six Kramers doublets. These interactions are represented by the spin-hamiltonian H:

$$H = H_{cf} + H_{so}, \tag{1}$$

$$H_{cf} = +\Delta_{ax}(2/3 - 1'^2_z), \tag{2}$$

$$H_{so} = +\alpha\lambda(1'_x s_x + 1'_y s_y + 1'_z s_z). \tag{3}$$

The hamiltonian will operate on the twelve basis states of the $^4T_1$ ground state in cubic symmetry. This state which is orbitally three-fold degenerate, will have an effective angular momentum $1'=1$. The associated orbital g-factor giving the magnetic moment has the theoretical value $\alpha=-3/2$. To find eigenstates and corresponding energies the 12x12 matrix $\langle ^4T_1|H|^4T_1\rangle$ has to be diagonalised. Due to the Kramers degeneracy there are actually two identical 6x6 matrices. These are found from table III by setting Δ_{rh} equal to zero. By forming suitable linear combinations the matrix can be decomposed into one cubic, one quadratic and one linear matrix, as specified in table IV. The solution of the associated eigenvalue equations does not present any special problem. The energies E_1 to E_6 of the doublets as a function of the axial field strength Δ_{ax}, both in units $\alpha\lambda$, are presented as figure 2. For iron in $3d^7$ configuration the spin-orbit coupling constant λ is negative. Having found these solutions the effect of a magnetic field $\vec{B}$ can now be evaluated. The hamiltonian of the Zeeman effect, treated as a small perturbation on the doublets, has the orbital and spin parts

$$H_{mf} = +\alpha\mu_B\vec{B}\cdot\vec{1}' + 2\mu_B\vec{B}\cdot\vec{S}. \tag{4}$$

Table III. Matrix elements of the crystalline field, axial and orthorhombic, and spin-orbit coupling in the $^4T_1$ state.

	$\begin{array}{l}\|-x,+3/2\rangle\\\|+x,-3/2\rangle\end{array}$	$\begin{array}{l}\|iy,+3/2\rangle\\\|iy,-3/2\rangle\end{array}$	$\begin{array}{l}\|-x,-1/2\rangle\\\|+x,+1/2\rangle\end{array}$	$\begin{array}{l}\|iy,-1/2\rangle\\\|iy,+1/2\rangle\end{array}$	$\begin{array}{l}\|+z,-3/2\rangle\\\|+z,+3/2\rangle\end{array}$	$\begin{array}{l}\|+z,+1/2\rangle\\\|+z,-1/2\rangle\end{array}$
$\begin{array}{l}\langle-x,+3/2\|\\\langle+x,-3/2\|\end{array}$	$-\Delta_{rh}-\Delta_{ax}/3$	$-3\,\alpha\lambda/2$	0	0	0	$-\sqrt{3}\,\alpha\lambda/2$
$\begin{array}{l}\langle-iy,+3/2\|\\\langle-iy,-3/2\|\end{array}$	$-3\,\alpha\lambda/2$	$+\Delta_{rh}-\Delta_{ax}/3$	0	0	0	$-\sqrt{3}\,\alpha\lambda/2$
$\begin{array}{l}\langle-x,-1/2\|\\\langle+x,+1/2\|\end{array}$	0	0	$-\Delta_{rh}-\Delta_{ax}/3$	$+\alpha\lambda/2$	$-\sqrt{3}\,\alpha\lambda/2$	$+\alpha\lambda$
$\begin{array}{l}\langle-iy,-1/2\|\\\langle-iy,+1/2\|\end{array}$	0	0	$+\alpha\lambda/2$	$+\Delta_{rh}-\Delta_{ax}/3$	$-\sqrt{3}\,\alpha\lambda/2$	$-\alpha\lambda$
$\begin{array}{l}\langle+z,-3/2\|\\\langle+z,+3/2\|\end{array}$	0	0	$-\sqrt{3}\,\alpha\lambda/2$	$-\sqrt{3}\,\alpha\lambda/2$	$+2\,\Delta_{ax}/3$	0
$\begin{array}{l}\langle+z,+1/2\|\\\langle+z,-1/2\|\end{array}$	$-\sqrt{3}\,\alpha\lambda/2$	$-\sqrt{3}\,\alpha\lambda/2$	$+\alpha\lambda$	$-\alpha\lambda$	0	$+2\,\Delta_{ax}/3$

The magnetic field will lift the remaining Kramers degeneracy in the doublets as shown in figure 1. Usually, the magnetic resonance experiment is carried out between the levels originating from the ground state doublet. Figure 2 shows that irrespective of the sign or strength of the axial field the doublet E_1 always has lowest energy. For this ground state, which is derived from the cubic equation, the results for the g-tensor are summarised by

$$\Delta_{ax}/\alpha\lambda = -(x-2)(x+1)(x+6)/2x(x+2), \qquad (5)$$

$$E/\alpha\lambda = -(x^3+5x^2+13x+6)/3x(x+2), \qquad (6)$$

$$g_{\parallel} = +2[x^4+4x^3+14x^2+72x+72+2\,\alpha(x^2-12x-12)]/(x^4+4x^3+18x^2+24x+24), \qquad (7)$$

$$g_{\perp} = +4[x^4+4x^3+16x^2+24x-2\,\alpha x^2(x+2)]/(x^4+4x^3+18x^2+24x+24). \qquad (8)$$

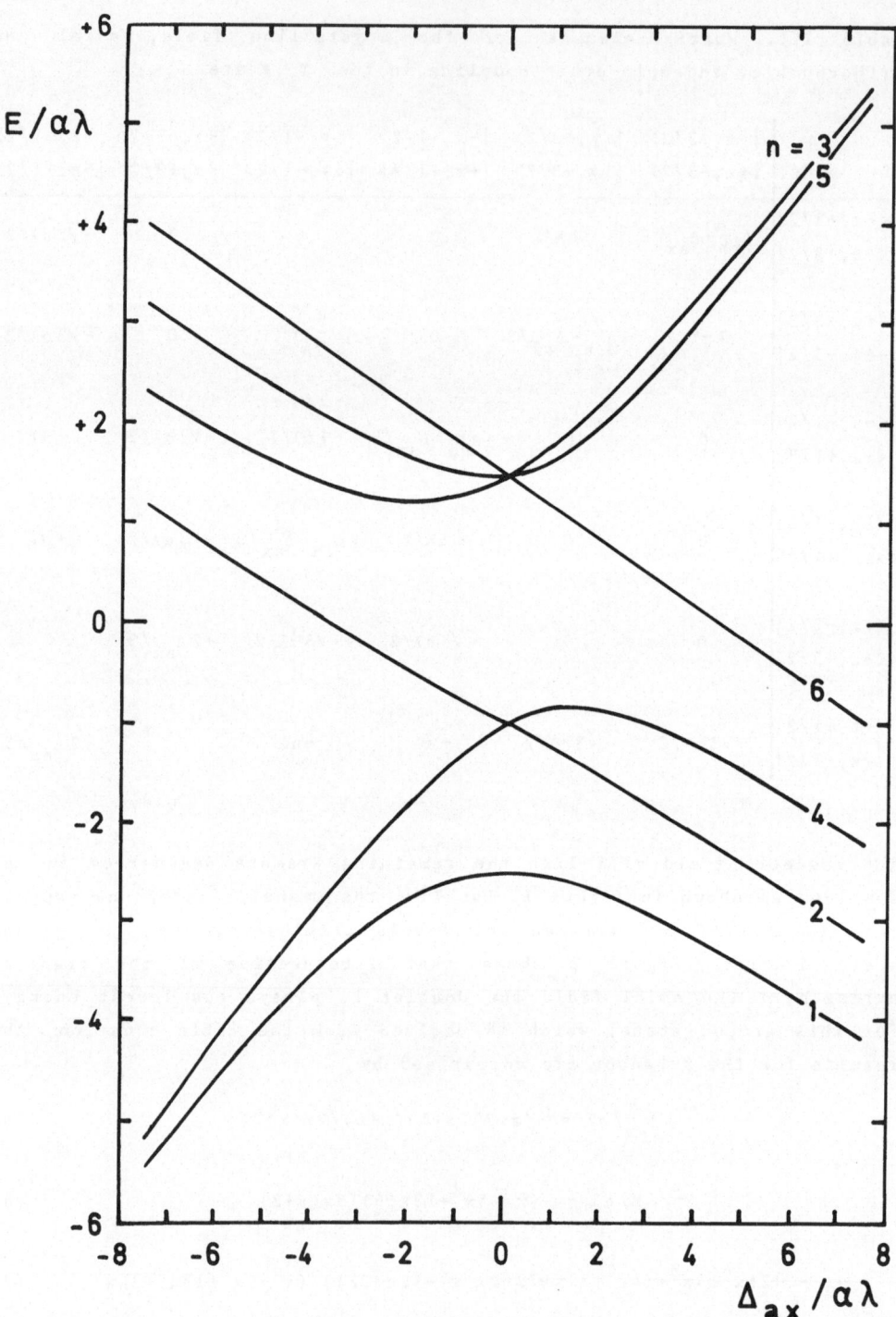

Fig. 2 Energy positions E_n, in units $\alpha\lambda$, of the six Kramers doublets, labelled $n=1,\ldots,6$, as a function of the trigonal crystal field, in reduced units $\Delta_{ax}/\alpha\lambda$.

Table IV. Matrix elements of the axial crystalline field and spin-orbit coupling in modified $^4T_1$ states.

	$\|-1,+3/2\rangle$ $\|+1,-3/2\rangle$	$\|0,+1/2\rangle$ $\|0,-1/2\rangle$	$\|+1,-1/2\rangle$ $\|-1,+1/2\rangle$
$\langle -1,+3/2\|$ $\langle +1,-3/2\|$	$-\Delta_{ax}/3 - 3\alpha\lambda/2$	$+\sqrt{6}\,\alpha\lambda/2$	0
$\langle\ 0,+1/2\|$ $\langle\ 0,-1/2\|$	$+\sqrt{6}\,\alpha\lambda/2$	$+2\,\Delta_{ax}/3$	$+\sqrt{2}\,\alpha\lambda$
$\langle +1,-1/2\|$ $\langle -1,+1/2\|$	0	$+\sqrt{2}\,\alpha\lambda$	$-\Delta_{ax}/3 - \alpha\lambda/2$

	$\|\ 0,+3/2\rangle$ $\|\ 0,-3/2\rangle$	$\|+1,+1/2\rangle$ $\|-1,-1/2\rangle$
$\langle\ 0,+3/2\|$ $\langle\ 0,-3/2\|$	$+2\,\Delta_{ax}/3$	$+\sqrt{6}\,\alpha\lambda/2$
$\langle +1,+1/2\|$ $\langle -1,-1/2\|$	$+\sqrt{6}\,\alpha\lambda/2$	$-\Delta_{ax}/3 + \alpha\lambda/2$

	$\|+1,+3/2\rangle$ $\|-1,-3/2\rangle$
$\langle +1,+3/2\|$ $\langle -1,-3/2\|$	$-\Delta_{ax}/3 + 3\alpha\lambda/2$

The dummy variable x relates the axial field Δ_{ax} to energy and g-values. A useful way to represent the result is the elimination of x between the equations (7) and (8), obtaining a direct relation between $g_\|$ and $g_\perp$. A graphical representation of the result is given in figure 3. The special case of cubic symmetry, with $\Delta_{ax}=0$ and x=2, leads to $g_\|=g_\perp=13/3$, and is applicable for isolated interstitial iron. In the upper left corner of the plot the solutions for $\Delta_{ax}/\alpha\lambda<0$, when the singulet level forms the ground state, are shown. The lower right part gives the solutions corresponding to $\Delta_{ax}/\alpha\lambda>0$ and the ^4E doublet ground state. Included in the figure are also the experimental data points, as taken from table II. Although a tendency of the measured g-values to follow the theoretical relationship is apparent, not an entirely satisfactory agreement is observed. The discrepancy may be explained by a reduction of the orbital contribution to the magnetism. Complete quenching, expressed by $\alpha=0$ in the formulas and in figure 3, leads to an underestimation of the g-values. Another curve, for $\alpha=-1$ can also be justified theoretically, as will be discussed in paragraph 2.4. A best agreement is obtained for the empirical value of the orbital g-factor $\alpha=-(0.3\pm0.05)$.

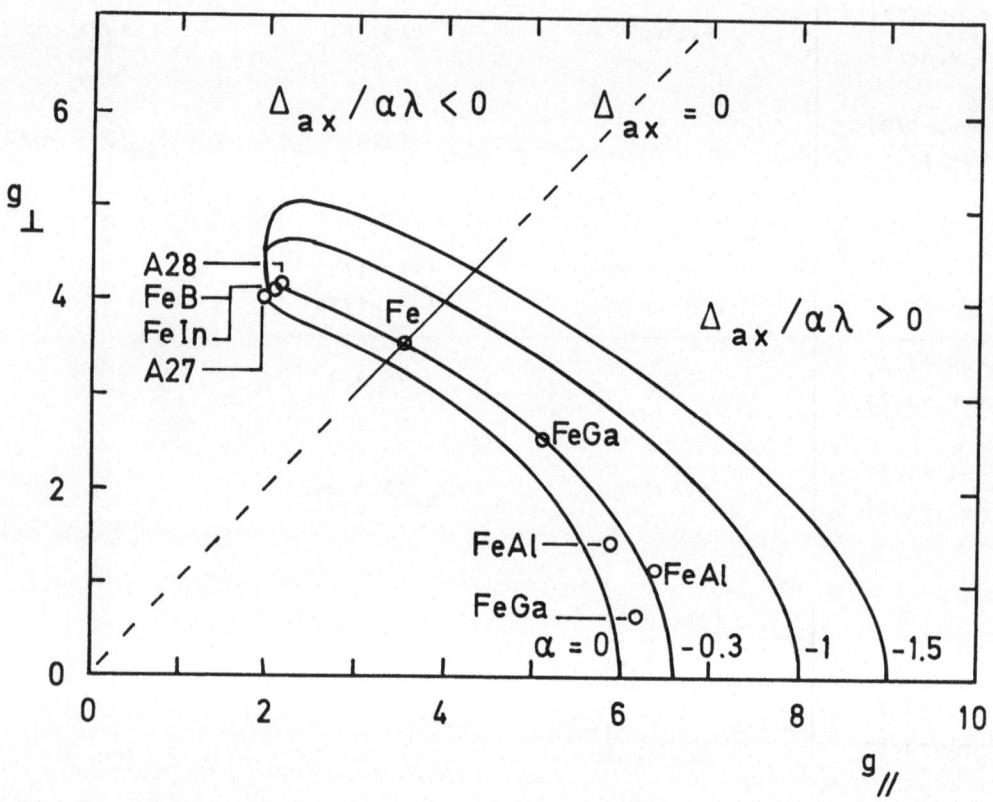

Fig. 3 Theoretical relations and experimental data for the Zeeman splitting factors $g_{\parallel}$ and $g_{\perp}$ for a $3d^7$ iron ion in an axial crystal field.

Some special attention may be given to the data points close to $g_{\parallel}=2$, $g_{\perp}=4$. The significance of these points may be questioned, as it is known that a spin quartet split by a strong axial field will always yield one doublet with $g_{\parallel} \approx g_e \approx 2$ and $g_{\perp} \approx 2g_e \approx 4$, in the J=1/2 formalism. To check more carefully on this aspect, the portion of figure 3 near $g_{\parallel}=2$ and $g_{\perp}=4$ is shown in close-up in figure 4. It shows clearly that also the data points for these cases require a value α=-0.3 for their interpretation.

Having selected the value α=-0.3 an alternative way of representing the results is shown as figure 5. Again, of course, the agreement is apparent. Values for $\Delta_{ax}/\alpha\lambda$ for which this agreement is obtained can

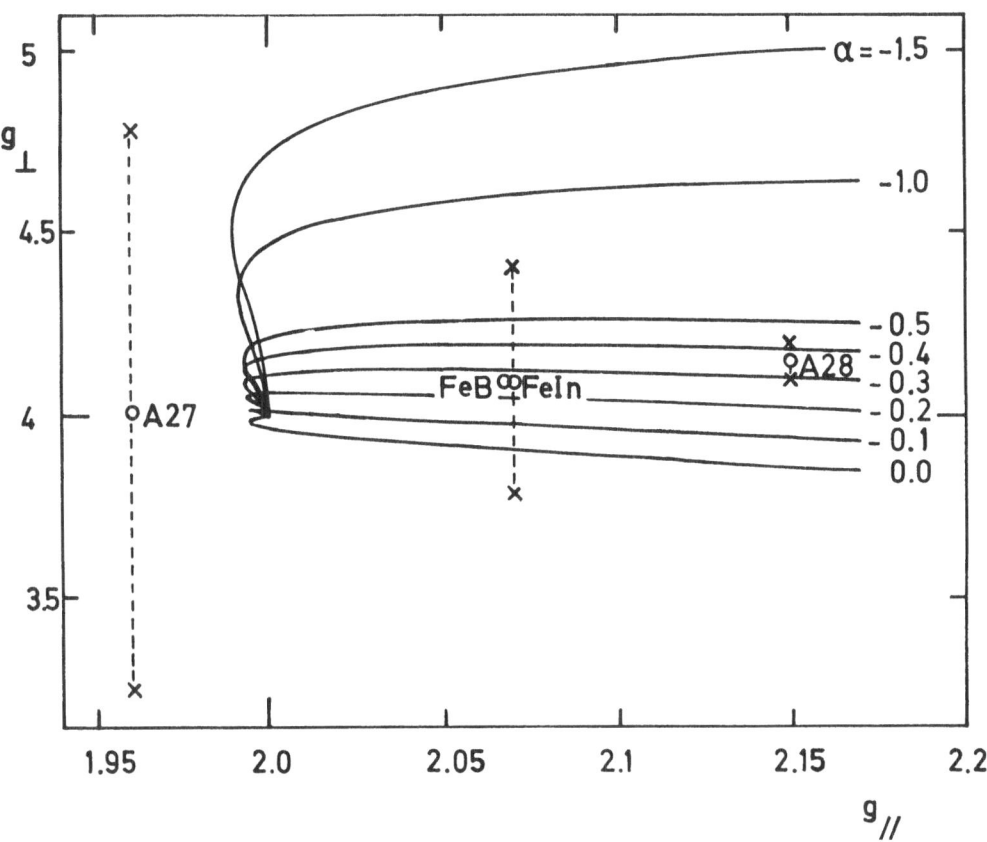

Fig. 4 Theoretical relations and experimental data for the Zeeman splitting factors $g_{\parallel}$ and $g_{\perp}$ for a $3d^7$ iron ion in a large axial crystal field with $\Delta_{ax}/\alpha\lambda<0$.

be read from the bottom scale. In table II the results of the analysis are summarised. To calculate the axial field Δ_{ax} the value λ=-14.3 meV was used for the spin-orbit coupling constant [15]. In the analysis as presented the two unknown quantities α and x, or alternatively α and Δ_{ax}, are calculated from the measured quantities $g_{\parallel}$ and $g_{\perp}$. If the equations allow for a solution, the agreement will then be exact.

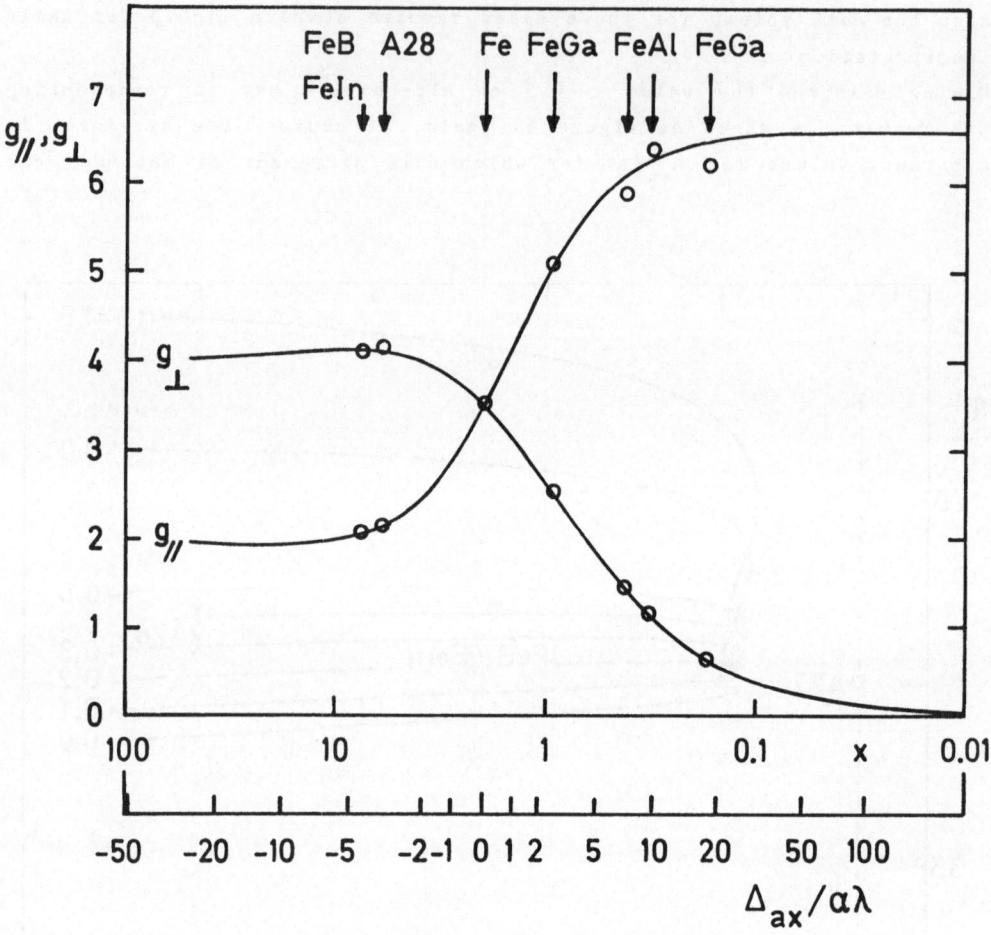

Fig. 5 Theoretical and experimental Zeeman splitting factors $g_{\parallel}$ and $g_{\perp}$ as a function of the axial crystal field, for an effective orbital g-factor $\alpha=-0.3$.

2.2 Orthorhombic crystal field

For this more general case the crystal field Hamiltonian H_{cf} takes the form

$$H_{cf} = +\Delta_{ax}(2/3-1_z^{'2}) + \Delta_{rh}(1_x^{'2}-1_y^{'2}),\qquad(9)$$

replacing equation (2). Operation of the orthorhombic hamiltonian on the basis states of $^4T_1$ will produce the matrixelements of table III. For solution of the eigenvalue equation numerical methods are required. A solution for the eigenstates can be written as

$$|+\rangle=a|-x,+3/2\rangle+b|iy,+3/2\rangle+c|-x,-1/2\rangle+d|iy,-1/2\rangle+e|+z,-3/2\rangle+f|+z,+1/2\rangle, \quad (10)$$

$$|-\rangle=a|+x,-3/2\rangle+b|iy,-3/2\rangle+c|+x,+1/2\rangle+d|iy,+1/2\rangle+e|+z,+3/2\rangle+f|+z,-1/2\rangle. \quad (11)$$

In terms of the coefficients a to f the principal g-values are derived as:

$$g_x = 2\langle+|(+\alpha l'_x+2S_x)|-\rangle=|-4c^2+4d^2+4f^2-4\sqrt{3}ac+4\sqrt{3}bd+4\sqrt{3}ef-\alpha(4be+4df)|, \quad (12)$$

$$g_y = 2i\langle+|(+\alpha l'_y+2S_y)|-\rangle=|+4c^2-4d^2+4f^2-4\sqrt{3}ac+4\sqrt{3}bd-4\sqrt{3}ef+\alpha(4ae+4cf)|, \quad (13)$$

$$g_z = 2\langle+|(+\alpha l'_z+2S_z)|+\rangle=|+6a^2+6b^2-2c^2-2d^2-6e^2+2f^2-\alpha(4ab+4cd)|. \quad (14)$$

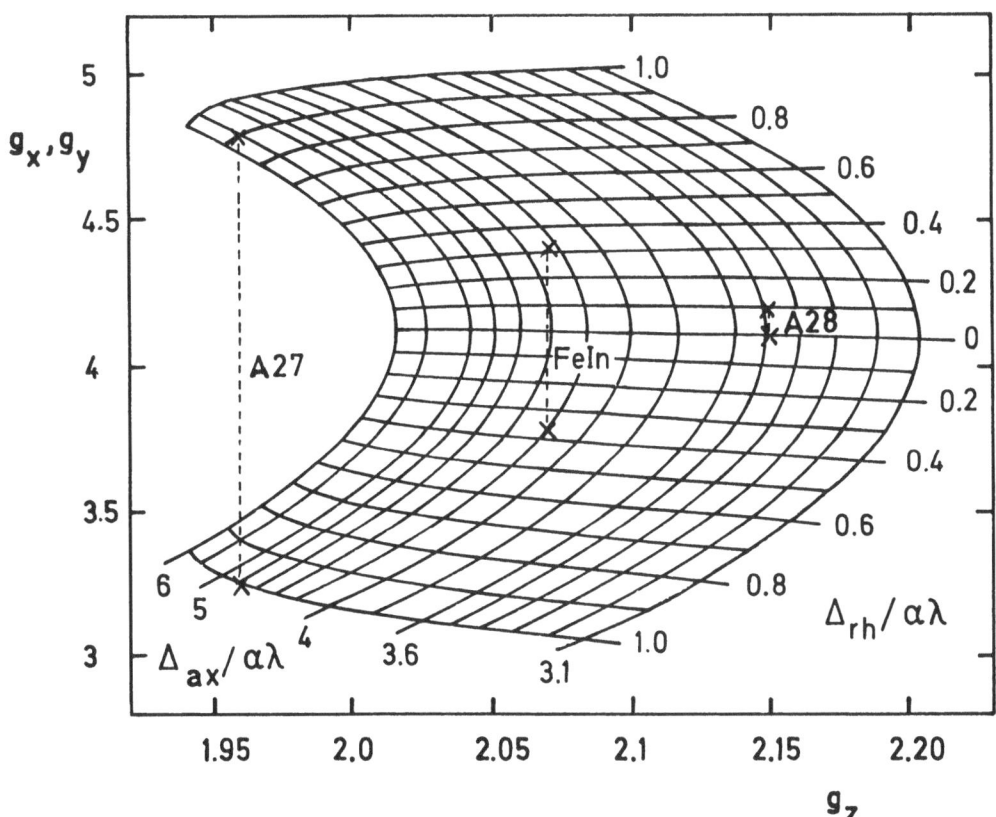

Fig. 6 Theoretical relations and experimental data for the Zeeman splitting factors near $(g_x,g_y,g_z) = (4,4,2)$, as a function of axial and orthorhombic crystal fields, for effective orbital g-factor $\alpha=-0.3$.

Again the number of unknowns to be determined, i.e. Δ_{ax}, Δ_{rh} and α equals the number of equations to be satisfied, i.e. for g_x, g_y and g_z. In the region near $(g_x, g_y, g_z) = (4, 4, 2)$ the solution is numerically stable. This can be illustrated by reference to the figures 4 and 6. From figure 6 it is concluded that Δ_{rh} is mainly determined by the difference $g_y - g_x$, rather independent of g_z. The average value of g_x and g_y essentially determines the factor α. The effects of α, Δ_{ax}, and Δ_{rh} are sufficiently orthogonal to allow the accurate determination of the parameters from the g-values. As a result the unambiguous solutions as given in table I are obtained for the centres A27, A28 and FeIn. The former centre for which it was not possible to find a solution in the axial approximation does not present any difficulty using the generalised crystal field of equation (9). The analysis of these orthorhombic and monoclinic centres confirms the reduced g-factor $\alpha = -0.3$. Unfortunately, in other regions of the parameter space the magnitude of α is strongly correlated with the crystal fields, and unambiguous solutions can not be obtained. For instance, the parameters $\alpha = -0.3$, $\Delta_{ax} = +66$ meV and $\Delta_{rh} = +13$ meV result in $g_x = 1.240$, $g_y = 1.609$ and $g_z = 5.885$. On the other hand, the quite different parameter selection $\alpha = -1.5$, $\Delta_{ax} = 1579$ meV and $\Delta_{rh} = 390$ meV gives $g_x = 1.238$, $g_y = 1.611$ and $g_z = 5.885$, representing a fit with comparable good agreement for the FeAl(2) spectrum. In these cases therefore the argument is reversed by requiring $\alpha = -0.3$ for consistency with other results. The crystal field parameters can then be determined.

2.3 Excited states

In figure 2 the energy levels are shown in their dependence on axial field and spin-orbit interaction. For positive values of the reduced axial field $\Delta_{ax}/\alpha\lambda$ the four doublets derived from the ^4E state are lowest in energy. For large values of the crystal field the levels approach equal separation by the amount $\alpha\lambda$. With the results of the previous analysis, $\alpha = -0.3$ and $\lambda = -14.3$ meV, this can be estimated as 4.3 meV, or T=50K in temperature units. Except at the lowest temperatures, the excited levels E_2, E_4 and E_6 may become appreciably populated thermally and resonance in them may be observable. For negative axial field the excited level E_4 even tends to coincide with the ground level E_1. Under such conditions the observation of resonances in both doublets derived from the $^4A_1$ spin quartet may be expected. Also, the assumption of small Zeeman energies compared to the doublet separation may then break down. This is more quantitatively discussed by Gehlhoff, et al. [13]. These authors actually report resonances in excited doublets for FeAl, FeGa and possibly FeIn.

2.4 Orbital g-factor

Obviously, the selection on empirical grounds of a strongly reduced effective orbital g-factor $\alpha=-0.3$ needs justification. The theoretical value $\alpha=-3/2$ reflects the transformation properties of a pure ${}^4T_1$ ground state in cubic symmetry. Any deviation from this description may lead to changes in α. In a first attempt the quenching of orbital momentum due to Jahn-Teller distortion may be considered. For isolated interstitial iron, which has cubic symmetry, with $\Delta_{ax}=0$ and $x=2$ in equation (5), the theoretical g-value' $g=(10/3)-(2/3)\alpha$ ranges from $g=13/3$ for $\alpha=-3/2$ without any quenching, to $\alpha=0$ and $g=10/3$ for complete quenching of the orbital moment. The experimental value $g=3.524$ thus corresponds to $\alpha=-0.286$ or 81% quenching. This reduction observed for iron in undistorted cubic symmetry was explained by Ham in a classical paper as a manifestation of a dynamical Jahn-Teller effect of the ${}^4T_1$ orbital triplet state [16]. Applying this concept to the pairs FeGa and FeAl one notes that also for these centres the successful analysis requires $\alpha=-0.3$. It is considered remarkable that these pairs with an E doublet ground state experience an equal Ham reduction factor as the triplet state for iron. Even more surprisingly, the pairs FeB, FeIn, and the A27 and A28 defects also have $\alpha=-0.3$. Since these centres have an A_2 orbital singulet ground state no Jahn-Teller instability and associated quenching is expected. On the basis of these results the Jahn-Teller mechanism as an explanation for the reduction of α appears unlikely. A recent theoretical analysis confirms this conclusion [17].

Modification of the ground state wave function by hybridisation offers another explanation for the reduction of α. Besides lifting the degeneracy of the ${}^4F$ free-ion ground state, the cubic field also has matrixelements between the ${}^4F$ and ${}^4P$ terms. Some p-character, derived from the ${}^4P$ term, will be admixed to the ${}^4T_1$ ground state [18]. An improved expression for the wave function is thus of the form

$$\psi = \eta_F \phi_F + \eta_P \phi_P, \qquad (15)$$

normalised by $\eta_F^2 + \eta_P^2 = 1$. The effective orbital g-factor associated with the hybridised wave function is

$$\alpha = -(3/2)\,\eta_F^2 + \eta_P^2. \qquad (16)$$

Admixture of p-functions through the cubic field has a maximum $\eta_P^2 = 0.2$.

Corresponding to this maximum, equation (16) gives a lower limit of -1 for α. Therefore, the effect of hybridisation, leading to $-3/2 \leq \alpha \leq -1$, may account for a reduction of α, but its possible effect is too small to fully explain the observed reduction to -0.3. In addition, the lower limit $\alpha = -1$ corresponds to an infinite cubic field. A more realistic estimate for the cubic field may give it a strength comparable to the ${}^4P - {}^4F$ splitting, which equals ≈ 1.4 eV [19]. This estimate raises the lower limit of α to ≈ -1.4.

Along similar lines the effect on the ground state wave function by covalent hybridisation with host atoms may be examined. Although the defect electrons will certainly be found in the impurity space, appreciable covalent delocalisation may occur. A recent experiment in which the electron distribution around the positive iron ion in silicon was measured by electron nuclear double resonance will be discussed in section 3 of this paper [20]. A suitable wave function may be constructed as a linear combination of 3d iron and ligand silicon orbitals:

$$\psi = \eta_{Fe} \phi_{Fe} + \eta_{Si} \phi_{Si}, \tag{17}$$

again normalised by $\eta_{Fe}^2 + \eta_{Si}^2 = 1$. In the impurity space the electrons will be described mainly by d-electrons with $\alpha = -3/2$ and, for iron, a spin-orbit coupling constant $\lambda = -14.3$ meV [15]. In the crystal around the impurity the 3p orbitals on the silicon atoms appearing in the expansion (17) will have $\alpha = +1$ and $\lambda = -20$ meV [21]. The effective orbital g-factor may be approximated by the weighted average

$$\alpha = -(3/2) \eta_{Fe}^2 + \eta_{Si}^2. \tag{18}$$

Agreement with the measured value $\alpha = -0.3$ is obtained for $\eta_{Si}^2 = 0.48$. This numerical example shows that the reduced orbital contribution can be understood by assuming considerable covalent delocalisation.

3. Hyperfine structure

3.1 LCAO analysis

To analyse the hyperfine interactions the one-electron wave function for the unpaired electrons is expanded in atomic orbitals, as indicated schematically in equation (17). At the central site the 3d orbitals of the iron impurity are included. The term $\eta_{Si} \phi_{Si}$ actually is a summation over hybrid 3s3p orbitals centered on silicon sites surrounding the impurity. For the analysis it is appropriate to decompose the measured hyperfine interaction tensor $\overset{\leftrightarrow}{A}$ into an isotropic part $a \cdot \overset{\leftrightarrow}{1}$, with $a=(1/3)\text{Tr}(\overset{\leftrightarrow}{A})$, and the remaining anisotropic tensor $\overset{\leftrightarrow}{B}$. The isotropic interaction a with a particular nucleus can then be related to contact spin density $|\psi(r)|^2$ on the site r of that nucleus. This is expressed by:

$$a = (2/3)\mu_o g_e g_N \mu_B \mu_N |\psi(r)|^2. \tag{19}$$

Anisotropic tensor $\overset{\leftrightarrow}{B}$ results from dipole-dipole interaction between the nucleus and electronic spin in the silicon 3p orbitals. The principal values of this axial tensor (2b,-b,-b) are related to the 3p-orbitals by

$$b = (2/5)(\mu_o/4\pi)g_e g_N \mu_B \mu_N \langle r^{-3} \rangle_{3p}. \tag{20}$$

Relations (19) and (20) allow the calculation of $|\psi(r)|^2$ and $\langle r^{-3} \rangle_{3p}$ valid for the defect electron wave function from the measured a and b. By comparison with tabulated atomic values for fully occupied orbitals [22] the coefficients in the expansion are obtained.

3.2 Impurity ENDOR

With the iron ion, i.e. with the 57-iron isotope which has nuclear spin I=1/2, the observed hyperfine interaction is isotropic. Its strength $a=2.985 \times 10^{-4}$ cm^{-1} has been measured accurately by ENDOR [23]. With equation (19) one finds $|\psi(0)|^2=0.25 \times 10^{30}$ m^{-3}, which is very small in comparison to the free-ion value $|\psi(0)|^2=5.65 \times 10^{30}$ m^{-3}. It leads to the conclusion that only about 5% of the electrons is accommodated in impurity orbitals. This result should, however, be considered with some reserve. The impurity d-orbitals themselves do not have any contact density. Their effect is only indirect through spin polarisation of the s-electrons in the core. Calculations of this process were performed by Watson and Freeman [24]. The accuracy of such calculations is not well known.

3.3 Ligand ENDOR

In a recent experiment the hyperfine interactions with ^{29}Si nuclei near the iron impurity were accurately measured by ENDOR [20]. The interactions with 98 atoms in 8 shells of symmetry related sites around the centre were resolved. Analysing these data in a one-electron model, using the equations (19) and (20), the coefficients of the expansion in silicon 3s and 3p orbitals are obtained. The total amount of spin density transferred to the surrounding crystal space is found to be 26%, indicating substantial covalency. More than 90% of this spin density is in silicon 3p orbitals. However, adding spin densities on impurity and silicon sites, one concludes that only just over 30% of the electrons have been revealed in the ENDOR experiment. This can be understood as ENDOR measures spin density whereas charge density is required. Spin, being a vector quantity, may cancel for different electrons, while charge directly adds. A measurement of spin density may therefore severely underestimate charge density. The figure of 26% reported above represents a lower limit.

4. Conclusion

An analysis of the fine structure, measured by electron spin resonance, and of the hyperfine structure, obtained from electron nuclear double resonance, for the positively charged interstitial iron impurity in silicon was presented. The g-values describing the Zeeman splitting of the impurity, either in isolated form or as part of a complex, could be understood by considering the action of crystal field and spin-orbit coupling on the $^4T_1$ ground state of a $3d^7$ configuration. The reduction of the orbital g-value is most likely not related to Jahn-Teller distortions, nor to intra-atomic hybridisation on the impurity. Covalent delocalisation of the defect electrons over silicon atoms in the vicinity of the impurity may have the predominant effect. The picture of substantial covalent character is supported by the low spin density found by ENDOR on the iron ion. In addition, the extended distribution of electrons over lattice atoms is evidenced by observations in ENDOR of hyperfine interactions with 98 silicon atoms. The accurate mapping of spatial extent of the wave function is unfortunately severely hampered by differences between the spin and charge distribution, which are apparent for this many-electron system.

References
1. M.H.L. Pryce, Proc. Phys. Soc. A 63, 25 (1950)
2. D.A. van Wezep, R. van Kemp, E.G. Sieverts and C.A.J. Ammerlaan, Phys. Rev. B 32, 7129 (1985)
3. D.A. van Wezep, T. Gregorkiewicz, E.G. Sieverts and C.A.J. Ammerlaan, Phys. Rev. B 34, 4511 (1986)
4. R. van Kemp, E.G. Sieverts and C.A.J. Ammerlaan, Phys. Rev. B 36, 3528 (1987)
5. E.G. Sieverts, D.A. van Wezep, R. van Kemp and C.A.J. Ammerlaan, Mat. Science Forum 10-12, 729 (1986)

6. H. Katayama-Yoshida and A. Zunger, Mat. Res. Soc. Symp. Proc. 46, 111 (1985)
7. L.V.C. Assali and J.R. Leite, Phys. Rev. B 36, 1296 (1987)
8. G.W. Ludwig and H.H. Woodbury, Phys. Rev. Lett. 5, 98 (1960)
9. H.H. Woodbury and G.W. Ludwig, Phys. Rev. 117, 102 (1960)
10. G.W. Ludwig and H.H. Woodbury, Solid State Physics 13, 223 (1962)
11. J.J. van Kooten, G.A. Weller and C.A.J. Ammerlaan, Phys. Rev. B 30, 4564 (1984)
12. W. Gehlhoff and K.H. Segsa, Phys. Stat. Sol. (b) 115, 443 (1983)
13. W. Gehlhoff, K. Irmscher and J. Kreissl, this volume
14. J.W. Corbett, private communication
15. T.M. Dunn, Transactions Faraday Soc. 57, 1441 (1961)
16. F.S. Ham, Phys. Rev. 138, A1727 (1965)
17. F.G. Anderson, Ph.D.-thesis, Lehigh University, 1987
18. A. Abragam and M.H.L. Pryce, Proc. Roy. Soc. (London) A206, 173 (1951)
19. C.E. Moore, Atomic Energy Levels, Natl. Bur. Stand. (U.S.) Circ. No. 467
20. J.J. van Kooten, E.G. Sieverts and C.A.J. Ammerlaan, to be published
21. G.D. Watkins, Phys. Rev. 155, 802 (1967)
22. J.R. Morton and K.F. Preston, J. Magn. Resonance 30, 577 (1978)
23. G.W. Ludwig and H.H. Woodbury, Phys. Rev. 117, 1286 (1960)
24. R.E. Watson and A.J. Freeman, in Hyperfine Interactions, edited by A.J. Freeman and R.B. Frankel (Academic, New York, 1967), p. 53

ELECTRONIC PROPERTIES OF PAIRS OF SHALLOW ACCEPTORS WITH IRON OR MANGANESE IN SILICON

W. Gehlhoff, K. Irmscher and J. Kreissl
Academy of Sciences of the GDR, Centre for Scientific Instruments
Rudower Chaussee 6, Berlin 1199, GDR

1. Introduction

Some fundamental statements concerning the chemical nature, geometry, and electronic structure of 3d transition metals (TM) and TM shallow acceptor pair centres were derived from the systematic EPR studies by Ludwig and Woodbury /1/ in the early sixties. Recently, theoretical considerations led to some doubt about the general validity of the so far accepted Ludwig-Woodbury model of TM /2,3/ as well as the pure ionic model of TM-acceptor pairs /4/ in silicon. Therefore, new experimental results in this field including TM-acceptor pairs are of great interest for the understanding of the electronic structure of these defects.Unambiguous identification and extensive characterization of defects require the combination of several measurement techniques.

In this work iron or manganese diffused p-type silicon was investigated by EPR and capacitance transient spectroscopy (DLTS). The EPR analysis of iron-acceptor pairs revealed new centres with orthorhombic symmetry. The DLTS results on these pairs clarify the severe problem to associate the measured defect properties to atomistically identified centres, because it is very difficult to ensure that actually the same centres are investigated with both methods. A successful centre identification and nearly complete electrical defect characterization is demonstrated on the example of the manganese-boron pair.

2. Iron-Shallow Acceptor Pairs

All up to now analysed EPR spectra of iron-shallow acceptor pairs can be phenomenologically described with an effective spin $S'= 1/2$ and their apparent g'-values are given in Table 1. Following /1/, it was believed for a long time that TM-acceptor pairs prefer a $\langle 111 \rangle$-axial arrangement with the only exception of the orthorhombic (FeIn) pair. But recently, also an orthorhombic configuration of the (FeAl) pair was detected at T= 4.2 K /6/.

Table 1: Effective g'-values of the EPR spectra of the Fe-acceptor
pairs in silicon (S'= 1/2)

trigonal ($\langle 111 \rangle$-axial)				orthorhombic		($x,y \parallel \langle 110 \rangle, z \parallel \langle 100 \rangle$)	
centre	$g'_\perp$	$g'_{\parallel}$		centre	g'_x	g'_y	g'_z
FeB	4.09	2.07	/1,5/	FeAl	1.25	1.61	5.88 /6,7/
FeAl	1.14	6.39	/6,7/		5.36	2.51	1.73 /7/
FeGa	2.53	5.09	/1/	FeGa	0.59	0.69	6.19 /+/
					4.65	3.37	2.02 /+/
				FeIn	4.40	3.78	2.07 /1/*

+ this work; * probably excited state

Own measurements /7/ confirmed this result (see Fig. 1, left) and
showed in addition the existence of a second set of lines at higher
temperature (see Fig. 1, right). Two similar orthorhombic line sets
with the g'-values given in Table 1 were found in iron-doped Si:Ga.

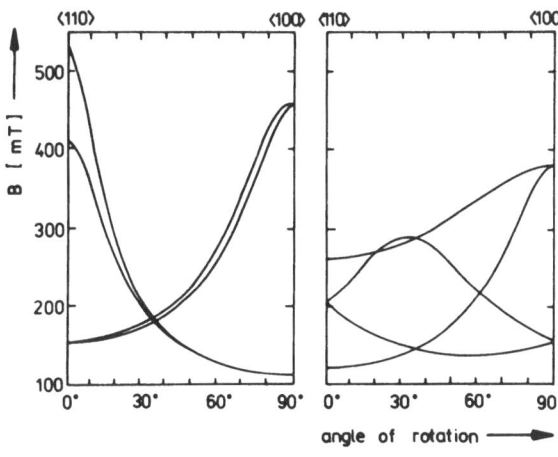

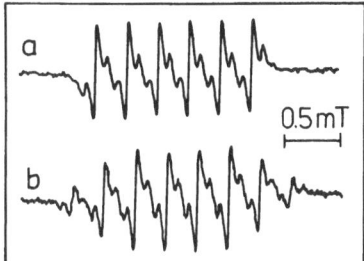

Fig.2:
Hyperfine structure of the
orthorhombic (FeAl) pair for
the EPR line at B= 263.3 mT
with $\vec{B} \parallel \langle 110 \rangle$ (excited state)
a) normal Fe-diffusion;
sixfold splitting due to ^{27}Al
(I= 5/2, 100 %)
b) enriched ^{57}Fe diffusion;
additional twofold splitting
due to ^{57}Fe (I= 1/2, $\approx$50%).
Further structure originate
from surrounding ^{29}Si nuclei
(I= 1/2, 4.7%).

Fig.1: Angular dependence of the X-band
EPR-line positions of the orthorhombic (FeAl) pair for rotation
of the magnetic field in a (110)
-plane. Transitions within the
ground doublet state on the left
and within the exited one on the
right.

A detailed EPR analysis reveals that they are due to the same FeAl and
FeGa centre, respectively. In the framework of crystal field theory the
observed apparent g'-values can be understood. The magnetic resonances
are mainly due to the Fe^+ ion $(3d^7, {}^4F)$ in a crystal field of orthorhombic symmetry, consistent with the assumption that the Fe^+ may be placed
at the next-nearest interstitial site. The observed two sets
of lines for the FeAl and FeGa pairs are caused by the spin transitions
within an isolated quartet ground state with a zero-field splitting
very large compared to the microwave quantum $h\nu$ $(2\sqrt{D^2+3E^2} \gg h\nu)$. The
observed g'-values $(S' = 1/2)$ correspond to the following actual g-
values and the fine structure parameters D and E

	g_x	g_y	g_z	E/D	D
FeAl	2.04	2.07	2.07	−0.253	≈ -7 cm^{-1}
FeGa	1.97	2.05	2.07	−0.09	

The pairing of one iron ion with one acceptor ion could be proved by
the well-resolved hyperfine structures due to the corresponding acceptor ion and by the additional twofold splitting of the whole spectrum
appeared in samples doped with enriched ^{57}Fe (Fig. 2,3).

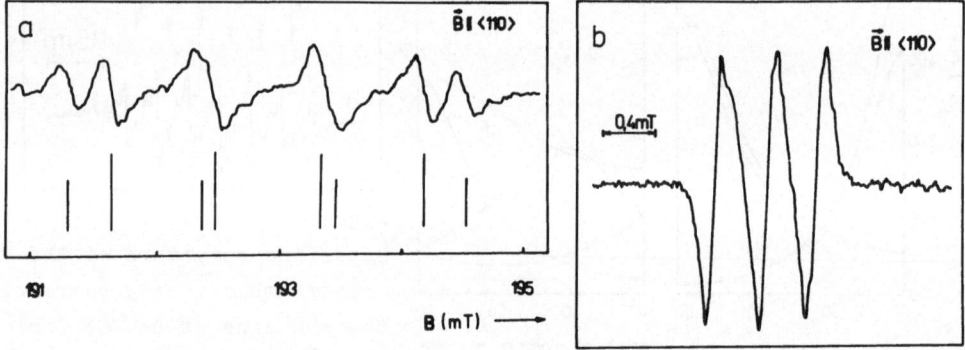

Fig.3: Hyperfine structure of the orthorhombic (FeGa) pair for $\vec{B} \parallel \langle 110 \rangle$
(excited state)
a) $\vec{B} \parallel x$; normal Fe diffusion; fourfold splittings due to ^{69}Ga
(I= 3/2, 60.4%) and ^{71}Ga (I= 3/2, 39.6%), respectively. The
ratio of the experimental hyperfine constants ^{69}Ga / ^{71}Ga
agrees with the corresponding one of the magnetic dipole mo-
ments.
b) $\vec{B} \parallel y$; enriched ^{57}Fe diffusion; additional twofold splitting
due to ^{57}Fe (I= 1/2, α 70%).

DLTS measurements of Chantre and Bois /8/ revealed a defect bistability
via the charge state controlled deep level concentrations of two levels
at E_v +0.13 eV and E_v +0.20 eV in Fe-doped Si:Al. They ascribed their
observations to a charge state dependent preference of the $Fe^{+/++}$ ion
on the nearest or next-nearest T_d-interstitial site to the
substitutional ion. For other Fe-shallow acceptor pairs Chantre and
Kimerling /9/ gave similar experimental evidences and explanations. Own
DLTS measurements of Fe doped Si:Al essentially confirmed the
experimental results of /8/. Though the EPR spectrum of the orthorhombic (FeAl) pair was detected in these samples an ultimate defect
identification can not yet be given. The observation of the configu-
rational bistability by EPR experiments failed up to now.

3. The Manganese-Boron Pair

Ludwig and Woodbury /1/ proposed a model, where an interstitially
incorporated Mn^{2+} ion is located in the immediate vicinity of a B^- ion
on a substitutional site, i. e. a $\langle 111 \rangle$-axial configuration. A more
detailed EPR analysis in /10/ supported this model and explained all
details of the complicated spectrum (for instance: isotopic effect in
the fine structure due to the boron isotopes ^{10}B and ^{11}B). Photo-EPR
hinted at a donor level at about E_v +0.6 eV.

DLTS measurements were accomplished with the same samples of Mn-
diffused p-type Si:B indicating the characteristic EPR spectrum of the
(MnB) pairs only. The only detected DLTS-peak originated from an electron trap near midgap (Fig. 4). It should be mentioned that the trap
was recharged by forward biasing of Schottky-diodes. This means that
there was a weak minority carrier injection. The concentration ratios
of (MnB) pairs determined by both methods for samples of different Mn-
dffusion temperatures or sample annealing at 375 °C agreed well.

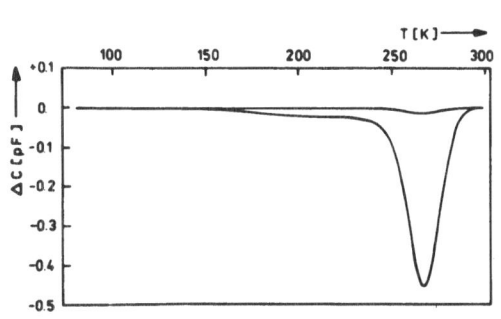

Fig.4:
DLT-spectrum of a Mn-doped
sample of 0.3 Ohm cm-p-Si:B
(emission rate of 500 s^{-1}).
No hole traps ($\Delta C > 0$) and
only one electron trap
($\Delta C < 0$) were detected.

Concerning the carrier capture and emission rates it was possible to
measure by DLTS the temperature dependence of the electron emission
rate and the hole capture coefficient, the latter from 80 K to 300 K
indicating a three order of magnitude variation. The carrier recombination parameters could be explained within the theory of non-radiative
multiphonon recombination processes. A quantitative fitting of the
measured data succeeded using Paessler's analytical expressions /11/.
The resultant configuration-coordinate diagram is depicted in Fig. 5.
The theoretical description especially explains the low value of the
hole capture coefficient and predicts a more than four orders of magnitude higher electron capture coefficient at the temperatures, naturally
interpreting the possibility of complete recharging of the (MnB) trap
by forward biasing Schottky-diodes.

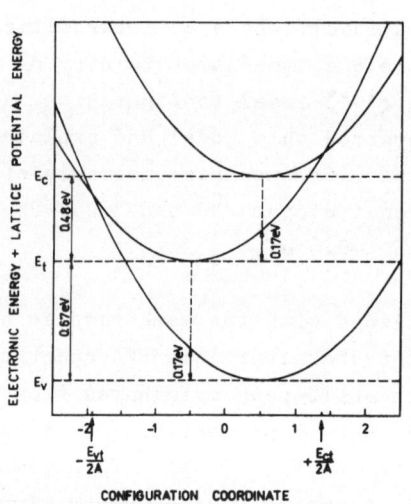

Fig.5:
Configuration coordinate diagram for the donor level of
the (MnB) pair in silicon as
deduced from the hole capture
and electron emission data.
($E_{vt} + E_{ct} = E_g$ (200 K) =
1.15 eV, lattice adjustment
energy of 0.17 eV, effective
phonon energy of 0.022 eV).

Lemke /12/ observed an electron trap using capacitance spectroscopy
exclusively which he assigned to the (MnB) pair. The own measurements
combining DLTS and EPR undoubtedly confirm his assumption. The unusual
DLTS features of this electron trap including the carrier capture
properties should be regarded as a unique "finger-print" of the (MnB)
pair.

References

/ 1/ G.W. Ludwig, H.H. Woodbury; Solid State Physics _13_, 223 (1962)

/ 2/ H. Katayama-Yoshida, A. Zunger; Phys. Rev. B31, 7877 (1985),

Phys. Rev. B31, 8317 (1985)

/ 3/ F. Beeler, O.K. Anderson, M. Scheffler; Phys. Rev. Lett.
_55_, 1498 (1985)

/ 4/ L.V.C. Assali, J.R. Leite; Phys. Rev. B36 (2), (1987)

/ 5/ W. Gehlhoff, K.H. Segsa; phys. stat. sol.(b) _115_, 443 (1983)

/ 6/ J.J. van Kooten, G.A. Weller, C.A.J. Ammerlaan; Phys. Rev.
B30, 4564 (1984)

/ 7/ K. Irmscher, W. Gehlhoff; Proc. Arbeitstagung "EPR-Spektroskopie"
Rohrbach (DDR) 1986, p. 205

/ 8/ A. Chantre, D. Bois; Phys. Rev. B31, 7979 (1985)

/ 9/ A. Chantre, L.C. Kimerling; Proc. ICDS-14, Materials Science
Forum _10-12_, 387 (1986)

/10/ J. Kreissl, W. Gehlhoff; phys. stat. sol. (b) _112_, 695 (1982)

/11/ R. Paessler; phys. stat. sol. (b) _85_, 203 (1978)

/12/ H. Lemke; phys. stat. sol. (a) _64_, 549 (1981)

MOCVD TECHNOLOGY

Lars Ledebo

Innovance AB

223 70 Lund, Sweden

Epitaxial growth of III-V semiconductors has been an important field of research for many years, and is now of significant commercial importance for the purpose of producing semiconductor devices. The techniques of Liquid Phase Epitaxy (LPE), Molecular Beam Epitaxy (MBE) and Metal-Organic Chemical Vapour Deposition (MOCVD) all have their special merits and it is highly unlikely that any one of them will ever totally replace any of the others.

Recent developments based on early experience in MOCVD has resulted in rapid progress, and a wide variation of materials and devices have been grown with MOCVD with excellent results. It is the intention in this paper to describe some new versions of epitaxy based on MOCVD in order to give an impression of the potential of the technique. However, this paper is not an attempt to give a review. Those more interested in the details of the background are referred to the review paper by Ludowise [1] for essentially III-V semiconductors, and papers by Mullin et al [2] and Cockayne and Wright [3] for information on MOCVD for II-VI materials. Also, the proceedings from the international conference series on MOCVD in Ajaccio 1981 [4], Sheffield 1984 [5] and Universal City 1986 [6] give valuable information in a concentrated form.

SOME BASIC FACTS

With GaAs as an example, the sum formula for epitaxy is:

$$(CH_3)_3Ga + AsH_3 \longrightarrow GaAs + 3 CH_4$$

The epitaxial GaAs is deposited on a substrate of GaAs, placed on a susceptor heated to something like 700 C. The Ga compound, trimethylgallium (TMGa) is a liquid at temperatures around room temperature with a vapour pressure of 65 torr at 0 C. It is transported as an undersaturated vapour by bubbling hydrogen gas through a steel cylinder containing the liquid, and is led by a carrier gas to a gas switching network (manifold), where TMGa and arsine are mixed and are rapidly transported towards the substrate by carrier gas flows. On their way to the substrate the gases hit the heated susceptor, which has a temperature that is high enough to make both arsine and TMGa disintegrate to As vapour and Ga-rich radicals. In a typical reactor cell the arsine will become only partially pyrolysed, whereas the pyrolysis of TMGa is more or less complete. The exact temperature of the substrate is normally not important, as the growth rate is determined by the availability of TMGa. Arsenic vapour should be present in excess. In a typical reactor well-behaved growth for GaAs will be obtained for an arsine/TMGa mole ratio of about 20, but this value is not critical. Depending on the exact geometry of the reactor, the carrier gas flow and the temperature distribution, different mole ratios give optimum results. Parameters from the literature are therefore of limited value when a different reactor geometry is to be tested.

Both hydrides and metal-organics can be made very pure. The strength of the MOCVD technique is that the reactivity and purity of the hydrides and the metal-organics are combined to epitaxially grow semiconducting materials. Hydrides of many group IV, V and group VI elements are readily available in high purity, even though their use in some cases is complicated due to high toxicity. Metal-organics of groups II and III, plus some group IV, group V and group VI elements are also readily available. Chemical engineering is used to produce molecules with a convenient vapour pressure for MOCVD at pratical bubbler temperatures in combination with a sufficiently high instability to readily pyrolyse at the growth temperature. Most metal-organics are pyrophoric, i.e. they react spontaneously in air. This is an unavoidable drawback caused by our condition for successful epitaxial growth that they should be unstable. Even though the pyrophoric nature of the metal-organics is a potential difficulty, in well-constructed reactors they can easily be handled.

In many laboratories in most industrialized countries of the world MOCVD has since some years been used to produce GaAs and AlGaAs heterostructures on GaAs substrates. Lasers, light emitting diodes and majority carrier devices all get high-quality properties. In a similar fashion InP, GaInAs and GaInAsP are grown on InP substrates. The results are generally good, even though the problem of lattice matching must be considered as with any alternative technique. A vast amount of other materials have been grown, and the abundance of scientific papers found in the literature on different alloys grown by MOCVD is a good illustration of the flexibility of the technique.

GaAs ON SILICON SUBSTRATES

Devices in GaAs, and also its alloy AlGaAs, have proven to be of high technical importance, and large volumes of such devices are now consumed in various sectors of importance for daily life in our society. Integration of such devices with each other, and with the intricate logical networks that today's silicon technology has made possible, could lead to new breakthroughs.

There are several reasons for developing the heteroepitaxy of GaAs on silicon. Firstly, the higher degree of integration requires larger areas of each substrate. Large-area GaAs substrates are not easy to grow with uniform properties. The edge of the wafer often has a high dislocation density, and a typical 4-inch diameter GaAs substrate has an average dislocation density of something like 10 000 per cm^2. This is in contrast to the more or less dislocation-free silicon 4-inch substrate which is routinely fabricated and consumed in large amounts within the silicon IC industry. Secondly, GaAs is more fragile than Si. The silicon industry has large amounts of money invested in process lines for silicon, and it is an advantage to use the same line for GaAs. However, when attempting this, it is found that the automatic handling and transport that works excellently for Si cause a large amount of the GaAs substrates to break. Thirdly, integration between the optoelectronics possible with III-V semiconductors and the logical network of silicon would be much simplified if it could all be made on the same substrate.

In the early days of III-V semiconductor epitaxy it would have been considered stupid to attempt to grow GaAs on Si, since the lattice mismatch is about 4 %. However both MBE and MOCVD can be used to deposite various semiconductor layers between the silicon substrate and the final GaAs. These layers contain misfit dislocations that gradually change the lattice constant to the right value for GaAs epitaxy.

Pulsed laser emission at 77 K from AlGaAs-GaAs double heterostructure lasers grown by MBE on a Ge-coated Si substate was reported in 1984 [7] by Windhorn et al. In 1985 laser emission at room temperature was reported [8] by Sakai et al. MOCVD was used to grow an initial nearly lattice-matched layer of GaP, followed by a double GaAs/GaAsP strained superlattice buffer region that accommodates the lattice mismatch. Finally the laser structure was grown. A similar strategy was employed by van der Ziel et al [9], also using MOCVD.

The first homoepitaxial GaAs lasers, which where made many years ago, had short lifetimes due to rapid degradation by dislocation multiplication induced by the high concentration of non-equilibrium carriers. Such lasers have now been developed to have probable lifetimes in excess of 10 years of continuous operation. Heteroepitaxial lasers made in GaAs on Si substrates still have the same problem as the early homoepitaxial lasers, and their lifetime is short due to dark-spot defects that originate in a dislocation density that is too high in the heteroepitaxial layers. It is, at this time, not clear whether it will be possible to bring down the dislocation density in GaAs/Si lasers to a sufficiently low level to permit longterm continuous operation. The same problem does not exist with majority carrier devices or devices like photodiodes where minority carriers have concentrations close to thermal equilibrium and for those devices the GaAs/Si heteroepitaxy is already sufficiently well developed to permit commercial products.

In any case, the strained superlattice of GaAsP/GaAs is of crucial importance for the successful growth of lattice-matched GaAs onto the silicon substrate. Since a phosphorous source is very difficult to use in MBE, the method that is now preferred for the fabrication of such superlattices is MOCVD.

INTERDIFFUSED MULTILAYER PROCESSING (IMP)

For some alloys, controlling the composition is very delicate in MOCVD. An example which is of great commercial interest and has also been studied extensively, mainly with the object of making infrared detectors, is $Cd(x)Hg(1-x)Te$ (CMT). The most favourable starting materials have been found to be Hg vapour from heated metallic Hg, Cd from DMCd and Te from DETe.

In CMT the bond between Hg and Te is much weaker than that between Cd and Te. Thus an excess of DMCd over DETe results in the formation of CdTe with little or no HgTe, irrespective of the concentration of Hg vapour. The composition x is in fact primarily controlled by the ratio DMCd/DETe. To obtain good morphology, certain constraints are imposed on this ratio, similar to the case of GaAs or AlGaAs for which V/III ratios of the order of 20 are required.

In addition, the rates of formation of HgTe and CdTe have different temperature dependences due to different pathways for the respective reactions.

In summary, the composition of the alloy $Cd(x)Hg(1-x)Te$ is found to be critically dependent on gas flow and temperature, making it difficult to achieve good lateral homogeneity in the layers [10]. To remedy this difficulty, the IMP process was devised [11,12].

In the IMP process, CdTe and HgTe are grown separately and succes-
sively as individual layers under the optimum conditions for each
compound. A practical layer thickness is of the order of 0.1 micro-
meter, which permits _interdiffusion_ to a sufficient homogeneity
in a time of the order of 10 minutes (at about 400 C). The alloy composition will simply be determined by the ratio of HgTe layer thickness to CdTe layer thickness. In order to ensure complete interdiffusion of the final alloy, a short period at the growth temperature
should be included at the end of each growth run.

Some disadvantages can be found with IMP. The lattice constant is
slightly different for HgTe and CdTe, which may lead to the formation
of misfit dislocations. Secondly, the interdiffusion will exclude
devices that are based on abrupt interfaces. In addition, repeated
gas switching is needed, which requires a reactor that is not only a
first order machine, but a little more advanced. However, in spite of
these potential disadvantages the IMP process has produced results
for CMT that are excellent when compared with other techniques [10].

In a good but not untypical sample from work by Whiffin et al.
[13] the composition x of the alloy varies by no more than 0.002 over
most of the sample area. Heterostructures do indeed lack abrupt
interfaces, but this could, in some cases, be argued to become an
advantage due to the disappearance of band discontinuity spikes. From
IMP CMT, a 32 x 32 starring array has been fabricated by Mullard.
This was the first focal plane array in MOCVD material [14]. However,
CMT infrared detectors of excellent performance have been made also
with non-IMP processes in MOCVD [15], but will probably require a
much closer control of flow patterns, temperatures and wafer screening than the IMP equivalent.

The IMP method is of course not restricted to CMT, but can also be
used with other alloys with high self-diffusion coefficients. It is
thus to be expected that it could be successfully applied to many
II-VIs, particularly the systems CdSeTe and CdZnTe, however not to
III-V alloys.

ALTERNATIVES TO ARSINE AS THE ARSENIC SOURCE

The use of arsine, AsH_3, for the As source, has, since the beginning of MOCVD, been the single most problematic issue with the technique. In the early days, the purity of the arsine was the major problem, presumably due to water adsorbed on the cylinder walls. Howe-

ver, for commercial arsine of the highest purity that can be found today this is normally not a problem. Instead, the major concern has become the toxicity of arsine. The arsine is stored in high-pressure cylinders, with a potential threat of widespread contamination and hazard if a leak or reactor error occurs. It is supposed that the maximum permissible concentration for arsine in air is 3 ppm for a single event of 1 hour human exposure [16]. Exposure in excess of this, or repeated exposure, will lead to serious symptoms. The problem is similar for phosphine, PH_3, however the corresponding maximum permissible concentration is supposed to be 7 ppm for a single event 1 hour exposure [16]. Another difficulty with both arsine and phosphine is the unsuitably slow pyrolysis rate at temperatures below 550 C [17], which are sometimes of interest for VPE growth processes.

For all of these reasons, significant effort is being devoted to the replacement of arsine with an alternative arsenic source. The simplest solution, the use of a solid As source, is not suitable for growth of device quality GaAs and AlGaAs by MOCVD [18]. The resulting material has morphological problems and is contaminated by carbon. The carbon presumably originates from the TMGa. It has been postulated [19] that atomic H from the arsine is responsible for the removal of CH_3 radicals adsorbed at the growth interface from the final stages of TMGa pyrolysis on the GaAs surface. One CH_3 radical and one hydrogen atom form methane, which escapes from the surface into the vapour. Since the elemental As provides no atomic hydrogen, C incorporation is dramatically increased. Similar problems, or worse, are apparently encountered with the use of trimethylarsenic (TMAs) or triethylarsenic (TEAs) even though some promising results have been obtained [20, 21].

Results using diethylarsine (DEAs) are more encouraging [22]. This molecule contains two ethyl radicals and a hydrogen atom, which seems to be efficient in removing the CH_3 radicals. Even more efficient radical removal is expected using tertiary-butylarsine (TBAs), with one butyl radical and two hydrogens bonded to the As [23]. The TBAs pyrolyses at lower temperatures, removing the need for extremly high V/III ratios at low growth temperatures when using arsine. Also, since TBAs is a liquid with a fairly low vapour pressure at room temperature, it is much safer to use than arsine. The toxicity as such is not known, but is expected to be considerably less than that of arsine [23]. Similar improvements have been obtained using tertiary-butylphosphine (TBP) for the growth of InP [24].

However, the initial phase of using a new source material is far from straightforward. Before device-quality wafers can be produced, data on vapour pressure, pyrolysis and toxicity must be obtained. In addition, before a source is consumed in large volumes there is no driving force for optimizing the purity, and early batches of sources may be contaminated by various dopants. Thus the evaluation of a new source material is complicated. It is therefore to be expected that arsine and phosphine will remain to be the most important group V sources for MOCVD growth for a long time to come.

GROWTH STIMULATED BY UV ILLUMINATION

The addition of energy in the form of photons to the region of growth can have marked effects and permit epitaxy at reduced temperatures. To observe the effect, normally UV light must be used. The reduced temperature decreases smear-out of compositional and dopant profiles. In addition, UV assisted growth opens up the possibility of localized deposition without masking, which may be important for pattern definition in integrated circuits and make possible the definition of three-dimensional hidden structures that can be produced without interrupting the growth cycle.

It is found that the reactants can conveniently be stimulated by an excimer laser with emission at 193 nm. In an experimental arrangement of the type shown in figure 1 the growth rate has been determined with and without UV broad-area illumination, and is shown in figure 2 [25].

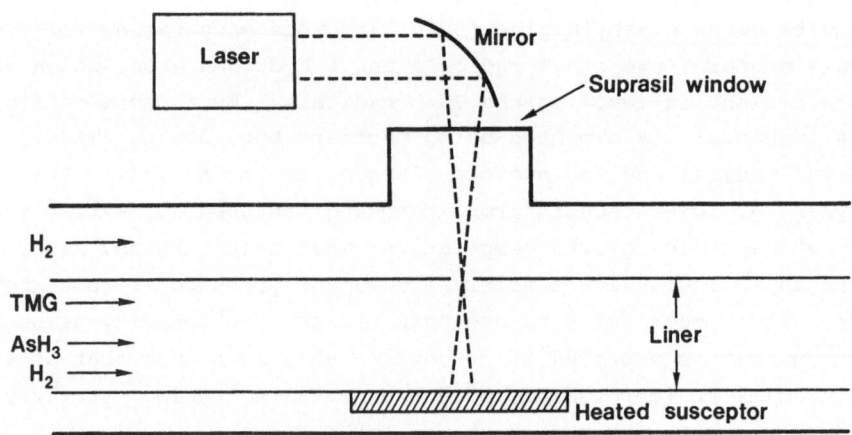

Fig. 1. Example of an experimental arrangement for UV-stimulated growth.

The contrast in growth rate, as presented in figure 2, between illuminated and non-illuminated areas becomes observable at low temperatures, where the decomposition of TMG is the rate-limiting step for
the growth. The contrast is seen not to be very high, but it is
important to note that the experiment reported is far from optimized
with respect to contrast. As an example, it is probable that most of
the UV radiation was absorbed at a considerable distance from the
substrate.

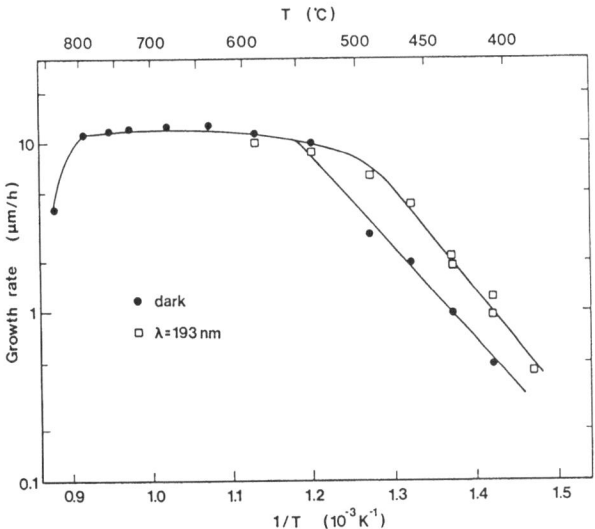

Fig. 2. Growth rate in the dark and with UV laser
illumination [25].

It is of course of interest to understand the detailed mechanisms
which are responsible for the increased growth rate with UV illumination. The following are the main possibilities:

1. A photon is absorbed by a molecule of the starting material,
which transfers the molecule to an excited state. The excited state
has a short lifetime, and the molecule dissociates to a final state
of importance for the growth. This is photolytic stimulation.

2. A photon is absorbed without producing an excited state of importance for the growth rate. The absorption can take place in the
gas phase or in the substrate. The absorption increases the temperature locally. Therefore the growth rate is increased for temperatures
at which a thermally activated step is controlling the growth rate.
This is pyrolytic stimulation.

3. The generation of free carriers in the substrate might change

the fermi level in the surface of the substrate. This may change the growth rate, particularly for steps in the reaction that increase or decrease the number of free carriers. Such a mechanism may be called fermi-level-controlled stimulation.

It is probable that photolytic stimulation is active in the experiment illustrated in figure 2. It is likewise probable that pure pyrolytic stimulation is active in the experiment with ALE growth described in figure 3, as also in recent experiments with direct laser writing for localized growth of GaAs on GaAs [26].

ATOMIC LAYER EPITAXY - ALE

ALE has been used with excellent results for several years for mainly II-VI semiconductors, and has commercial applications for electro-luminescent displays based on ZnS. For ZnS the ratio between the sticking coefficients for say Zn on a S or a Zn surface is large. Therefore, with a fresh Zn surface, a pulse of S vapour gives exactly one monolayer, given enough time. Also, with some simplification, the fresh S surface when exposed to a Zn pulse is covered by exactly one monolayer of Zn. Growth with excellent control of thickness and homogeneity can thus be attained with alternating pulses of S and Zn vapour.

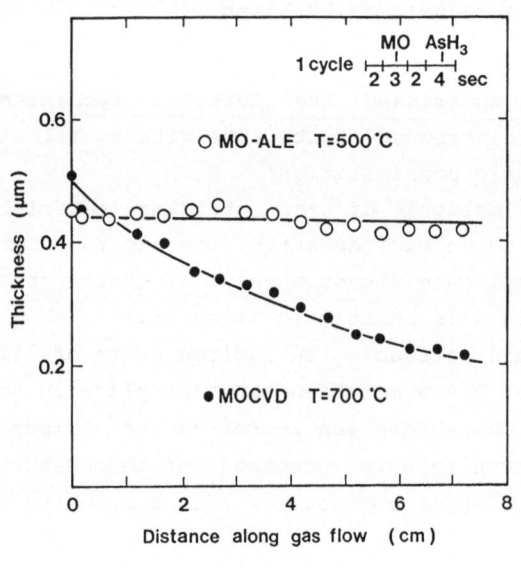

Fig. 3. MO-ALE growth at 500 C with reduced pressure to 100 torr gives excellent homogeneity in thickness [27].

With GaAs the ratio between the sticking coefficients is not that large, and at the normal growth temperatures in MOCVD Ga is deposited at about the same rate irrespective of whether the fresh surface is Ga-rich or As-rich. However, at low enough temperatures a region of ALE-mode growth can be found. Figure 3 demonstrates this, with an excellent thickness homogeneity as a result over the full diameter of a 75 mm diameter GaAs wafer. For comparison, the thickness distribution is also shown for the same reactor with normal MOCVD. Obviously the ALE method has potential for large batch production of GaAs epitaxial wafers, which is one of the reasons for the amount of work being put into the technique at present.

The ALE mode type of growth gives an interesting possibility of planar doping, since the dopants can be introduced in a chemical pulse with the result that the dopant is included only into a single atomic layer. Since the temperature of growth is low, diffusion of dopants is low, and very abrupt doping profiles can be obtained [28].

It is possible to further decrease the temperature of growth in the ALE mode by stimulating the decomposition of TMG by illumination with laser light [29, 30], which could add interesting possibilities within selective area growth.

CHEMICAL BEAM EPITAXY

Chemical Beam Epitaxy (CBE) is a rapidly developing mixture of the MBE and MOCVD processes, and combines the high vacuum and analysis facilities of MBE technology with the advantages of using MOCVD type sources. In the CBE modification of MBE the source elements are partially or completely replaced by chemical compound sources [31, 32]. The technique is alternatively with CBE referred to as MOMBE (Metal-Organic Molecular Beam Epitaxy), CMBE (Compound Molecular Beam Epitaxy) or gas-MBE, with the choice of name sometimes made depending on which traditional sources that are replaced with MOCVD-type sources.

One of the most obvious difficulties in MOCVD is the control of the gas flow pattern in the reaction chamber in order to avoid stagnant volumes and to obtain a homogeneity in thickness as well as composition and doping. By using low pressure, typically in the range of 20 to 200 torr, rather than atmospheric pressure, these disadvantages are reduced, but still remain.

In MBE one of the dominating difficulties is to find chemically suitable source materials of sufficiently high purity. It is difficult to find a phosphorous source at all. Phosphorous is reactive and loading of a traditional effusion cell is highly impractical. Very little MBE work has thus been done with semiconductor alloys containing phosphorous. Generally, the limited source capability is a major concern, particularly in production environments since source replacement means long down-time due to the high-vacuum requirements of MBE.

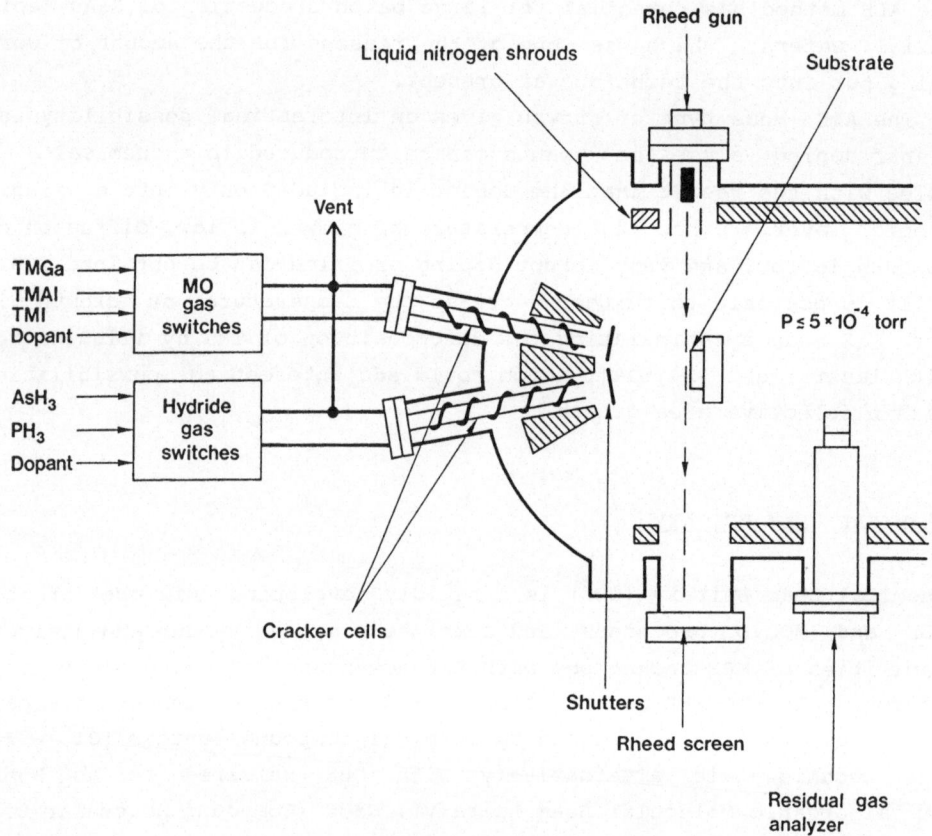

Fig. 4. A schematic diagram of a CBE reactor.

The above difficulties in MOCVD and MBE are resolved in CBE, which is schematically presented in figure 4. A clean substrate surface is obtained by using high vacuum rather than hot hydrogen gas as in MOCVD. The vacuum chamber looks much like a traditional MBE chamber, but the different metal-organics are introduced through the same port, and the hydrides through another port, thus greatly reducing the need in MBE of large-diameter multiport systems for the growth of alloys. The chemical beams are directed towards the substrate where the growth takes place at a total pressure in the 10^{-4} mbar range. Practical substrate temperatures range from about 500 to 700 C for GaAs, and for InP from about 400 to 600 C. These temperatures are below the temperatures needed for efficient cracking of arsine and phoshine. Thus a cracker cell for the hydrides is needed, operating at something like 850 C. The decomposition rate for the metal-organics is normally sufficiently rapid at the surface of the substrate at practical growth temperatures. The growth rate in such cases is limited by the arrival rate of the metal-organics to the surface of the substrate. To permit the growth of alloy systems with abrupt interfaces either mechanical shutters or a pressure balanced vent/run system is needed. Both in combination may be desirable. Hydrogen or some inert gas can be used as the carrier gas.

Some excellent results have been obtained with the CBE technique [33, 34], indicating the usefulness of the concept for research as well as production. However, the disadvantages of high cost and impracticability of operation due to the need for a high vacuum remain, both inherited from the pure MBE.

REFERENCES

1. M J Ludowise, J Appl Phys 58, R31 (1985)
2. J B Mullin, S J C Irvine and J Tunnicliffe, J Crystal Growth 68, 214 (1984)
3. B Cockayne and P J Wright, J Crystal Growth 68, 223 (1984)
4. J Crystal Growth 55, no 1 (1981)
5. J Crystal Growth 68, no 1 (1984)
6. J Crystal Growth 77, no 1-3 (1986)
7. T H Windhorn, G M Metze, B Y Tsaur and J C C Fan, Appl Phys Lett 45, 309 (1984)
8. S Sakai, T Soga, M Takeyasu and M Umenu, Jpn J Appl Phys 24, L666 (1985)
9. J P van der Ziel, R D Dupuis and J C Bean, Appl Phys Lett 48, 1713 (1986)
10. J B Mullin, J Giess, S J C Irvine, J S Gough and A Royle, in MRS Symposia Proceedings, 90, 367 (1987), editors R F C Farrow, J F Schetzina and J T Cheung
11. S J C Irvine, J Tunnicliffe and J B Mullin, Mat Lett 2, 305 (1984)
12. J Tunnicliffe, S J C Irvine, O D Dosser and J B Mullin, J Crystal Growth 68, 245 (1984)

13. P A C Whiffin, B C Easton, P Capper and C D Maxey, J Crystal Growth 79, 935 (1986)

14. J B Mullin, S J C Irvine, J Giess and A Royle, J Crystal Growth 72, 1 (1985)

15. L T Specht, W E Hoke, S Oguz, P J Lemonias, V G Kreismanis and R Korenstein, Appl Phys Lett 48, 417 (1986)

16. K L Hess and R J Riccio, J Crystal Growth 77, 95 (1986)

17. J Nishizawa and T Kurabayashi, J Electrochem Soc 130, 413 (1983)

18. R Bhat, J Electron Mater 14, 433 (1985)

19. T F Keuch and E Veuhoff, J Crystal Growth 68, 148 (1984)

20. D W Vook, S Reynolds and J F Gibbons, Appl Phys Lett 50, 1386 (1987)

21. D M Speckman and J P Wendt, Appl Phys Lett 50, 676 (1987)

22. R Bhat, M A Koza and B J Skromme, Appl Phys Lett 50, 1194 (1987)

23. C H Chen, C A Larsen and G B Stringfellow, Appl Phys Lett 50, 218 (1987)

24. C H Chen, C A Larsen, G B Stringfellow, D W Brown and A J Robertson, J Crystal Growth 77, 11 (1986)

25. P Balk, M Fischer, D Grundmann, R Lyckerath, H Lyth and W Richter, to be published 1987

26. N H Karam, N A El-Masry and S M Bedair, Appl Phys Lett 49, 880 (1986)

27. K Mori, A Ogura, M Yoshida and H Terao, Solid State Device and Materials Conference 1986, Tokyo

28. N Kobayashi, T Makimoto and Y Horikoshi, Appl Phys Lett 50, 1435 (1987)

29. A Doi, Y Aoyagi and S Namba, Appl Phys Lett 48, 1787 (1986)

30. A Doi, Y Aoyagi and S Namba, Appl Phys Lett 49, 785 (1986)

31. W T Tsang, Appl Phys Lett 45, 1234 (1984)

32. M B Panish, H Temkin and S Sumski, J Vac Sci Techn B3, 657 (1985)

33. M Frei, D C Tsui, W T Tsang, Appl Phys Lett 50, 606 (1987)

34. H Temkin, D Gershoni and M B Panish, Appl Phys Lett 50, 1776 (1987)

EPITAXIAL GROWTH OF PbTe DOPING SUPERLATTICES
ON (111) BaF$_2$ AND (100) GaAs

B.Tranta and H.Clemens
Institut für Physik, Montanuniversität,
A-8700 Leoben, Austria

1.)   Introduction

Doping superlattices have been the subject of intensive investigation during the past years. Esaki and Tsu /1/ have suggested that a modulation of the conduction and valence band edges can be achieved either
by compositional superlattices or by a periodic sequence of doping layers
in a homogeneous semiconductor. Döhler /2/ published several theoretical
papers on this subject. The first realization of this idea was performed
by Ploog /3/ in 1980 using GaAs. Subsequently, PbTe was used as a host
material for this type of superlattices. So far, the Hot Wall technique
was used to grow a periodic sequence of n- and p-doped PbTe layers /4,5/.
The purpose of this paper is to present data on the doping mechanism in
PbTe for this application as well as RHEED (Reflection High Energy E-
lectron Diffraction) studies in order to get information about the surface properties by an in-situ control.

2.)   Realization of PbTe doping superlattices

N- and p-doped region in PbTe can be realized by two different methods:
Deviation from stochiometry, either by Pb or Te vacancies causes p-
and n-type conductivities, Fig.1. Each of these resonant vacancy levels
releases either two holes or two electrons into the valence and conduc-
tive band, respectivily.

Type of diffusion	Diffusand	D_{PbTe} at 650°C (cm^2/sec)
p/n junction	p into n	$4*10^{-11}$
	n into p	$4*10^{-9}$

Table I: Diffusion coefficient for Pb and Te diffusion
in PbTe /6/

However, according to published data on the interdiffusion (see Table I.)

it would be difficult to produce abrupt interfaces between the n- and p-regions with this method. The data by Walpole et al./6/ were obtained on /6/ highly doped ($>10^{18}$cm^{-3}) n- and p-type PbTe in form of single p-n junction. Due to the interdiffusion mechanism involving vacancies a dependence of diffusion constant on vacancy concentration and vacancy compensation has to be anticipated.

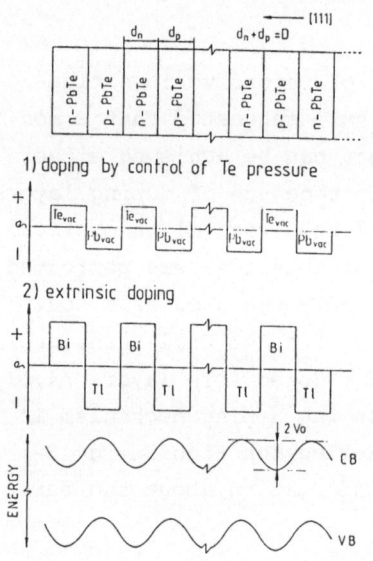

Fig.1. Top : PbTe doping superlattice ('nipi') of period D, schematically. The p- and n-layers contain donors and acceptors.

Middle: The positive and negative space charge can be realized through doping by control of Te pressure or for higher doping levels by extrinsic doping with Bi and Tl.

Bottom: Modulation of the conduction and valence band edges by the periodic space charge.

From our experimental data /4/ and those by Kinoshita et al. /5/, PbTe nipi's can be grown with individual periods of 500-1000Å and doping levels of the order of $5*10^{17}$cm^{3} without severe interdiffusion problems if the growth temperature is kept close to 350°C. This method of doping by a control of stochiometry using an additional Te-furnace, by which the Te partial pressure is adjusted, has the additional advantage that mobilities in excess of 10^{5}cm^{2}/Vs at T≈5K are possible in the nipi-structures. /4/.

The second method, shown also in Fig.1 uses an extrinsic doping by Bi and Tl for n- and p- type conduction. Bi$_2$Te$_3$ and Tl$_2$Te are used as the doping agents at temperatures both between 300-350°C /7/.

As demonstrated by Partin /8/, abrupt interfaces and doping profiles on a length scale of about 10 Å are possible with these dopants. With Bi and Tl doping concentrations of the order of $2*10^{18}$cm^{3} are readily achievable, thus causing a higher quasiparabolic band edge modulation $2V_0$ for the same layer width. The mobilities are, however, lower then those acheived by intrinsic doping approximatly by a factor of 2.

The deposition of PbTe on BaF$_2$ substrate is effected by cleavage steps of the (111) cleavage planes. Especially for multilayer growth these steps cause problem with electronic measurements, since the overgrowth follows the BaF$_2$ surface as shown by Clemens et al. /9/. So, it is always necessary to carefully choose areas within the nipi surface are free from cleavage steps in order to obtain reasonable electronic data.

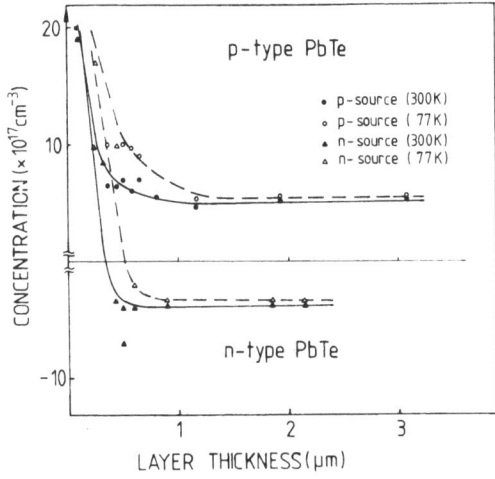

Fig.2.: Dependence of the electron and hole concentration from the film thickness. Note: only for a film thickness >4000 Å occurs n-type conduction.

As shown in Fig.2, the concentration of the electrons and holes in films deposited on BaF$_2$ depends critically on the thickness. For films which are less thick then of about 4000 Å, the initial island growth apparently does not allow n-type conduction. For films with thicknesses of about 1000 Å, where the islands merge together, the carrier concentration is more or less independent on the status of the source materials of the order of 10^{18}cm^3 (p-type conduction).

3.)   Control of the PbTe growth by in-situ RHEED investigation

A typical example of a sequence of RHEED patterns for the growth of PbTe on BaF$_2$ is shown in Fig.3. For the [$\bar{1}$10] azimuth was observed initially island growth with a three-dimensional pattern. A single orientation of the PbTe layer on BaF$_2$ is observed and for the [111] growth direction any evidence of twinning is absent. For larger thicknesses a streaked pattern can be observed, which indicates a rather smooth PbTe surface. For an interruption of the PbTe growth by closing the corresponding shutter for 2-3 seconds the RHEED pattern does not change.

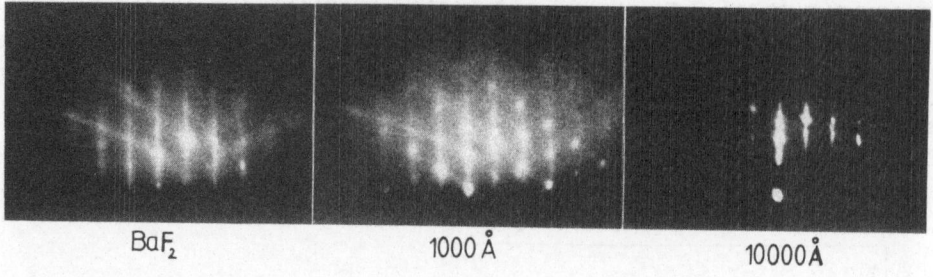

Fig.3.: RHEED patterns during deposition of PbTe on (111) BaF₂ [01Ī]-azimuth.
    Left : BaF₂ substrate after preheating.
    Middle: initial stage of PbTe overgrowth.
    Right : streak patterns appear for a 1μm thick PbTe
            layer indicating a smooth surface.

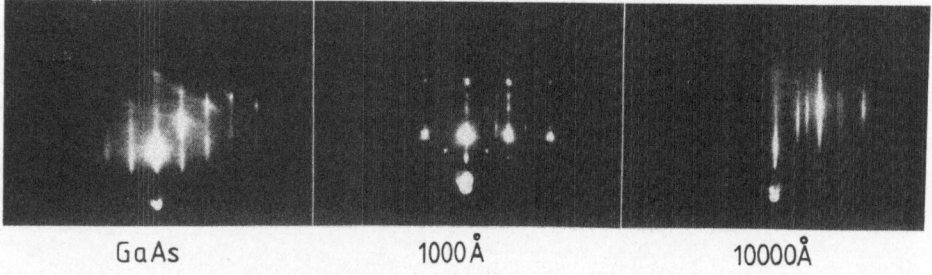

Fig.4.: RHEED patterns during deposition of PbTe on (100) GaAs [01Ī]-azimuth.
    Left : GaAs substrate prior to deposition.
    Middle: initial stage of growth. Two different orientations,
            (100) and (111) can be observed.
    Right : 1um thick PbTe layer. Only the (100) orientations
            remain, [01Ī] and [001] || [01Ī].

For the deposition of PbTe on (100) GaAs the RHEED pattern is much more
complicated, Fig.4. At substrate temperature of about 300°C both (111)
as well as (100) overlayer growth can be observed. For larger thickness
when already streaks appear and the growth is along a [001] direction
still two different orientations normally [01Ī] and [001] PbTe || [01Ī]
GaAs were observed. Despite of the fact that for thicker PbTe layers one
orientation remains and a parallel overgrowth is possible, there are serious complicates. The (100) GaAs layer offers of course perfect flatness
but the difference in the thermal expansion coefficients between the
growth temperature and room temperature is so large between PbTe and GaAs
that the mechanical strain is released by generation of microcracks.
Since for nipi structures total thicknesses of about 2-3 μm are necessary (10-15 periods , D≈1000 Å), the build-up of the elastic energy

with such thicknesses is to high. Therefore other growth modes like initial deposition at 300°C subsequent growth at 100°C would be necessary to avoid the formation of cracks.On the other hand, a deposition of PbTe nipi's on CdTe layer deposited epitaxially on GaAs is another choice.

Acknoledgements

We thank Prof.G.Bauer, H.Krenn and P.Ofner for helpful and stimulating discussions and H.Ulrich for expert technical assistance. This work is supported by "Fonds zur Förderung der wissenschaftlichen Forschung" (project P5321), Vienna, Austria.

References

1. L.Esaki and R.Tsu, IBM J.Res.Develop. 14, 61(1970).
2. G.H.Döhler, Phys.stat.sol. (b) 52, 79 (1972), (b) 52, 533 (1972).
    G.H.Döhler, Surface Science 73, 97 (1978).
3. K.Ploog and Döhler, Advances in Physics 32, 285(1983).
4. W.Jantsch, G.Bauer, P.Pichler and H.Clemens,
    Appl.Phys.Lett. 47 738(1985).
5. H.Kinoshita, T.Sakashita and H.Fujiyasu, J.Appl.Phys. 52,
    2869(1981).
6. J.N.Walpole and R.L.Guldi, in "Physics of IV-VI-Compounds
    and Alloys", Ed.: S.Rabii, Gordon and Breach Sciencie
    Publishers, New York (1974) p.167.
7. H.Clemens, Ph.D.thesis, unpublished
8. D.L.Partin, Appl.Phys.Lett., 45 487(1984)
9. H.Clemens, E.J.Fantner, W.Ruhs and G.Bauer
    J.Crystal Growth 66, 251(1984)

ENERGY DISTRIBUTION OF INTERFACE STATES IN GaAs-Cr/Au SCHOTTKY CONTACTS OBTAINED FROM I-V CHARACTERISTICS

Zs. J. Horváth

Research Institute for Technical Physics

of the Hungarian Academy of Sciences,

Budapest, P.O.Box 76. H-1325, Hungary

Abstract

The effect of the presence of an interfacial layer and interface states on the I-V characteristics of Schottky contacts is analyzed. A simple evaluation of the interface state energy distribution and the interfacial layer thickness to its relative dielectric constant ratio (the relative interfacial layer thickness) is presented for three different models concerning the quasi Fermi level positions in the Schottky contacts. The validity and usability of these models are discussed and the first experimental results evaluated from the I-V characteristics of n-type (100) GaAs-Cr/Au mesa Schottky structures are presented.

Introduction

Recently the effect of the presence of an interfacial layer and interface states on the I-V characteristics of Schottky contacts was analyzed, and a simple evaluation of the interface state energy distribution and the interfacial layer thickness to its relative dielectric constant ratio (the rela-

tive interfacial layer thickness) was proposed [1,2]. In this
paper the possible models, concerning the quasi Fermi level
positions in Schottky contacts, suitable for the explanation
of experimental I-V characteristics, are treated, and their
usability for the evaluation of the above parameters are discussed. The first experimental results are presented.

Analysis

The I-V characteristics of the most Schottky contacts may
be described by the thermionic emission theory [3-6]:

$$J = A^{***} T^2 \exp\left(-\frac{q\phi_b}{kT}\right) \left[\exp\left(\frac{qV}{kT}\right) - 1\right] \tag{1}$$

where J is the current density, A^{***} the effective Richardson
constant including the transmission probability across the
interfacial layer, T the temperature, q the electron charge,
ϕ_b the barrier height, k the Boltzmann constant and V the
applied voltage.

Assuming that A^{***} does not depend on the applied voltage, for the derivatives of the lnI-V characteristics one can
get:

$$\frac{d\ln J}{dV} = \frac{1}{J}\frac{dJ}{dV} = \frac{q}{kT}\left\{1 - \frac{d\phi_b}{dV} + \left[\exp\left(\frac{qV}{kT}\right) - 1\right]^{-1}\right\} \tag{2}$$

The slope for forward direction may be characterized with
the ideality factor n (V>3kT/q) [3,4] which on the basis of
(2) may be expressed as

$$n = -\frac{q}{kT}\frac{1}{\dfrac{d\ln J}{dV}} = \frac{1}{1-\beta} \tag{3}$$

where

$$\beta = d\phi_b / dV \tag{4}$$

For the slope of the reverse characteristics (V<3kT/q)
one obtains:

$$s = \frac{d\ln J}{dV} = \frac{q\beta}{kT} \tag{5}$$

In the presence of an interfacial layer the applied
voltage is divided among it and the semiconductor:

$$V = V_I + V_S \tag{6}$$

where V_I and V_S are the voltage drop on the interfacial layer and the semiconductor respectively. V_I is connected with the maximum electric field in the semiconductor E_{MAX}:

$$V_I = \frac{\delta}{\mathcal{E}_I} \left(\mathcal{E}_S E_{MAX} - \frac{Q_{SS}}{\mathcal{E}_0} \right) \tag{7}$$

where δ is the interfacial layer thickness, $\mathcal{E}_I$ and $\mathcal{E}_S$ the relative dielectric constants of the interfacial layer and the semiconductor respectively, Q_{SS} the charge in the interface states located at the interfacial layer - semiconductor interface (including fixed charge) and $\mathcal{E}_0$ the dielectric constant of vacuum.

Using the model proposed by Card and Rhoderick [3], i.e. assuming that a part of the interface states with charge Q_{SSA} and density D_{SA} is in equilibrium with the metal, and the other part with charge Q_{SSB} and density D_{SB} with the semiconductor, the contribution of the interfacial layer and the interface states to the barrier height lowering with the applied voltage may be expressed on the basis of (6) and (7) as

$$\beta_I = \frac{dV_I}{dV} = \frac{\dfrac{\mathcal{E}_S}{w} + \dfrac{qD_{SB}}{\mathcal{E}_0}}{\dfrac{\mathcal{E}_I}{\delta} + \dfrac{\mathcal{E}_S}{w} + \dfrac{q(D_{SA}+D_{SB})}{\mathcal{E}_0}} \tag{8}$$

where w is the depletion depth in the semiconductor:

$$w = (dE_{MAX}/dV_S)^{-1} = [2\mathcal{E}_0 \mathcal{E}_S (\phi_K - V_S)/(qN)]^{1/2} \tag{9}$$

Here N is the free carrier concentration and ϕ_K the built-in potential.

Multiplying both the numerator and denominator in (8) with $\mathcal{E}_0$ one gets β_I as a function of the depletion, the interfacial layer and the interface state capacitances.

Possible models

The investigated n-type GaAs-Cr/Au Schottky contacts have shown much higher β_I values for forward direction than for reverse one [7]. This behavior may be explained by any of the following three models:

Model A:

For forward direction all interface states are in equilibrium with the semiconductor, while for reverse direction with the metal ($Ds_F = Ds_B$, $Ds_R = Ds_A$). This model is considered to be valid for semiconductors with high mobility, and corresponds to the thermionic emission current mechanism [5,6].

In this case changing the applied voltage both for the forward and reverse direction the same interface states located over the metal Fermi level (n-type semiconductor) will be filled or emptied, therefore the interface state density at a given energy level must be the same.

On the basis of (3) and (8) for forward direction one obtains:

$$Ds_F = Ds_B = \frac{\varepsilon_0}{q} \left[\frac{\varepsilon_I}{\delta} (n_I - 1) - \frac{\varepsilon_s}{w_F} \right] \tag{10}$$

The same interface state density obtained for reverse direction using (5) and (8):

$$Ds_R = Ds_A = \frac{\varepsilon_0}{q} \left[\frac{\varepsilon_s}{w_R} \left(-\frac{q}{s_I \, kT} - 1 \right) - \frac{\varepsilon_I}{\delta} \right] \tag{11}$$

From Equ. (10) and (11) both the interface state density and the relative interfacial layer thickness may be determined:

$$Ds = \frac{\varepsilon_0}{n_I q} \left[\frac{\varepsilon_s}{w_R} \left(-\frac{q}{s_I \, kT} - 1 \right) (n_I - 1) - \frac{\varepsilon_s}{w_F} \right] \tag{12}$$

$$\frac{\delta}{\varepsilon_I} = n_I \left[\frac{\varepsilon_s}{w_R} \left(-\frac{q}{s_I \, kT} - 1 \right) + \frac{\varepsilon_s}{w_F} \right]^{-1} \tag{13}$$

Model B:

For forward direction all interface states are in equilibrium with the semicundoctor, while in reverse direction their common density is negligible ($Ds_F = Ds_B$, $Ds_R = 0$).

On the basis of (3), (5) and (8) the wanted parameters for this model may be given by the following expressions:

$$Ds = \frac{\varepsilon_0}{q} \left[\frac{\varepsilon_s}{w_R} \left(-\frac{q}{s_I \, kT} - 1 \right) (n_I - 1) - \frac{\varepsilon_s}{w_F} \right] \tag{14}$$

$$\frac{\delta}{\varepsilon_I} = \left[\frac{\varepsilon_s}{w_R} \left(-\frac{q}{s_I \, kT} - 1 \right) \right]^{-1} \tag{15}$$

Model C:

Either both for forward and reverse direction all interface states are in equilibrium with the metal (Case X), or for forward direction the common interface state density is negligible, while for reverse direction they are in equilibrium with the metal (Case Y) ($D_{SF}=0$, $D_{SR}=D_{SA}$). Case X corresponds to the diffusion current mechanism.'

In Case X for forward direction D_{SA} may be neglected, because the change of the voltage drop on the interfacial layer, i.e. the shift of the electron quasi Fermi level for forward direction, is a small part of kT/q only, therefore the change of the interface charge must be negligible.

The interface state density and the relative interfacial layer thickness for this model may be expressed as

$$D_s = \frac{\mathcal{E}_0}{q} [\frac{\mathcal{E}_s}{w_R} (\frac{q}{s_I kT} -1) - \frac{\mathcal{E}_s}{w_F (n_I -1)}] \qquad (16)$$

$$\frac{\delta}{\mathcal{E}_I} = \frac{w_F}{\mathcal{E}_s} (n_I -1) \qquad (17)$$

For all three models the corresponding energy level may be found on the basis of (6) integrating β_I according to the applied voltage.

Discussion of models

To obtain the conditions of the negligibility of D_{SA} and D_{SB} for Models B and C one must rearrange Equ. (8):

$$\frac{\mathcal{E}_I}{\delta} = (\frac{1}{\beta_I} -1) (\frac{\mathcal{E}_s}{w} + \frac{qD_{SB}}{\mathcal{E}_0}) - \frac{qD_{SA}}{\mathcal{E}_0} \qquad (18)$$

This expression gives the following conditions:

$$D_{SA} << (\frac{1}{\beta_I} -1) \frac{\mathcal{E}_0 \mathcal{E}_s}{qw} \qquad (19)$$

and

$$D_{SB} << \frac{\mathcal{E}_0 \mathcal{E}_s}{qw} \qquad (20)$$

Here condition (20) is much stronger than (19), because β_I much less than unity. Moreover if (19) is not fulfilled, instead of Model B Model A will be got. For Model C condition (19) a priori is fulfilled (Case X).

Comparing Equ. (12) and (13) with (14) and (15) it is seen that the highest deviation between the results obtained for Models A and B is $(n_I -1)/n_I$. In our investigations n<1.5, therefore the deviation of results obtained for these two models is less than 50%.

The same experimental I-V characteristics give much higher interface state density and relative interfacial layer thickness for Model C than for Models A and B. The $\delta/\mathcal{E}_I$ values evaluated for Model C has been in the range of 0.5-2.5 nm [7] which is not consistent with Model C: for so thick interfacial layers it is not likely the equilibrium of the interface states with the metal.

Models A and B give acceptable $\delta/\mathcal{E}_I$ values (0.2-0.4 nm), but in the case of Model B the condition of the negligibility of D_{SB} (19) for reverse direction is very hard. The depletion depth for reverse direction is in the order of 10^{-3}-10^{-6} cm for concentrations of 10^{14}-10^{18} cm^{-3} respectively. This means that - if Model A is not valid, i.e. the interface states are in equilibrium with the semiconductor - the interface state density below the metal Fermi level must be much less than 10^9-10^{12} cm^{-2}eV^{-1} respectively, which for the lower concentrations is not likely to be fulfilled. Therefore at low concentrations - in the case of validity of Model B - the error of the obtained results will be high. Indeed, the preliminary results obtained for Model A on n-type GaAs-Cr/Au devices with low concentrations (below $2*10^{15}$ cm^{-3} including) have shown higher interface state density than those for higher concentrations. This phenomenon may be connected with the possible fact that for reverse direction a part of the interface states is in equilibrium also with the semiconductor.

Experimental results

Fig. 1 shows typical energy distribution of interface states evaluated from I-V characteristics of n-type (100) GaAs-Cr/Au mesa Schottky contacts [8] (N=$7*10^{15}$ cm^{-3}) using Model A. In the evaluation the image force lowering has been taken into account. The Schottky barrier height ϕ_{bo} determining the energy position below the conduction band edge is

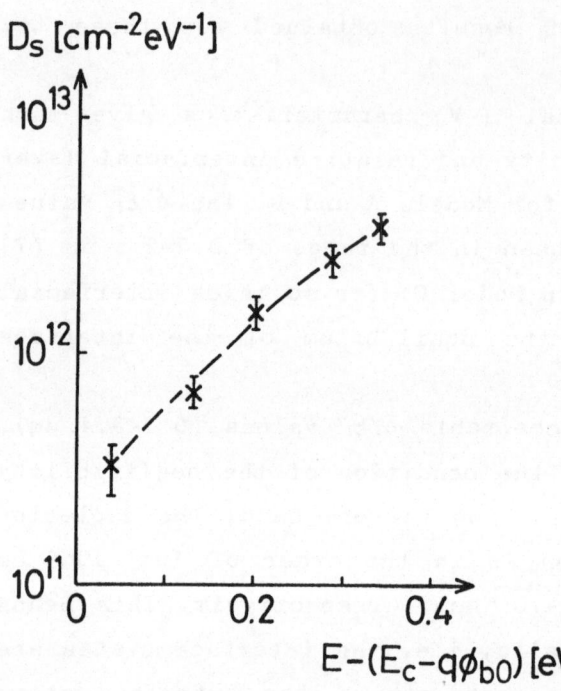

Fig. 1: Interface state energy distribution obtained from the I-V characteristics of an n-type GaAs-Cr/Au Schottky contact

about 0.8-0.9 eV [7,8]. The energy distribution is in agreement with those obtained for n-type GaAs contacts with different metallization by capacitance spectroscopy [9]. The obtained relative interfacial layer thickness for this diode is 0.28+0.03 nm.

Note

Independently of this work Tseng and Wu [4] developed a systematic selfconsistent iteration method to evaluate the interface state energy distribution and the relative interfacial layer thickness (the interfacial layer capacitance) from the Schottky I-V characteristics. The model used by them for the evaluation is identical to Model B in this work.

Acknowledgements

The author is indebted to I. Mojzes, B. Pödör, B. Szentpáli and P. Tüttö for their support and the valuable discussions and advices. He also thanks A. Csonka, J. Földesi, I. Gyúró and M. Németh-Sallay for the device preparation.

References:

1. Zs.J.Horváth, Proc.Symp.Electron.Technol.'87, Sept.15-18, 1987, Budapest, Hungary, in press.
2. Zs.J.Horváth, to be published.
3. H.C.Card, E.H.Rhoderick, J.Phys.D: Appl.Phys. 4, 1589, (1971).
4. H.-H.Tseng, C.-Y.Wu, Solid-State Electron. 30, 383 (1987).
5. C.R.Crowell, M.Beguwala, Solid-State Electron. 14, 1149 (1971).
6. E.H.Rhoderick, J.Phys.D: Appl.Phys. 5, 1920 (1972).
7. Zs.J.Horváth, B.Pécz, P.Tüttö, E.Jároli, M.Németh-Sallay, I.Gyúró, Proc. 2nd Conf. on Physics and Technology of GaAs and other III-V Semiconductors, Sept.8-11, 1986, Budapest, Hungary, in press.
8. Zs.J.Horváth, I.Gyúró, M.Németh-Sallay, B.Szentpáli, K.Kazi, phys.stat.sol.(a) 94, 719 (1986).
9. F.Chekir, C.Barret, A.Vapaille, J.Appl.Phys. 54, 6476 (1983).

SURFACE WORK FUNCTION TRANSIENTS OF TUNNEL SIO$_2$-SI STRUCTURES

J.Mizsei and I.Zolomy

Technical University of Budapest,Department of Electron Devices,
Budapest,H-1521

The surface potential of the external surface of silicon-dioxide layers upon silicon substrate was measured by the vibrating capacitor /Kelvin/ method.This method is very sensitive for surface charge and/or work function change of the surface[1].According to this method,the electric field between the vibrating reference electrode and the surface to be investigated is compensated to a zero value/Fig.1./.

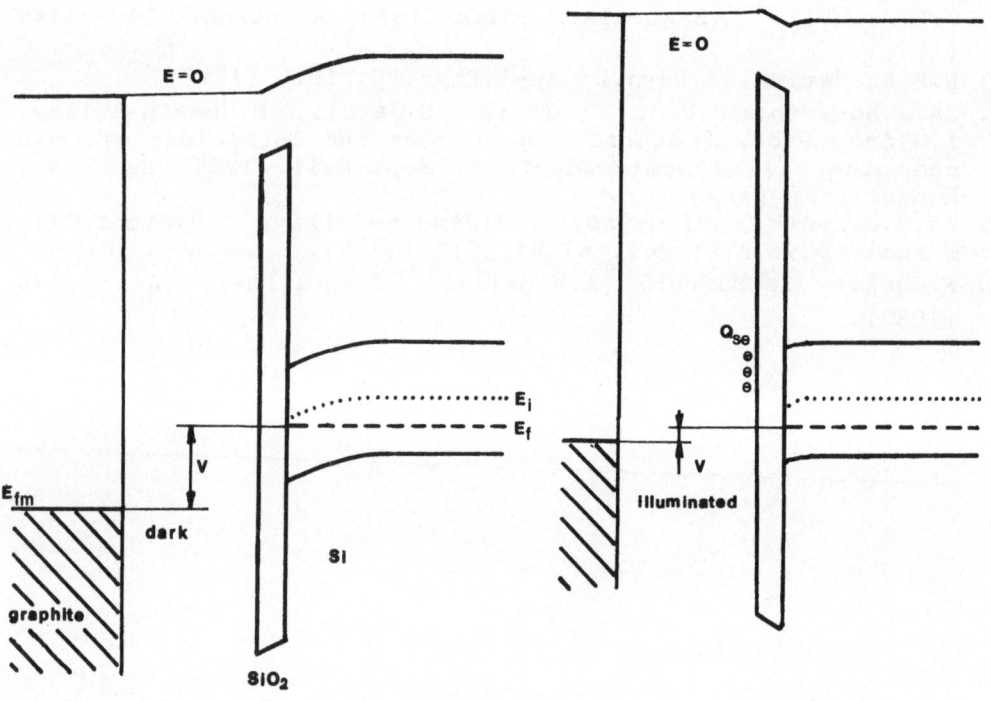

Figure 1.

The electric field can be compensated to a zero value by adjusting the dc. bias voltage /V/ till the alternating current in the connecting wire between the electrodes becomes zero.The method is simple,and the compensation can be automatized by a feedback system /Fig.2./ [2] .

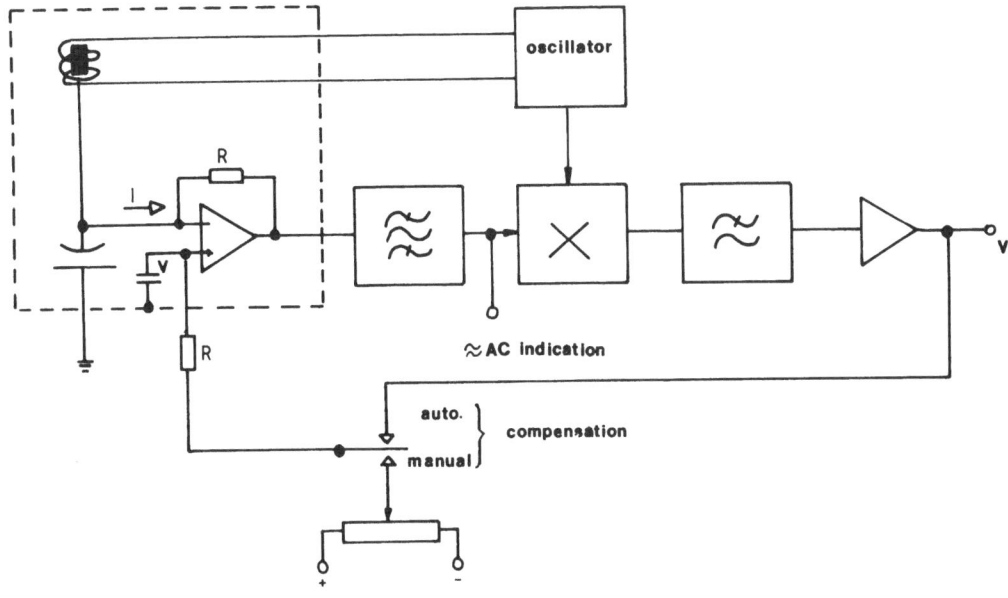

Figure 2.

As the electric field between the reference electrode and the surface to be measured is zero,the field lines originating from the charge on the external surface of the oxide will not enter into the external space, but will penetrate the oxide and end on the charges in the semiconductor. Thus there is no charge on the vibrating reference electrode,and therefore in spite of the change of the capaticance between this electrode and the surface under investigation,no alternating current will flow in the external circuit.

Illuminating the surface of the ultrathin SiO_2-Si system,a change of some hundreds of millivolts in the necessary bias voltage /the work function difference/was observed.After switching off the light, the voltage difference V /bias voltage/ returned to its original /dark/ value through a transient process.Brattain and Bardeen investigated the trap charging and discharging processes on germanium surface[3] with a similar experimental setup.The transient process reported in this paper seems to be composed of two sections,the first one is more rapid,the

second one is slower.The time constant of the second section increased
very strongly with increasing oxide thicknesses.Analising this transient
process it was possible to observe the native oxide growth ,other oxides
have also been investigated/oxides prepared in boiling nitric acid and
thermally grown oxides/.

Experimental results

Let's consider at first the native oxides.On the freshly HF etched
silicon surface the measured bias V was about 6oo-8oo mV,depending on
doping concentration and surface state charge.On the freshly etched
surface there was not any sign of the transient process mentioned above.
As the native oxide was growing,the dark value of V decreased, and the
transient process developed/Fig.3./ ,that is the change of V denoted

by ΔV increased.

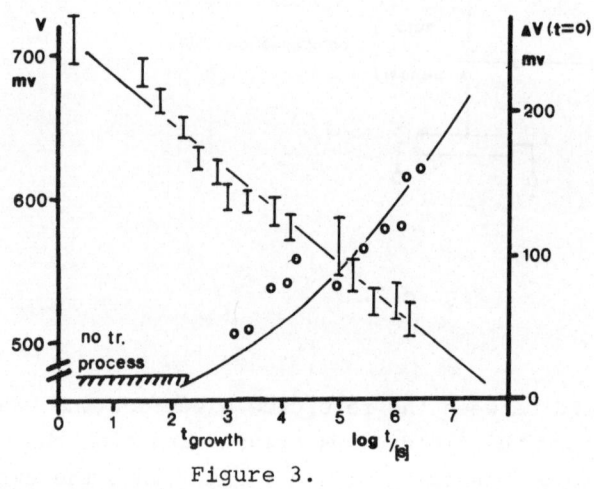

Figure 3.

Fig.4. shows typical measured V-t functions for different oxides,
demonstrating different time constants.In Fig.5. the measured τ time
constants are given versus the oxide thickness.Fig.6. shows the measured
dV/dt and the calculated electron tunnel current density J_{nt} as function
of time.

The deviation from simple exponential function is demonstrated in
Fig.4.as well as the strong time dependence of the transient process
on the SiO_2 thickness and quality.It seems,that the whole process consists of two parts,the first one being faster than the second one.

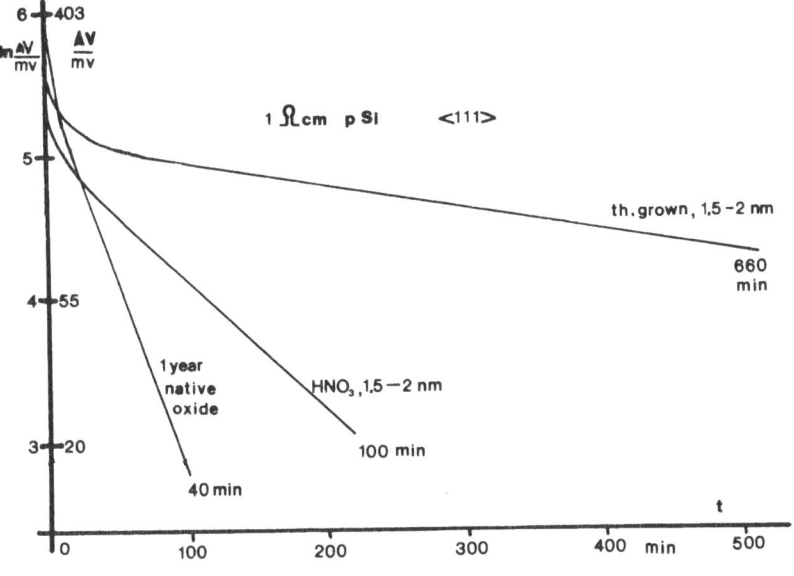

Figure 4.

In Fig.5. the time constants of this second part are plotted in logarithmic scale against the oxide thickness. The time constant τ is increasing exponentially with oxide thickness, as it can be seen in the case of the native oxide. The relation between the native oxide thickness and the growth time is taken from ref. [4]. There is a two or more orders of magnitude deviation in time constants between the native oxide and the thermally grown/nitric acid oxides.

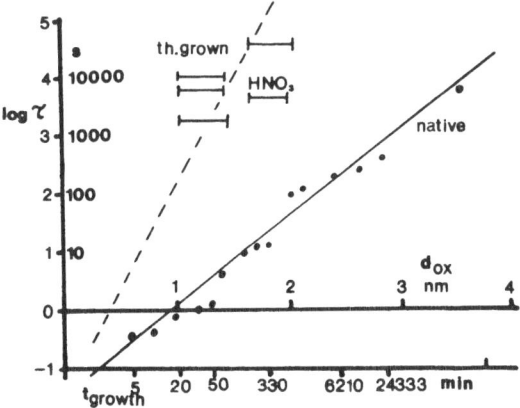

Figure 5.

This deviation can be caused by two possible reasons.The first one is, that the native oxide may have a different chemical composition and different structure /SiO$_x$/ as the thermally grown oxide,and according to this a potential barrier with different shape and height is formed with respect to the thermally grown SiO$_2$.The second possible reason could be, that the native oxides were thinner than it was given in ref. [4] .We have not found yet native oxides thicker than 2.5 nm,even among ten years old samples.The discrepancy can be reduced,if it is supposed, that the native oxides are thinner,corresponding to the dashed line in Fig.5.,as our Talystep measurements have shown a thickness of about 2 nm for 10 years old native oxides.

Discussion

For creating a physical picture let's consider a p-type substrate. In darkness there is an equilibrium of tunnel currents between the charge carriers in the semiconductor and the carriers captured in the traps on the external surface of the oxide.There is also a fix charge /usually positive/at the oxide-silicon interface/or on the silicon surface,if there is no oxide on it/,due to dangling bonds.As the native oxide grows,the number of dangling bonds and thus this fix charge decreases.A thin oxide layer has however another effect:electrons and holes are tunneling to the outer traps on the surface of the oxide, creating a surface charge.The electron tunnel current flowing from the interface to the surface electron charge and the one flowing in opposite direction are equal.As the electron tunnel currents depend upon the electron contrentration at the interface and upon the trapped surface electron charge,a higher surface electron contrentration results a higher electron charge.The same holds for holes.

In equilibrium at p-type material the surface can not be in the flat-band condition even if the fix charge were zero,because in this case the interface hole concentration would be much higher than the electron concentration/resulting a high hole tunnel current to the surface/,thus the net surface charge would be positive,which would bend the energy bands/near the interface/downward/to more positive potential/. The fix /positive/ interface charge has also similar effect upon the band bending.Thus in dark and equilibrium the semiconductor energy bands are bent toward positive voltages with a Ψ_{so} value.This value depends upon many factors,such as doping concentration,fixed interface charge density,oxide thickness and quality/surface trap density/,but its value is usually near to

$$\Psi_{so} \approx \frac{kT}{q} \ln \frac{N_a}{n_i} \quad , \qquad\qquad /1/$$

where N_a is the acceptor concentration. That is the surface is nearly in intrinsic condition. A strong deviation from this condition would increase either the electron or the hole concentrations, thus the tunnel currents, therefore the surface charge would change, which shifts the surface potential back near to its initial /near intrinsic/ value.

Under strong illumination the /minority/ electron concentration increases strongly and at the same time there is a relatively small change in the /majority/ hole concentration. Thus the resultant surface charge will be more negative, shifting the band upward /to more negative potentials/, approaching the flat band condition at sufficiently strong illumination [5] , and at a high concentration of surface traps.

After finishing the illumination the carrier concentrations in the silicon quickly return to their equilibrium values, and the electron charge at the external oxide surface starts to decrease as a result of tunneling back to the silicon. The net positive charge and the band bending increase toward the equilibrium or rather steady-state values. If Q_{se} is the surface electron charge and J_{nt} the electron tunnel current from the oxide surface to the semiconductor, the tunnel current changes the surface electron charge, thus

$$\frac{dQ_{se}}{dt} = -J_{nt} = -AQ_{se} \exp/-bd_{ox}/ \qquad /2/$$

as the electron tunnel current is proportional to the total electron charge and depends exponentially upon the thickness of the oxide.

The solution of equ./2/ is:

$$Q_{se} = Q_{se}/0/ \exp/- \frac{t}{\tau}/ \qquad /3/$$

where

$$\tau = \frac{\exp/bd_{ox}/}{A} \qquad /4/$$

is the time constant of the transient process, and b and A are constants.

Let's denote the total surface and interface charge by Q_s and its value in darkness by Q_{so}. The total charge can be expressed as:

$$Q_s = Q_{so} - Q_{se}/0/ \exp/- \frac{t}{\tau}/ = Q_{so}/1 - \exp/- \frac{t}{\tau}// \qquad /5/$$

As at t=0, in the case of high illumination level, the flat-band condition is approximated [5], the initial resultant surface charge is zero, that is $Q_s/0/=0$, and therefore $Q_{so}=Q_{se}/0/$.

The total surface potential Ψ_b is the sum of the voltage drops in the oxide and in the semiconductor. The first one is proportional to the maximum field, the second one to the square of the field. On the other hand, the maximum field is proportional to the total surface charge, thus

$$\Psi_b = V_{ox} + \Psi_{si} = C_1 Q_s + C_2 Q_s^2 \quad , \qquad /6/$$

where $C_1 = d_{ox}/\epsilon_{ox}$ and $C_2 = /2qN_a\epsilon_s/^{-1}$.

The change of the surface potential in respect to its dark value can be expressed with equs. /5/ and /6/ as

$$\Delta\Psi_b = \Delta V = \Psi_b/\infty/ - \Psi_b/t/ = Q_{so}/C_1 + 2C_2 Q_{so}/\exp/-\frac{t}{\tau}/ - C_2 Q_{so}^2 \exp/-\frac{2t}{\tau}/ \quad , \qquad /7/$$

$\Delta\Psi_b = \Delta V$ the change of the bias voltage.

According to equ. /7/ the transient process has two time constants, τ and $\tau/2$. After determining $d\Delta V/dt = dV/dt$, it can be shown, that it has a maximum value at

$$t' = \tau \ln\frac{4C_2 Q_{so}}{C_1 + 2C_2 Q_{so}} \quad . \qquad /8/$$

At this time the second member of equ. /6/ is only one forth of the first one, thus after t' the transient process has practically only one time constant, τ, and therefore from this section the τ can be determined. The electron tunnel current can also be obtained with equation /2/ and considering, that $Q_{se} = Q_{so} - Q_s$:

$$J_{nt} = \frac{dQ_s}{dt} \quad . \qquad /9/$$

Expressing Q_s from equ. /6/ and inserting it into equ. /9/, for J_{nt} results

$$J_{nt} = \frac{dV/dt}{/C_1^2 + 4C_2\Psi_b/^{1/2}} \quad . \qquad /10/$$

The tunnel current calculated from equ. /10/ is shown in Fig. 6. Measuring t', τ, and $\Delta V/0/$ the oxide thickness can be determined as follows: equ. /9/ with $t=0$ and equ. /8/ contain C_1 and Q_{so} as unknowns, thus they can be expressed from these equations as

$$Q_{so}=\sqrt{\frac{\Delta V/0/}{C_2/4\exp/-\frac{t'}{\tau}/-1/}} \qquad\qquad /11/$$

and

$$d_{ox}=C_1\epsilon_{ox}=\epsilon_{ox}\sqrt{\frac{\Delta V/0/C_2}{4\exp/-\frac{t'}{\tau}/-1}}\left|2-\exp/-\frac{t'}{\tau}/\right| \quad . \quad /12/$$

However, as $t' \ll \tau$, the accuracy of these expressions is rather poor, as d_{ox} depends exponentially upon the value of t', which also can not be determined accurately from the measurement.

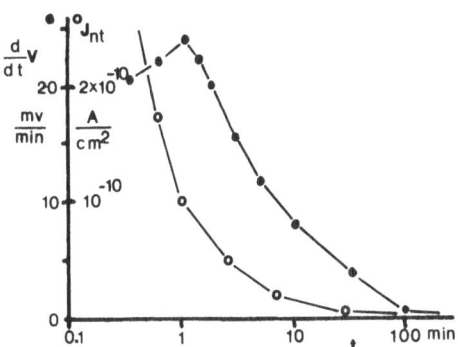

Figure 6.

References

[1] J.Bonnet and J-M.Palau,Proc.of the 4th Int. Conf. on Solid Surfaces and 3rd Europ. Conf. on Surf. Sci./Cannes,1980/ p.1129.

[2] J.Mizsei and I.Házman,Mérés és Automatika 31 /1983/ p.136.

[3] W.Brattain and J.Bardeen,Bell Syst. Techn.J.32/1953/p.1.

[4] FLukes,Surface Sci. 30 /1972/ p.91.

[5] S.R.Morrison,The Chemical Physics of Surfaces, Plenum Press,New York and London,1977.

INDEX OF CONTRIBUTORS